T0327639

Inverse Synthetic Aperture Radar Imaging with MATLAB Algorithms

WILEY SERIES IN MICROWAVE AND OPTICAL ENGINEERING

KAI CHANG, Editor
Texas A&M University

A complete list of the titles in this series appears at the end of this volume.

Inverse Synthetic Aperture Radar Imaging with MATLAB Algorithms

CANER ÖZDEMİR, PhD
Mersin University
Mersin, Turkey

A JOHN WILEY & SONS, INC., PUBLICATION

Published by John Wiley & Sons, Inc., Hoboken, New Jersey.
Published simultaneously in Canada.

For general information on our other products and services or for technical support, please
contact our Customer Care Department within the United States at (800) 762-2974, outside the
United States at (317) 572-3993 or fax (317) 572-4002.

Wiley also publishes its books in a variety of electronic formats. Some content that appears in
print may not be available in electronic formats. For more information about Wiley products,
visit our web site at www.wiley.com.

Library of Congress Cataloging-in-Publication Data:

Özdemir, Caner.
   Inverse synthetic aperture radar imaging with MATLAB / Caner Özdemir.
      p. cm. – (Wiley series in microwave and optical engineering ; 210)
   Includes bibliographical references.
   ISBN 978-0-470-28484-1 (hardback)
   1. Synthetic aperture radar.   2. MATLAB.   I. Title.
   TK6592.S95O93 2011
   621.3848'5–dc23

                              2011031430

Printed in Singapore.

10 9 8 7 6 5 4 3 2 1

To:

My wife,
My three daughters,
My brother,
My father,
and the memory of my beloved mother

Contents

Preface

Inverse synthetic aperture radar (ISAR) has been proven to be a powerful signal processing tool for imaging moving targets usually on the two-dimensional (2D) down-range cross-range plane. ISAR imagery plays an important role especially in military applications such as target identification, recognition, and classification. In these applications, a critical requirement of the ISAR image is to achieve sharp resolution in both down-range and cross-range domains. The usual way of obtaining the 2D ISAR image is by collecting the frequency and aspect diverse backscattered field data from the target. For synthetic aperture radar (SAR) and ISAR scenarios, there is always a trade-off between the down-range resolution and the frequency bandwidth. In contrast to SAR, the radar is usually fixed in the ISAR geometry and the cross-range resolution is attained by target's rotational motion, which is generally unknown to the radar engineer.

In order to successfully form an ISAR image, the target's motion should contain some degree of rotational component with respect to radar line of sight (RLOS) direction during the coherent integration time (or dwell time) of the radar system. But in some instances, especially when the target is moving along the RLOS direction, the target's viewing angle width is insufficient to be able to form an ISAR image. This restriction can be eliminated by utilizing bistatic or multistatic configurations that provide adequate look-angle diversity of the target. Another challenging problem occurs when the target's rotational velocity is sufficiently high such that the target's viewing angle width is not small during the dwell time of the radar. The target's translational movement is another issue that has to be addressed before displaying the final motion-free ISAR image. Therefore, motion effects have to be removed or mitigated with the help of motion compensation algorithms.

This book is devoted to the conceptual description of ISAR imagery and the explanation of basic ISAR research. Although the primary audience will be graduate students and other interested researchers in the fields of electrical engineering and physics, I hope that colleagues working in radar research and development or in a related industry may also benefit from the book. Numerical or experimental examples in Matlab technical language are provided for the presented algorithms with the aim of improving the understanding of the algorithms by the reader.

The organization of the book is as follows. In the first chapter, an overview of Fourier theory, which plays an important and crucial role in radar imaging, is presented to provide a fair knowledge of Fourier-based signal processing basics. Noting that the ISAR imaging can also be treated as a signal processing tool, an understanding of signal processing and Fourier theory will be required to get the full benefit from the chapters within the book. The next chapter is devoted to radar fundamentals. Since ISAR itself is a radar, the key parameters of the radar concept that is related to ISAR research are revisited. These include electromagnetic scattering, radar cross section, the radar equation, and the radar waveforms. Then, before stepping into inverse problem of ISAR, the forward problem of SAR is reviewed in Chapter 3. SAR and ISAR provide dual problems and share dual algorithms with similar difficulties. Therefore, understanding the ISAR imagery could not be complete without understanding the SAR concepts. In the SAR chapter, therefore, important concepts of SAR such as resolution, pulse compression, and image formation are given together with associated Matlab codes. Furthermore, some advanced concepts and trends in SAR imaging are also presented.

After providing the fundamentals for SAR imaging, we provide the detailed imaging procedure for conventional ISAR imaging and the basic ISAR concepts with associated Matlab codes in Chapter 4. The topics include range profile concept, range/cross-range resolutions, small-angle small-bandwidth ISAR imaging, large-angle wide-bandwidth ISAR imaging, polar reformatting, and three-dimensional ISAR imaging. In Chapter 5, we provide some design aspects that are used to improve the quality of the ISAR image. Down sampling/up sampling, image aliasing, point spread function and smoothing are covered in this chapter. Several imaging tricks and fine-tuning procedures such as zero-padding and windowing that are used for enhancing the image quality are also presented.

In Chapter 6, range-Doppler ISAR image processing is given in detail. ISAR waveforms, ISAR receiver for these waveforms, quadrature detection, Doppler shift phenomena, and range-Doppler ISAR imaging algorithms are presented. The design examples with Matlab codes are also provided. In Chapter 7, scattering center representation, which has proven to be a sparse but an effective model of ISAR imaging, is presented. We provide algorithms to reconstruct both the image and the field data from the scattering centers with good fidelity. In Chapter 8, motion compensation (MOCOMP), one of the most important and challenging problems of ISAR imagery, is taken up in detail. The concepts include Doppler effect due to target motion, translational and motion compensation routines, range tracking, and Doppler tracking subjects. Algorithms and numerical examples with Matlab codes are provided for the most popular MOCOMP techniques, namely, cross-correlation method, minimum entropy method, and joint-time frequency (JTF)-based motion compensation. In the final chapter, applications of the ISAR imaging concept to different but related engineering problems are presented. The employment of ISAR imagery to the antenna scattering problem (i.e., antenna SAR) and also to the antenna coupling problem (i.e., antenna coupling SAR) are explained. The imaging algorithms together with numerical examples are given. In addition, the application of the SAR/ISAR concept to the ground penetrating radar application is presented.

CANER ÖZDEMİR

Acknowledgments

I would like to address special thanks to the people below for their help and support during the preparation of this book. First, I am thankful to my wife and three children for their patience and continuous support while writing this book. I am very thankful to Dr. Hao Ling of the University of Texas at Austin for being a valuable source of knowledge, ideas, and also inspiration. He has been a great advisor since I met him.

I would like to express my sincere thanks to my former graduate students Betül Yılmaz, Deniz Üstün, Enes Yiğit, Şevket Demirci, and Özkan Kırık, who carried out some of the research detailed in this book.

Last but not least, I would like to show my special thanks to Dr. Kai Chang for inviting me to write this book. Without his kind offer, this study would not have been possible.

C.Ö.

CHAPTER ONE

Basics of Fourier Analysis

1.1 FORWARD AND INVERSE FOURIER TRANSFORM

Fourier transform (FT) is a common and useful mathematical tool that is utilized in numerous applications in science and technology. FT is quite practical, especially for characterizing nonlinear functions in nonlinear systems, analyzing random signals, and solving linear problems. FT is also a very important tool in radar imaging applications as we shall investigate in the forthcoming chapters of this book. Before starting to deal with the FT and inverse Fourier transform (IFT), a brief history of this useful linear operator and its founders is presented.

1.1.1 Brief History of FT

Jean Baptiste Joseph Fourier, a great mathematician, was born in 1768 in Auxerre, France. His special interest in heat conduction led him to describe a mathematical series of sine and cosine terms that can be used to analyze propagation and diffusion of heat in solid bodies. In 1807, he tried to share his innovative ideas with researchers by preparing an essay entitled "On the Propagation of Heat in Solid Bodies." The work was examined by Lagrange, Laplace, Monge, and Lacroix. Lagrange's oppositions caused the rejection of Fourier's paper. This unfortunate decision caused colleagues to wait for 15 more years to read his remarkable contributions on mathematics, physics, and, especially, signal analysis. Finally, his ideas were published in the book *The Analytic Theory of Heat* in 1822 [1].

Discrete Fourier transform (DFT) was developed as an effective tool in calculating this transformation. However, computing FT with this tool in the

Inverse Synthetic Aperture Radar Imaging with MATLAB Algorithms, First Edition.
Caner Özdemir.
© 2012 John Wiley & Sons, Inc. Published 2012 by John Wiley & Sons, Inc.

19th century was taking a long time. In 1903, Carl Runge studied the minimization of the computational time of the transformation operation [2]. In 1942, Danielson and Lanczos utilized the symmetry properties of FT to reduce the number of operations in DFT [3]. Before the advent of digital computing technologies, James W. Cooley and John W. Tukey developed a fast method to reduce the computation time in DFT. In 1965, they published their technique that later on became famous as the fast Fourier transform (FFT) [4].

1.1.2 Forward FT Operation

The FT can be simply defined as a certain linear operator that maps functions or signals defined in one domain to other functions or signals in another domain. The common use of FT in electrical engineering is to transform signals from time domain to frequency domain or vice versa. More precisely, forward FT decomposes a signal into a continuous spectrum of its frequency components such that the time signal is transformed to a frequency-domain signal. In radar applications, these two opposing domains are usually represented as "spatial frequency" (or wave number) and "range" (distance). Such use of FT will be examined and applied throughout this book.

The forward FT of a continuous signal $g(t)$ where $-\infty < t < \infty$ is described as

$$
\begin{aligned}
G(f) &= \mathcal{F}\{g(t)\} \\
&= \int_{-\infty}^{\infty} g(t) \cdot e^{-j2\pi ft} dt.
\end{aligned}
\tag{1.1}
$$

To appreciate the meaning of FT, the multiplying function $e^{-j2\pi ft}$ and operators (multiplication and integration) on the right side of Equation 1.1 should be investigated carefully: The term $e^{-j2\pi f_i t}$ is a complex phasor representation for a sinusoidal function with the single frequency of f_i. This signal oscillates only at the frequency of f_i and does not contain any other frequency component. Multiplying the signal in interest, $g(t)$, with the term $e^{-j2\pi f_i t}$ provides the similarity between each signal, that is, how much of $g(t)$ has the frequency content of f_i. Integrating this multiplication over all time instances from $-\infty$ to ∞ will sum the f_i contents of $g(t)$ over all time instants to give $G(f_i)$; that is, the amplitude of the signal at the particular frequency of f_i. Repeating this process for all the frequencies from $-\infty$ to ∞ will provide the frequency spectrum of the signal; that is, $G(f)$. Therefore, the transformed signal represents the continuous spectrum of frequency components; that is, representation of the signal in "frequency domain."

1.1.3 IFT

This transformation is the inverse operation of the FT. IFT, therefore, synthesizes a frequency-domain signal from its spectrum of frequency components to its time-domain form. The IFT of a continuous signal $G(f)$ where $-\infty < f < \infty$ is described as

$$g(t) = \mathcal{F}^{-1}\{G(f)\}$$
$$= \int_{-\infty}^{\infty} G(f) \cdot e^{j2\pi ft} df. \tag{1.2}$$

1.2   FT RULES AND PAIRS

There are many useful Fourier transform rules and pairs that can be very helpful when applying the FT or IFT to different real-world applications. We will briefly revisit them to remind the reader of the properties of FT. Provided that FT and IFT are defined as in Equations 1.1 and 1.2, respectively, FT pair is denoted as

$$g(t) \overset{\mathcal{F}}{\leftrightarrow} G(f), \tag{1.3}$$

where $\mathcal{F}$ represents the forward FT operation from time domain to frequency domain. The IFT operation is represented by $\mathcal{F}^{-1}$ and the corresponding alternative pair is given by

$$G(f) \overset{\mathcal{F}^{-1}}{\leftrightarrow} g(t). \tag{1.4}$$

Here, the transformation is from frequency domain to time domain. Based on these notations, the properties of FT are listed briefly below.

1.2.1   Linearity

If $G(f)$ and $H(f)$ are the FTs of the time signals $g(t)$ and $h(t)$, respectively, the following equation is valid for the scalars a and b:

$$a \cdot g(t) + b \cdot h(t) \overset{\mathcal{F}}{\leftrightarrow} a \cdot G(f) + b \cdot H(f). \tag{1.5}$$

Therefore, the FT is a linear operator.

1.2.2   Time Shifting

If the signal is shifted in time with a value of t_o, then its frequency domain signal is multiplied with a phase term as listed below:

$$g(t - t_o) \overset{\mathcal{F}}{\leftrightarrow} e^{-j2\pi ft_o} \cdot G(f) \tag{1.6}$$

1.2.3   Frequency Shifting

If the time signal is multiplied by a phase term of $e^{j2\pi f_o t}$, then the FT of this time signal is shifted in frequency by f_o:

$$e^{j2\pi f_o t} \cdot g(t) \overset{\mathcal{F}}{\leftrightarrow} G(f - f_o) \qquad (1.7)$$

1.2.4   Scaling

If the time signal is scaled by a constant a, then the spectrum is also scaled with the following rule:

$$g(at) \overset{\mathcal{F}}{\leftrightarrow} \frac{1}{|a|} G\left(\frac{f}{a}\right), \quad a \in \mathbb{R},\, a \neq 0. \qquad (1.8)$$

1.2.5   Duality

If the spectrum signal $G(f)$ is taken as a time signal $G(t)$, then the corresponding frequency-domain signal will be the time reversal equivalent of the original time-domain signal, $g(t)$:

$$G(t) \overset{\mathcal{F}}{\leftrightarrow} g(-f). \qquad (1.9)$$

1.2.6   Time Reversal

If the time is reversed for the time-domain signal, then the frequency is also reversed in the frequency-domain signal:

$$g(-t) \overset{\mathcal{F}}{\leftrightarrow} G(-f). \qquad (1.10)$$

1.2.7   Conjugation

If the conjugate of the time-domain signal is taken, then the frequency-domain signal is conjugated and frequency-reversed:

$$g^*(t) \overset{\mathcal{F}}{\leftrightarrow} G^*(-f). \qquad (1.11)$$

1.2.8   Multiplication

If the time-domain signals $g(t)$ and $h(t)$ are multiplied in time, then their spectrum signals $G(f)$ and $H(f)$ are convolved in frequency:

$$g(t) \cdot h(t) \overset{\mathcal{F}}{\leftrightarrow} G(f) * H(f). \qquad (1.12)$$

1.2.9   Convolution

If the time-domain signals $g(t)$ and $h(t)$ are convolved in time, then their spectrum signals $G(f)$ and $H(f)$ are multiplied in the frequency domain:

$$g(t)*h(t)\overset{\mathcal{F}}{\leftrightarrow}G(f)\cdot H(f). \qquad (1.13)$$

1.2.10   Modulation

If the time-domain signal is modulated with sinusoidal functions, then the frequency-domain signal is shifted by the amount of the frequency at that particular sinusoidal function:

$$
\begin{aligned}
g(t)*cos(2\pi f_o t)&\overset{\mathcal{F}}{\leftrightarrow}\frac{1}{2}(G(f+f_o)+G(f-f_o))\\[4pt]
g(t)*sin(2\pi f_o t)&\overset{\mathcal{F}}{\leftrightarrow}\frac{j}{2}(G(f+f_o)-G(f-f_o)).
\end{aligned}
\qquad (1.14)
$$

1.2.11   Derivation and Integration

If the derivative or integration of a time-domain signal is taken, then the corresponding frequency-domain signal is given as below:

$$
\begin{aligned}
\frac{d}{dt}g(t)&\overset{\mathcal{F}}{\leftrightarrow}2\pi f\cdot G(f)\\[4pt]
\int_{-\infty}^{t}g(\tau)d\tau&\overset{\mathcal{F}}{\leftrightarrow}\frac{1}{j2\pi f}G(f)+\pi G(0)\cdot\delta(f).
\end{aligned}
\qquad (1.15)
$$

1.2.12   Parseval's Relationship

A useful property that was claimed by Parseval is that since the FT (or IFT) operation maps a signal in one domain to another domain, the signals' energies should be exactly the same as given by the following relationship:

$$\int_{-\infty}^{\infty}|g(t)|^2\,dt\overset{\mathcal{F}}{\leftrightarrow}\int_{-\infty}^{\infty}|G(f)|^2\,df. \qquad (1.16)$$

1.3   TIME-FREQUENCY REPRESENTATION OF A SIGNAL

While the FT concept can be successfully utilized for the stationary signals, there are many real-world signals whose frequency contents vary over time. To be able to display these frequency variations over time, joint time-frequency (JTF) transforms/representations are used.

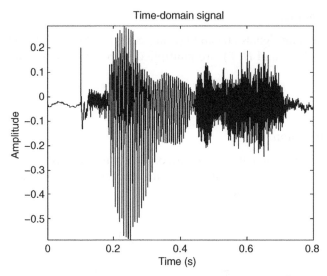

FIGURE 1.1   The time-domain signal of "prince" spoken by a lady.

1.3.1   Signal in the Time Domain

The term "time domain" is used while describing functions or physical signals with respect to time either continuous or discrete. The time-domain signals are usually more comprehensible than the frequency-domain signals since most of the real-world signals are recorded and displayed versus time. A common equipment to analyze time-domain signals is the *oscilloscope*. In Figure 1.1, a time-domain sound signal is shown. This signal is obtained by recording an utterance of the word "prince" by a lady [5]. By looking at the occurrence instants in the x-axis and the signal magnitude in the y-axis, one can analyze the stress of the letters in the word "prince."

1.3.2   Signal in the Frequency Domain

The term "frequency domain" is used while describing functions or physical signals with respect to frequency either continuous or discrete. Frequency-domain representation has been proven to be very useful in numerous engineering applications while characterizing, interpreting, and identifying signals. Solving differential equations and analyzing circuits and signals in communication systems are a few applications among many others where frequency-domain representation is much more advantageous than time-domain representation. The frequency-domain signal is traditionally obtained by taking the FT of the time-domain signal. As briefly explained in Section 1.1, FT is generated by expressing the signal onto a set of basis functions, each of which is a sinusoid with the unique frequency. Displaying the measure of the similarities of the original time-domain signal to those particular unique

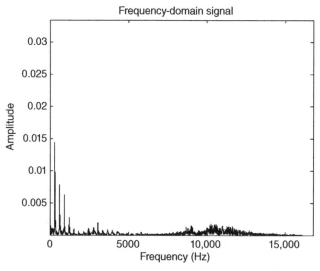

FIGURE 1.2   The frequency-domain signal (or the spectrum) of "prince."

frequency bases generates the Fourier transformed signal, or the frequency-domain signal. Spectrum analyzers and network analyzers are the common equipment which analyze frequency-domain signals. These signals are not as quite perceivable when compared to time-domain signals. In Figure 1.2, the frequency-domain version of the sound signal in Figure 1.1 is obtained by using the FT operation. The signal intensity value at each frequency component can be read from the *y*-axis. The frequency content of a signal is also called the spectrum of that signal.

1.3.3   Signal in the (JTF) Plane

Although FT is very effective for demonstrating the frequency content of a signal, it does not give the knowledge of frequency variation over time. However, most of the real-world signals have time-varying frequency content such as speech and music signals. In these cases, the single-frequency sinusoidal bases are not suitable for the detailed analysis of those signals. Therefore, JTF analysis methods were developed to represent these signals both in time and frequency to observe the variation of frequency content as the time progresses.

There are many tools to map a time-domain or frequency-domain signal onto the JTF plane. Some of the most well-known JTF tools are the short-time Fourier transform (STFT) [6], the Wigner–Ville distribution [7], the Choi-Willams distribution [8], the Cohen's class [9], and the time-frequency distribution series (TFDS) [10]. Among these, the most appreciated and commonly used is the STFT or the spectrogram. The STFT can easily display the variations in the sinusoidal frequency and phase content of local moments of a signal over time with sufficient resolution in most cases.

The spectrogram transforms the signal onto two-dimensional (2D) time-frequency plane via the following famous equation:

$$STFT\{g(t)\} \triangleq G(t,f)$$
$$= \int_{-\infty}^{\infty} g(\tau) \cdot w(\tau - t) e^{-j2\pi f \tau} d\tau. \tag{1.17}$$

This transformation formula is nothing but the short-time (or short-term) version of the famous FT operation defined in Equation 1.1. The main signal, $g(t)$, is multiplied with a shorter duration window signal, $w(t)$. By sliding this window signal over $g(t)$ and taking the FT of the product, only the frequency content for the windowed version of the original signal is acquired. Therefore, after completing the sliding process over the whole duration of the time-domain signal $g(t)$ and putting corresponding FTs side by side, the final 2D STFT of $g(t)$ is obtained.

It is obvious that STFT will produce different output signals for different duration windows. The duration of the window affects the resolutions in both domains. While a very short-duration time window provides a good resolution in the time domain, the resolution in the frequency domain becomes poor. This is because of the fact that the time duration and the frequency bandwidth of a signal are inversely proportional to each other. Similarly, a long duration time signal will give a good resolution in the frequency domain while the resolution in the time domain will be bad. Therefore, a reasonable compromise has to be attained about the duration of the window in time to be able to view both domains with fairly good enough resolutions.

The shape of the window function has an effect on the resolutions as well. If a window function with sharp ends is chosen, there will be strong sidelobes in the other domain. Therefore, smooth waveform type windows are usually utilized to obtain well-resolved images with less sidelobes with the price of increased main lobe beamwidth; that is, less resolution. Commonly used window types are Hanning, Hamming, Kaiser, Blackman, and Gaussian.

An example of the use of spectrograms is demonstrated in Figure 1.3. The spectrogram of the sound signal in Figure 1.1 is obtained by applying the STFT operation with a Hanning window. This JTF representation obviously demonstrates the frequency content of different syllables when the word "prince" is spoken. Figure 1.3 illustrates that while the frequency content of the part "prin ..." takes place at low frequencies, that of the part "... ce" occurs at much higher frequencies.

JTF transformation tools have been found to be very useful in interpreting the physical mechanisms such as scattering and resonance for radar applications [11–14]. In particular, when JTF transforms are used to form the 2D image of electromagnetic scattering from various structures, many useful physical features can be displayed. Distinct time events (such as scattering from

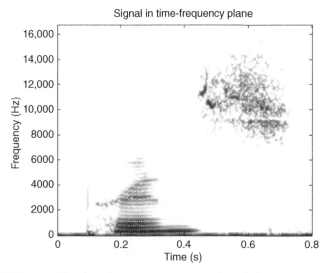

FIGURE 1.3   The time-frequency representation of the word "prince."

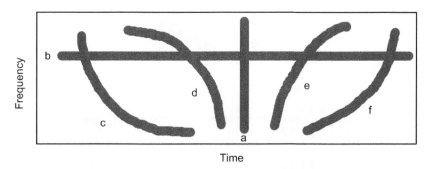

FIGURE 1.4   Images of scattering mechanisms in the joint time-frequency plane: (a) scattering center, (b) resonance, (c and d) dispersion due to material, (e and f) dispersion due to geometry of the structure.

point targets or specular points) show up as vertical line in the JTF plane as depicted in Figure 1.4a. Therefore, these scattering centers appear at only one time instant but for all frequencies. A resonance behavior such as scattering from an open cavity structure shows up as horizontal line on the JTF plane. Such mechanisms occur only at discrete frequencies but over all time instants (see Fig. 1.4b). Dispersive mechanisms, on the other hand, are represented on the JTF plane as slanted curves. If the dispersion is due to the material, then the slope of the image is positive as shown in Figure 1.4c,d. The dielectric coated structures are the good examples of this type of dispersion. The reason for having a slanted line is because of the modes excited inside such materials. As frequency increases, the wave velocity changes for different modes inside these materials. Consequently, these modes show up as slanted curves in the JTF plane. Finally, if the dispersion is due to the geometry of the structure, this type of mechanism appears as a slanted line with a negative slope. This style of behavior occurs for such structures such as waveguides where there exist different modes with different wave velocities as the frequency changes as seen in Figure 1.4e,f.

An example of the use of JTF processing in radar application is shown in Figure 1.5 where spectrogram of the simulated backscattered data from a dielectric-coated wire antenna is shown [14]. The backscattered field is collected from the Teflon-coated wire ($\varepsilon_r = 2.1$) such that the tip of the electric field makes an angle of 60° with the wire axis as illustrated in Figure 1.5. After the incident field hits the wire, infinitely successive scattering mechanisms occur. The first four of them are illustrated on top of Figure 1.5. The first return comes from the near tip of the wire. This event occurs at a discrete time that

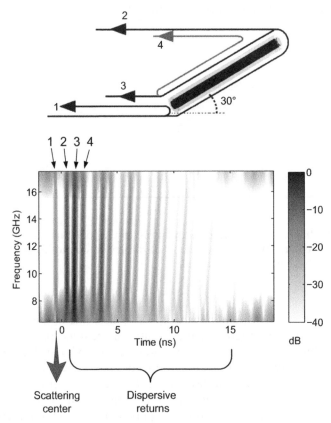

FIGURE 1.5   JTF image of a backscattered measured data from a dielectric-coated wire antenna using spectrogram.

exists at all frequencies. Therefore, this return demonstrates a scattering center-type mechanism. On the other hand, all other returns experience at least one trip along the dielectric-coated wire. Therefore, they confront a dispersive behavior. As the wave travels along the dielectric-coated wire, it is influenced by the dominant dispersive surface mode called Goubau [15]. Therefore, the wave velocity decreases as the frequency increases such that the dispersive returns are tilted to later times on the JTF plane. The dominant dispersive scattering mechanisms numbered 2, 3, and 4 are illustrated in Figure 1.5 where the spectrogram of the backscattered field is presented. The other dispersive returns with decreasing energy levels can also be easily observed from the spectrogram plot. As the wave travels on the dielectric-coated wire more and more, it is slanted more on the JTF plane, as expected.

1.4   CONVOLUTION AND MULTIPLICATION USING FT

Convolution and multiplication of signals are often used in radar signal processing. As listed in Equations 1.12 and 1.13, convolution is the inverse operation of multiplication as the FT is concerned, and vice versa. This useful feature of the FT is widely used in signal and image processing applications. It is obvious that the multiplication operation is significantly faster and easier to deal with when compared to the convolution operation, especially for long signals. Instead of directly convolving two signals in the time domain, therefore, it is much easier and faster to take the IFT of the multiplication of the spectrums of those signals as shown below:

$$
\begin{aligned}
g(t)*h(t) &= \mathcal{F}^{-1}\{\mathcal{F}\{g(t)\} \cdot \mathcal{F}\{h(t)\}\} \\
&= \mathcal{F}^{-1}\{G(f) \cdot H(f)\}.
\end{aligned}
\tag{1.18}
$$

In a dual manner, convolution between the frequency-domain signals can be calculated in a much faster and easier way by taking the FT of the product of their time-domain versions as formulated below:

$$
\begin{aligned}
G(f)*H(f) &= \mathcal{F}\{\mathcal{F}^{-1}\{G(f)\} \cdot \mathcal{F}^{-1}\{H(f)\}\} \\
&= \mathcal{F}\{g(t) \cdot h(t)\}.
\end{aligned}
\tag{1.19}
$$

1.5   FILTERING/WINDOWING

Filtering is the common procedure that is used to remove undesired parts of signals such as noise. It is also used to extract some useful features of the signals. The filtering function is usually in the form of a window in the frequency domain. Depending on the frequency inclusion of the window in the frequency axis, the filters are named low-pass (LP), high-pass (HP), or band-pass (BP).

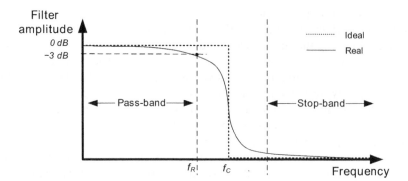

FIGURE 1.6   An ideal and real LP filter characteristics.

The frequency characteristics of an ideal LP filter are depicted as the dotted line in Figure 1.6. Ideally, this filter should pass frequencies from DC to the cutoff frequency of f_c and should stop higher frequencies beyond. In real practice, however, ideal LP filter characteristics cannot be realized. According to the Fourier theory, *a signal cannot be both time limited and band limited.* That is to say, to be able to achieve an ideal band-limited characteristic as in Figure 1.6, then the corresponding time-domain signal should theoretically extend from minus infinity to plus infinity which is of course impossible for realistic applications. Since all practical human-made signals are time limited, that is, they should start and stop at specific time instances, the frequency contents of these signals normally extend to infinity. Therefore, an ideal filter characteristic as the one in Figure 1.6 cannot be realizable, but only the approximate versions of it can be implemented in real applications. The best implementation of practical LP filter characteristic was achieved by Butterworth [16] and Chebyshev [17]. The solid line in Figure 1.6 demonstrates a real LP filter characteristic of Butterworth type.

Windowing procedure is usually applied to smooth a time-domain signal, therefore filtering out higher frequency components. Some of the popular windows that are widely used in signal and image processing are Kaiser, Hanning, Hamming, Blackman, and Gaussian. A comparative plot of some of these windows is given in Figure 1.7.

The effect of a windowing operation is illustrated in Figure 1.8. A time-domain signal of a rectangular window is shown in Figure 1.8a, and its FT is provided in Figure 1.8b. This function is in fact a *sinc (sinus cardinalis)* function and has major sidelobes. For the *sinc* function, the highest sidelobe is approximately 13 dB lower than the apex of the main lobe. This much of contrast, of course, may not be sufficient in some imaging applications. As shown in Figure 1.8c, the original rectangular time-domain signal is Hanning windowed. Its corresponding spectrum is depicted in Figure 1.8d where the sidelobes are highly suppressed, thanks to the windowing operation. For this example, the

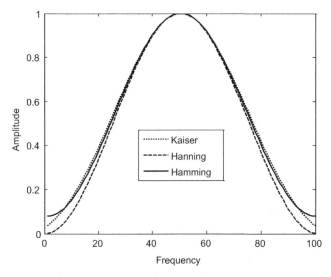

FIGURE 1.7   Some common window characteristics.

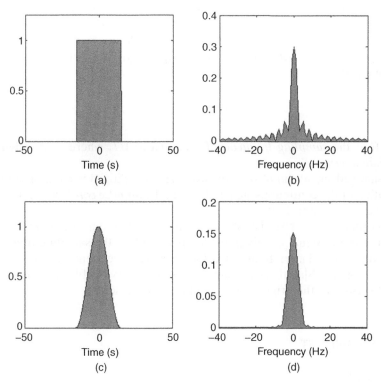

FIGURE 1.8   Effect of windowing: (a) rectangular time signal, (b) its Fourier spectrum: a *sinc* signal, (c) Hanning windowed time signal, (d) corresponding frequency-domain signal.

highest sidelobe level is now 32 dB below the maximum value of the main lobe, which provides better contrast when compared to the original, nonwindowed signal.

A main drawback of windowing is the resolution decline in the frequency signal. The FT of the windowed signal has worse resolution than the FT of the original time domain signal. This feature can also be noticed from the example in Figure 1.8. By comparing the main lobes of the figures on the right, the resolution after windowing is almost twice as bad when compared to the original frequency domain signal. A comprehensive examination of windowing procedure will be presented later on, in Chapter 5.

1.6 DATA SAMPLING

Sampling can be regarded as the preprocess of transforming a continuous or analog signal to a discrete or digital signal. When the signal analysis has to be done using digital computers via numerical evaluations, continuous signals need to be converted to the digital versions. This is achieved by applying the common procedure of sampling. Analog-to-digital (A/D) converters are common electronic devices to accomplish this process. The implementation of a typical sampling process is shown in Figure 1.9. A time signal $s(t)$ is sampled at every T_s seconds such that the discrete signal, $s[n]$, is generated via the following equation:

$$s[n] = s(nT_s), \quad n = 0, 1, 2, 3, \ldots \tag{1.20}$$

Therefore, the *sampling frequency*, f_s, is equal to $1/T_s$ where T_s is called the *sampling interval*.

A sampled signal can also be regarded as the digitized version of the multiplication of the continuous signal, $s(t)$, with the *impulse comb* waveform, $c(t)$, as depicted in Figure 1.10.

According to the Nyquist–Shannon sampling theorem, the perfect reconstruction of the signal is only possible provided that the sampling frequency, f_s, is equal to or larger than twice the maximum frequency content of the sampled signal [18]. Otherwise, signal aliasing is unavoidable, and only a distorted version of the original signal can be reconstructed.

1.7 DFT AND FFT

1.7.1 DFT

As explained in Section 1.1, the FT is used to transform continuous signals from one domain to another. It is usually used to describe the continuous

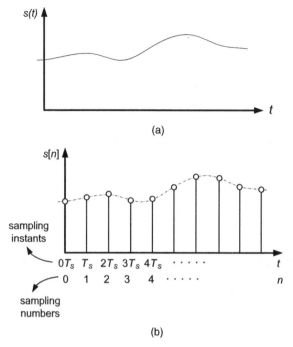

FIGURE 1.9   Sampling: (a) continuous time signal, (b) discrete time signal after the sampling.

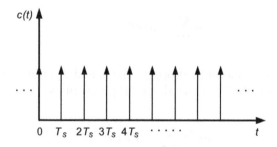

FIGURE 1.10   Impulse comb waveform composed of ideal impulses.

spectrum of an aperiodic time signal. To be able to utilize the FT while working with digital signals, the digital or DFT has to be used.

Let $s(t)$ be a continuous periodic time signal with a period of $T_o = 1/f_o$. Then, its sampled (or discrete) version is $s[n] \triangleq s(nT_s)$ with a period of $NT_s = T_o$ where N is the number of samples in one period. Then, the Fourier integral in Equation 1.1 will turn to a summation as shown below:

$$S(kf_o) = \sum_{n=0}^{N-1} s(nT_s) \cdot e^{-j2\pi(kf_o)(nT_s)}$$

$$= \sum_{n=0}^{N-1} s(nT_s) \cdot e^{-j2\pi\left(\frac{k}{NT_s}\right)(nT_s)} \qquad (1.21)$$

$$= \sum_{n=0}^{N-1} s(nT_s) \cdot e^{-j2\pi\frac{k}{N}n}.$$

Dropping the f_o and T_s inside the parenthesis for the simplicity of nomenclature and therefore switching to discrete notation, DFT of the discrete signal $s[n]$ can be written as

$$S[k] = \sum_{n=0}^{N-1} s[n] \cdot e^{-j2\pi\frac{n}{N}k} \qquad (1.22)$$

In a dual manner, let $S(f)$ represent a continuous periodic frequency signal with a period of $Nf_o = N/T_o$ and let $s[k] \triangleq s(kf_o)$ be the sampled signal with the period of $Nf_o = f_s$. Then, the inverse discrete Fourier transform (IDFT) of the frequency signal $S[k]$ is given by

$$s\left(\frac{n}{f_s}\right) = \sum_{k=0}^{N-1} S(kf_o) \cdot e^{j2\pi(kf_o)\left(\frac{n}{f_s}\right)}$$

$$s(nT_s) = \sum_{k=0}^{N-1} S(kf_o) \cdot e^{j2\pi(kf_o)\left(\frac{n}{Nf_o}\right)} \qquad (1.23)$$

$$= \sum_{k=0}^{N-1} S(kf_o) \cdot e^{j2\pi\frac{n}{N}k}.$$

Using the discrete notation by dropping the f_o and T_s inside the parenthesis, the IDFT of a discrete frequency signal $S[k]$ is given as

$$s[n] = \sum_{k=0}^{N-1} S[k] \cdot e^{j2\pi\frac{k}{N}n} \qquad (1.24)$$

1.7.2 FFT

FFT is the efficient and fast way of evaluating the DFT of a signal. Normally, computing the DFT is in the order of N^2 arithmetic operations. On the other hand, fast algorithms like Cooley–Tukey's FFT technique produce arithmetic operations in the order of Nlog (N) [4, 19, 20]. An example of DFT is given in Figure 1.11 where a discrete time-domain ramp signal is plotted in Figure 1.11a, and its frequency-domain signal obtained by an FFT algorithm is given in Figure 1.11b.

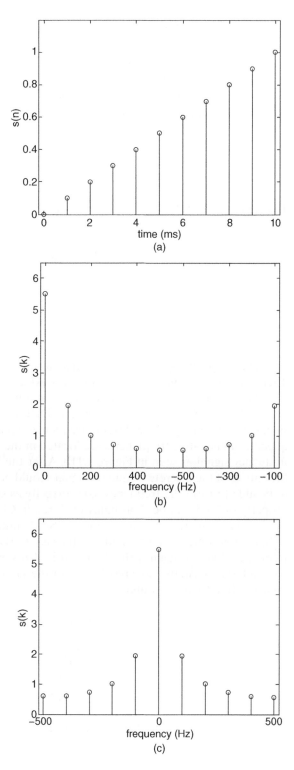

FIGURE 1.11 An example of DFT operation: (a) discrete time-domain signal, (b) discrete frequency-domain signal without FFT shifting, (c) discrete frequency-domain signal with FFT shifting.

1.7.3 Bandwidth and Resolutions

The duration, the bandwidth, and the resolution are important parameters while transforming signals from time domain to frequency domain or vice versa. Considering a discrete time-domain signal with a duration of $T_o = 1/f_o$ sampled N times with a sampling interval of $T_s = T_o/N$, the frequency resolution (or the sampling interval in frequency) after applying the DFT can be found as

$$\Delta f = \frac{1}{T_o}. \tag{1.25}$$

The spectral extend (or the frequency bandwidth) of the discrete frequency signal is

$$\begin{aligned} B &= N \cdot \Delta f \\ &= \frac{N}{T_o} \\ &= \frac{1}{T_s}. \end{aligned} \tag{1.26}$$

For the example in Figure 1.11, the signal duration is 1 ms with $N = 10$ samples. Therefore, the sampling interval is 0.1 ms. After applying the expressions in Equations 1.25 and 1.26, the frequency resolution is 100 Hz, and the frequency bandwidth is 1000 Hz. After taking the DFT of the discrete time-domain signal, the first entry of the discrete frequency signal corresponds to zero frequency, and negative frequencies are located in the second half of the discrete frequency signal as seen in Figure 1.11b. After the DFT operation, therefore, the entries of the discrete frequency signal should be swapped from the middle to be able to form the frequency axis correctly as shown in Figure 1.11c. This property of DFT will be thoroughly explored in Chapter 5 to demonstrate its use in inverse synthetic aperture radar (ISAR) imaging.

Similar arguments can be made for the case of IDFT. Considering a discrete frequency-domain signal with a bandwidth of B sampled N times with a sampling interval of Δf, the time resolution (or the sampling interval in time) after applying IDFT can be found as

$$\begin{aligned} \Delta t &= T_s \\ &= \frac{1}{B} \\ &= \frac{1}{N\Delta f}. \end{aligned} \tag{1.27}$$

The time duration of the discrete time signal is

$$T_o = \frac{1}{\Delta f}. \tag{1.28}$$

For the frequency-domain signal in Figure 1.11b or c, the frequency bandwidth is 1000 Hz with $N = 10$ samples. Therefore, the sampling interval in frequency is 100 Hz. After applying IDFT to get the time-domain signal as in Figure 1.11a, the formulas in Equations 1.27 and 1.28 calculate the resolution in time as 0.1 ms and the duration of the signal as 1 ms.

1.8 ALIASING

Aliasing is a type of signal distortion due to undersampling the signal of interest. According to Nyquist–Shannon sampling theorem [18], the sampling frequency, f_s, should be equal to or larger than twice the maximum frequency content, f_{max}, of the signal to be able to perfectly reconstruct the signal:

$$f_s \geq 2f_{max}. \tag{1.29}$$

Since processing the analog radar signal requires sampling of the received data, the concept of aliasing should be taken into account when dealing with radar signals.

1.9 IMPORTANCE OF FT IN RADAR IMAGING

The imaging of a target using electromagnetic waves emitted from radars is mainly based on the phase information of the scattered waves from the target. This is because of the fact that the phase is directly related to the range distance of the target. In the case of monostatic radar configuration as shown in Figure 1.12a, let the scattering center on the target be at R distance away from the radar.

The scattered field E^s from this scattering center on the target has a complex scattering amplitude, A, and a phase factor that contains the distance information of the target as follows:

$$E^s(k) \cong A \cdot e^{-j2kR}. \tag{1.30}$$

As is obvious from Equation 1.30, there exists a Fourier relationship between the wave number, k, and the distance, R. Provided that the scattered field is collected over a bandwidth of frequencies (Fig. 1.12b), it is possible to pinpoint the distance R by Fourier transforming the scattered field data as depicted in Figure 1.12c. The plot of scattered field versus range is called the *range profile*, which is an important phenomenon in radar imaging. Range profile is in

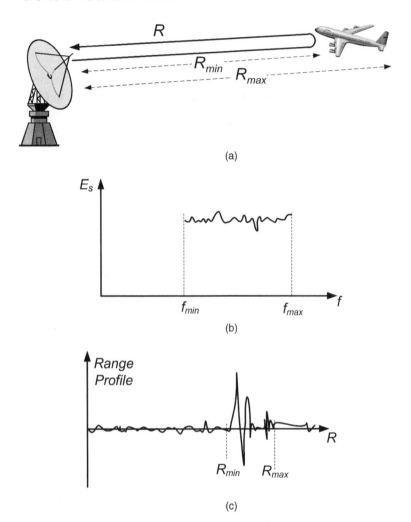

FIGURE 1.12 (a) Monostatic radar configuration, (b) scattered field versus frequency, (c) range profile of the target.

fact nothing but the one-dimensional range image of the target. An example is illustrated in Figure 1.13 where the range profile of an airplane is shown. The concept of range profiling will be thoroughly investigated in Chapter 4, Section 4.3.

Another main usage of FT in radar imaging is the ISAR imaging. In fact, ISAR can be regarded as the 2D range and cross-range profile image of a target. While the range resolution is achieved by utilizing the frequency diversity of the backscattered signal, the cross-range resolution is gathered by collecting the backscattered signal over different look angles of the target. An example of ISAR imaging for the same airplane is demonstrated in Figure

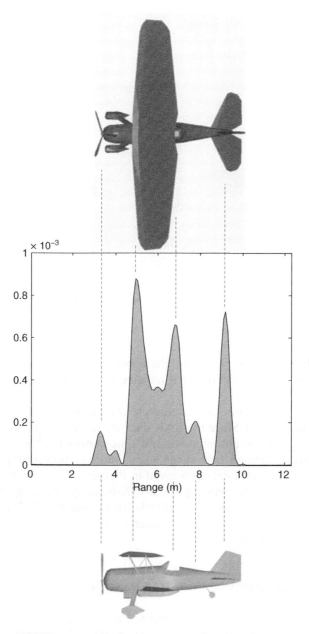

FIGURE 1.13   Simulated range profile of an airplane.

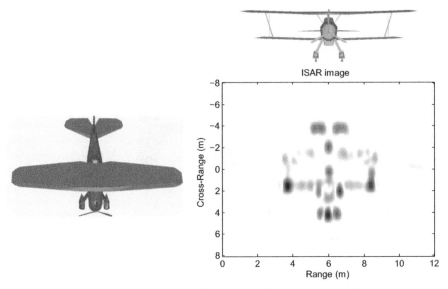

FIGURE 1.14 Simulated two-dimensional ISAR image of an airplane.

1.14 where both the CAD view and the constructed ISAR image for the airplane is shown. The concept of ISAR imaging will be examined in great detail in Chapter 4.

The FT operations are also extensively used in synthetic aperture radar (SAR) imaging as well. Since the SAR data are usually huge and processing this amount of data is an extensive and time-consuming task, the FTs are usually utilized to achieve both range and azimuth compression procedures.

An example of SAR imagery is given in Figure 1.15 where the image was acquired by spaceborne imaging radar-C/X-band synthetic aperture radar (SIR-C/X-SAR) onboard the space shuttle Endeavour in 1994 [21]. This SAR image covers an area of Cape Cod, Massachusetts. The details of SAR imagery will be explored in Chapter 3.

1.10 EFFECT OF ALIASING IN RADAR IMAGING

In radar applications, the data are collected within a finite bandwidth of frequencies. According to the sampling theory, if the radar signal is $g(t)$ and its spectrum is $G(f)$, and the frequency components beyond a specific frequency B are zero, that is,

$$G(f) = \begin{cases} \neq 0 & ;|f| < B \\ 0 & ;|f| \geq B, \end{cases} \qquad (1.31)$$

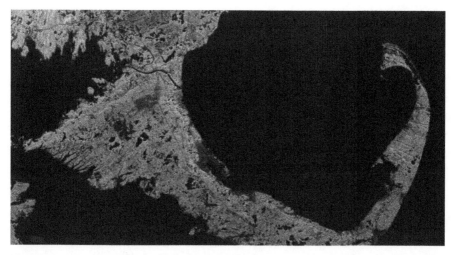

FIGURE 1.15   SAR image of the famous "hook" of Cape Cod, Massachusetts, USA [21].

then the time-domain signal $g(t)$ should be sampled at least twice the bandwidth in frequency as

$$f_s = 2B, \tag{1.32}$$

where f_s stands for the sampling frequency.

When the radar imaging is concerned, the scattered electric field has the form as given in Equation 1.30. The sampling theorem can be applied in the following manner:

Suppose that target to be imaged lies in the range direction within the range width or range extend of R_{max} such that

$$-\frac{R_{max}}{2} \leq r \leq \frac{R_{max}}{2} \tag{1.33}$$

as depicted in Figure 1.16b. In the case of imaging radar, this figure represents the range profile of the target. The FT of $g(r)$ represents its spectrum that theoretically extends to infinity in the frequency axis (see Fig. 1.16a). The main problem is to get the digitized (or sampled) versions of $g(r)$ from digitized (or sampled) versions of $G(k)$ or $G(f)$ with adequate samples so that no aliasing occurs. Here, k stands for the wave number and is related to the operating frequency as

$$k = 2\pi f / c \tag{1.34}$$

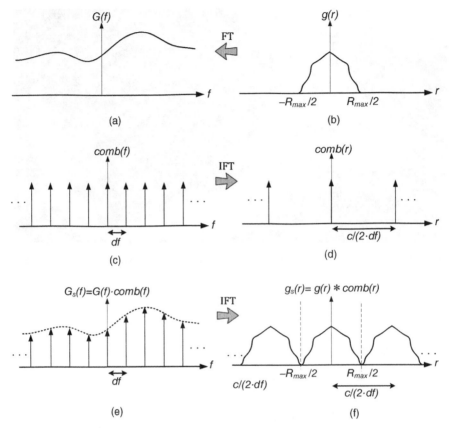

FIGURE 1.16 The Nyquist sampling procedure for getting unaliased range image: (a) frequency-domain radar signal for a range windowed data, (b) range domain signal, (c) sampling comb signal in frequency domain, (d) its range domain equivalent, (e) critically sampled version of the frequency-domain signal, and (f) its range domain equivalent.

where c is the speed of light. Utilizing the relationship between the wave number, k, and the distance, R, in Equation 1.30, the sampling in the wave-number domain should satisfy the following inequality:

$$dk \leq \frac{2\pi}{2R_{max}}$$
$$= \frac{\pi}{R_{max}}. \tag{1.35}$$

This inequality is forced by the famous Nyquist sampling condition. The minimum sampling frequency, then, should be equal to

$$(df)_{min} = \frac{c}{2\pi}(dk)_{min}$$

$$= \frac{c}{2\pi} \cdot \frac{\pi}{R_{max}} \qquad (1.36)$$

$$= \frac{c}{2R_{max}}.$$

Then, the sampled version of frequency-domain signal $G(f)$ is obtained by the multiplication of the $G(f)$ with the following impulse comb function:

$$comb(f) = \sum_n \delta(f - n \cdot df). \qquad (1.37)$$

The plot of this comb function is shown in Figure 1.16c. By taking the IFT of this comb function, we can get another comb function in the range domain as

$$comb(r) = \mathcal{F}^{-1}\{comb(f)\}$$
$$= \mathcal{F}^{-1}\left\{\sum_n \delta(f - n \cdot df)\right\} \qquad (1.38)$$
$$= \frac{c}{2df} \sum_n \delta\left(r - n \cdot \frac{c}{2df}\right)$$

as depicted in Figure 1.16d. The sampled version of frequency-domain signal, $G_s(f)$, can be obtained by multiplying the original frequency-domain signal, $G(f)$, with the impulse comb function in Figure 1.16c as

$$G_s(f) = G(f) \cdot comb(f)$$
$$= \sum_n G(f) \cdot \delta(f - n \cdot df), \qquad (1.39)$$

which is shown in Figure 1.16e. The range domain equivalent of the frequency-domain sampled signal can be found via inverse Fourier transformation as

$$g_s(r) = \mathcal{F}^{-1}\{G_s(f)\}$$
$$= \mathcal{F}^{-1}\{G(f) \cdot comb(f)\}$$
$$= g(r) * comb(r)$$
$$= g(r) * \frac{c}{2df} \sum_n \delta\left(r - n \cdot \frac{c}{2df}\right) \qquad (1.40)$$
$$= \frac{c}{2df} \sum_n g\left(r - n \cdot \frac{c}{2df}\right),$$

where "*" is the convolution operation. Therefore, the resultant range domain signal is periodic with $(c/2df)$ intervals. If df is chosen to be equal to

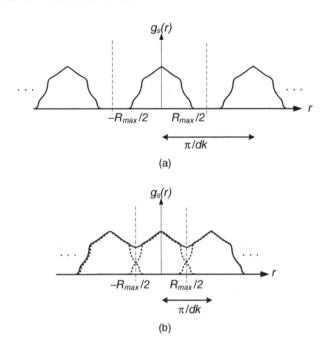

FIGURE 1.17 The effect of sampling rate: (a) no aliasing due to oversampling and (b) aliased or distorted range domain waveform due to undersampling.

$(df)_{min} = c/2R_{max}$, the period of $g_s(r)$ becomes R_{max} as illustrated in Figure 1.16f. If $G(f)$ is sampled and found to be finer than $(df)_{min}$, that is, over-sampled, no aliasing occurs, and the resultant $g_s(r)$ signal will be similar to the one in Figure 1.17a. Therefore, when the DFT is used, the original range domain signal falls within one period of $g_s(r)$ and can be recovered without any distortion. When a sampling rate of $df \geq (df)_{min}$ is used, $g_s(r)$ signal is aliased as demonstrated in Figure 1.17b, and the original range domain signal $g(r)$ is distorted within one period of $(c/2df)$. Therefore, the recovery of $g(r)$ is not possible due to undersampling of $G(f)$. The effect of aliasing in ISAR imaging will be covered in Chapter 5, Section 5.2.

1.11   MATLAB CODES

Below are the Matlab source codes that were used to generate all of the Matlab-produced Figures in Chapter 1. The codes are also provided in the CD that accompanies this book.

Matlab code 1.1: Matlab file "Figure1-1.m"

```
%---------------------------------------------------------------
% This code can be used to generate Figure 1_1
```

```
%---------------------------------------------------------------
% This file requires the following files to be present in the
same
% directory:
%
%prince.wav

clear all
close all

% Read the sound signal "prince.wav"
[y,Fs,bits] = wavread('prince.wav');
sound(y,Fs);
N = length(y);

% TIME DOMAIN SIGNAL
t = 0:.8/(N-1):.8;
plot(t,y,'k'); %downsample for plotting
set(gca,'FontName', 'Arial', 'FontSize',14,'FontWeight','Bold');
axis tight;
xlabel('Time [s]');
ylabel('Amplitude');
title('time domain signal');
```

Matlab code 1.2: Matlab file "Figure1-2.m"
```
%---------------------------------------------------------------
% This code can be used to generate Figure 1.2
%---------------------------------------------------------------
% This file requires the following files to be present in the
same
% directory:
%
%prince.wav

clear all
close all

% Read the sound signal "prince.wav"
[y,Fs,bits] = wavread('prince.wav');
sound(y,Fs); %play the sound
N = length(y);

t = 0:.8/(N-1):.8; %form time vector

% FREQUENCY DOMAIN SIGNAL
Y = fft(y)/N;
% Calculate the spectrum of the signal
df = 1/(max(t)-min(t)); % Find the resolution in frequency
f = 0:df:df*(length(t)-1); % Form the frequency vector
```

```
plot(f(1:2:N),abs(Y(1:(N+1)/2)),'k') %downsample for plotting
set(gca,'FontName', 'Arial', 'FontSize',14,'FontWeight','Bold');
axis tight;
xlabel('Frequency [Hz]');
ylabel('Amplitude');
title('frequency domain signal');
```

Matlab code 1.3: Matlab file "Figure1-3.m"
```
%----------------------------------------------------------------
% This code can be used to generate Figure 1.3
%----------------------------------------------------------------
% This file requires the following files to be present in the
same
% directory:
%
%prince.wav

clear all
close all

% Read the sound signal "prince.wav"
[y,Fs,bits] = wavread('prince.wav');
sound(y,Fs);
N = length(y);

t = 0:.8/(N-1):.8; %form time vector

% TIME FREQUENCY PLANE SIGNAL
A = spectrogram(y,256,250,400,1e4); % Calculate the
spectrogram
matplot(t,f, (abs(A)),30); % Display the signal in T-F
domain
colormap(1-gray); % Change the colormap to grayscale
set(gca,'FontName', 'Arial', 'FontSize',14,'FontWeight','Bold');
xlabel('Time [s]');
ylabel('Frequency [Hz]');
title('signal in time-frequency plane');
```

Matlab code 1.4: Matlab file "Figure1-5.m"
```
%----------------------------------------------------------------
% This code can be used to generate Figure 1.5
%----------------------------------------------------------------
% This file requires the following files to be present in the
same
% directory:
%
% tot30.mat
clear all
close all
```

```
load tot30; % load the measured back-scattered E-field

% DEFINITION OF PARAMETERS
f = linspace(6,18,251)*1e9; %Form frequency vector
BW = 6e9; % Select the frequency window size
d = 2e-9; %Select the time delay

% DISPLAY THE FIELD IN JTF PLANE
[B,T,F] = stft(tot30,f,BW,50,d);
xlabel('--->Time (nsec)');
ylabel('--> Freq. (GHz)');
colorbar;
colormap(1-gray)
set(gca,'FontName', 'Arial', 'FontSize',14,'FontWeight','Bold
');
axis tight;
xlabel('Time [ns]');
ylabel('Frequency [GHz]');
```

Matlab code 1.5: Matlab file "Figure1-8.m"

```
%----------------------------------------------------------------
% This code can be used to generate Figure 1.8
%----------------------------------------------------------------

clear all
close all

%% DEFINE PARAMETERS
t = linspace(-50,50,1001); % Form time vector
df = 1/(t(2)-t(1)); %Find frequency resolution
f = df*linspace(-50,50,1001);% Form frequency vector

%% FORM AND PLOT RECTANGULAR WINDOW
b(350:650) = ones(1,301);
b(1001) = 0;
subplot(221);
h = area(t,b);
set(gca,'FontName', 'Arial', 'FontSize',14,'FontWeight','Bold
');
xlabel('Time [s]');
axis([-50 50 0 1.25])
set(h,'FaceColor',[.5 .5 .5])

subplot(222);
h = area(f,fftshift(abs(ifft(b))));
set(gca,'FontName', 'Arial', 'FontSize',14,'FontWeight','Bold
');
xlabel('Frequency [Hz]')
axis([-40 40 0 .4])
```

```
set(h,'FaceColor',[.5 .5 .5])

%% FORM AND PLOT HANNING WINDOW
bb = b;
bb(350:650) = hanning(301)';

subplot(223);
h = area(t,bb);
set(gca,'FontName', 'Arial', 'FontSize',14,'FontWeight','Bold
');
xlabel('Time [s]');
axis([-50 50 0 1.25])
set(h,'FaceColor',[.5 .5 .5])

subplot(224);
h = area(f,fftshift(abs(ifft(bb))));
set(gca,'FontName', 'Arial', 'FontSize',14,'FontWeight','Bold
');
xlabel('Frequency [Hz]')
axis([-40 40 0 .2])
set(h,'FaceColor',[.5 .5 .5])
```

Matlab code 1.6: Matlab file "Figure1-11.m"
```
%---------------------------------------------------------------
% This code can be used to generate Figure 1.11
%---------------------------------------------------------------
clear all
close all

%--- Figure 1.11(a) --------------------------------------------
% TIME DOMAIN SIGNAL
a = 0:.1:1;
t = (0:10)*1e-3;
stem(t*1e3,a,'k','Linewidth',2);
set(gca,'FontName', 'Arial', 'FontSize',14,'FontWeight','Bold
');
xlabel('time [ms]'); ylabel('s[n]');axis([-0.2 10.2 0 1.2]);

% FREQUENCY DOMAIN SIGNAL
b = fft(a);
df = 1./(t(11)-t(1));
f = (0:10)*df;
ff = (-5:5)*df;
%--- Figure 1.11(b) --------------------------------------------
Figure;
stem(f,abs(b),'k','Linewidth',2);
set(gca,'FontName', 'Arial', 'FontSize',14,'FontWeight','Bold
');
```

```
xlabel('frequency [Hz]'); ylabel('S[k]');axis([-20 1020 0
6.5]);
set(gca,'FontName', 'Arial', 'FontSize',12,'FontWeight',
'Bold');

%--- Figure 1.11(c)------------------------------------------------
Figure;
stem(ff,fftshift(abs(b)),'k','Linewidth',2);
set(gca,'FontName', 'Arial', 'FontSize',14,'FontWeight','Bold
');
xlabel('frequency [Hz]'); ylabel('S[k]');axis([-520 520 0
6.5]);
set(gca,'FontName', 'Arial', 'FontSize',12,'FontWeight',
'Bold');
```

REFERENCES

1 J. Fourier. *The analytical theory of heat.* Dover Publications, New York, 1955.

2 C. Runge. *Zeit für Math und Physik* 48 (1903) 433.

3 G. C. Danielson and C. Lanczos. Some improvements in practical Fourier analysis and their application to X-ray scattering from liquids. *J Franklin Inst* 233 (1942) 365.

4 J. W. Cooley and J. W. Tukey. An algorithm for the machine calculation of complex Fourier series. *Math Comput* 19 (1965) 297–301.

5 From http://www.ling.ohio-state.edu/~cclopper/courses/prince.wav (accessed at 08.10.2007).

6 J. B. Allen. Short term spectral analysis, synthesis, and modification by discrete Fourier transform. *IEEE Trans Acoust* ASSP-25 (1977) 235–238.

7 A. T. Nuttall. Wigner distribution function: Relation to short-term spectral estimation, smoothing, and performance in noise, Technical Report 8225, Naval Underwater Systems Center, 1988.

8 L. Du and G. Su. Target number detection based on a order Choi-Willams distribution, signal processing and its applications. Proceedings, Seventh International Symposium on Volume 1, Issue, 1–4 July 2003, pp. 317–320, vol.1, 2003.

9 L. Cohen. Time frequency distribution—A review. *Proc IEEE* 77(7) (1989) 941–981.

10 S. Qian and D. Chen. *Joint time-frequency analysis: Methods and applications.* Prentice Hall, New Jersey, 1996.

11 V. C. Chen and H. Ling. *Time-frequency transforms for radar imaging and signal processing.* Artech House, Norwood, MA, 2002.

12 L. C. Trintinalia and H. Ling. Interpretation of scattering phenomenology in slotted waveguide structures via time-frequency processing. *IEEE Trans Antennas Propagat* 43 (1995) 1253–1261.

13 A. Filindras, U. O. Larsen, and H. Ling. Scattering from the EMCC dielectric slabs: Simulation and phenomenology interpretation. *J Electromag Waves Appl* 10 (1996) 515–535.

14 C. Özdemir and H. Ling. Joint time-frequency interpretation of scattering phenomenology in dielectric-coated wires. *IEEE Trans Antennas Propagat* 45(8) (1997) 1259–1264.

15 J. H. Richmond and E. H. Newman. Dielectric-coated wire antennas. *Radio Sci* 11 (1976) 13–20.

16 R. W. Daniels. *Approximation methods for electronic filter design*. McGraw-Hill, New York, 1974.

17 A. B. Williams and F. J. Taylors. *Electronic filter design handbook*. McGraw-Hill, New York, 1988.

18 C. E. Shannon. Communication in the presence of noise. *Proc Inst Radio Eng* 37(1) (1949) 10–21.

19 N. Brenner and C. Rader. A new principle for fast Fourier transformation. *IEEE Trans Acoust* 24 (1976) 264–266.

20 P. Duhamel. Algorithms meeting the lower bounds on the multiplicative complexity of length-2n DFTs and their connection with practical algorithms. *IEEE Trans Acoust* 38 (1990) 1504–1511.

21 From internet, http://www.jpl.nasa.gov/radar/sircxsar/capecod2.html (accessed at 01.07.2008).

CHAPTER TWO

Radar Fundamentals

2.1 ELECTROMAGNETIC (EM) SCATTERING

Scattering is the physical phenomenon that occurs when an EM wave hits a discontinuity/nonuniformity or an object. The deviation of the wave trajectory or path is generally known as *scattering*.

The classification of scattering phenomena can be made according to the size of the scattering object (or the scatterer) with respect to the wavelength of the EM wave. Radar signals reflect or scatter in different ways depending on the wavelength of the EM wave and the shape of the object (scatterer). If the wavelength of the EM wave is much smaller than the size of the scatterer, the EM wave bounces back in a similar way to how light reflects from a large surface. This type of scattering is often called *scattering in the optical region* [1]. If the EM wave's wavelength is of comparable size of the scatterer, such as within a few wavelengths, some resonances may occur, and the scattering intensity may fluctuate considerably for different frequencies. The scattering direction is mainly affected by the incident wave direction in this region that is usually called *Mie region* (or *resonant region*) [2, 3]. This particular scattering type is therefore called *Mie scattering*. If the wavelength of the EM wave is much longer than the size of the scatterer, the wave is dispersed around the scatterer. This type of reflection is named as *Rayleigh scattering* [2, 3].

Scattering types can also be classified according to the wave's trajectory from different structures that may have planar surfaces, curved surfaces, corners, edges, or tips. Below is the list of such scattering types that we may observe in most radar applications:

Inverse Synthetic Aperture Radar Imaging with MATLAB Algorithms, First Edition.
Caner Özdemir.
© 2012 John Wiley & Sons, Inc. Published 2012 by John Wiley & Sons, Inc.

1. *Specular Scattering.* When the EM wave experiences a mirror-like reflection from a planar surface, this phenomenon is usually called *specular reflection* or *perfect reflection* (illustrated in Fig. 2.1a). When the incoming wave makes an angle θ_i with the surface normal, the direction of the outgoing scattered wave makes the same angle $\theta_r = \theta_i$ with the surface normal, obeying Snell's law of reflection. Scattering of the EM wave from sufficiently large perfectly conducting plates is a good example of this type of scattering.

2. *Multiple Scattering.* When the EM wave experiences multiple bounces around the object, it is called *multiple scattering*. The reflections of EM waves from dihedral (Fig. 2.1b) and trihedral (Fig. 2.1c) corner reflectors are good examples of this type of scattering. If the reflectors are perfect electric conductors (PECs), the wave undergoes the laws of specular reflection at each plate and leaves the object in one outgoing direction. Most of the energy is reflected back in the opposite direction of the incoming wave if the angle between the plates is 90°.

3. *Surface Scattering.* This type of scattering mechanism is demonstrated in Figure 2.1d. Since the surface of the object is not flat in general, the wave scatters in various directions. For PECs, the direction of the outgoing ray of wave is again determined by Snell's law. Since the normal direction of the surface changes for different points on the surface of the object, the direction of the reflection ray also changes such that the EM wave scatters in various directions.

4. *Edge/Tip Scattering.* Another type of EM scattering occurs when an EM wave impinges with an edge or a corner (Fig. 2.1e) or a tip (Fig. 2.1f) of an object. For such situations, the wave scatters in all directions, and this type of reflection is often called *diffraction*. When an EM wave hits an edge, some energy reflects in the backward direction, some scatters in the direction that makes Snell's angle with one plate, some scatters in the direction that makes Snell's angle with the other plate, and the rest of the energy reflects in all other directions.

5. *Traveling Wave Scattering.* When the wavelength is of comparable order with the object size, traveling waves occur along the object. This wave travels along the object, and the scattering occurs when the wave impinges upon the discontinuity at the edge of the target as illustrated in Figure 2.1g. The multiple scattering from wires is a good example of this type of scattering.

EM scattering is, of course, the key inspiration for radar theory and radar imaging. In fact, what imaging radar displays is nothing but the scattered energy from a target or a scene. With this understanding, imaging radar can be thought of as analogous to an optical imaging system. While the optical image displays the light reflectivity of an object or a scene, the radar image displays the EM reflectivity or scattering from a target or a scene. In radar

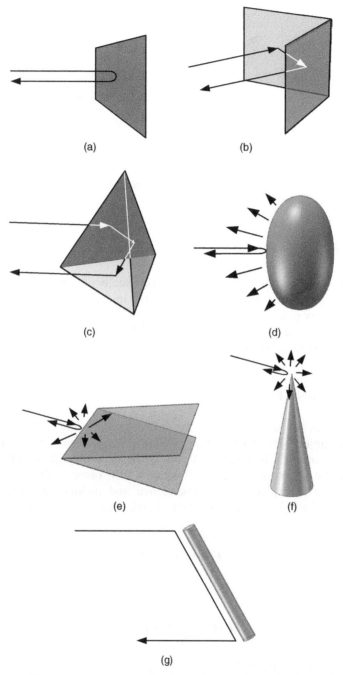

FIGURE 2.1 Different scattering mechanisms: (a) specular reflection, (b) scattering from dihedral, (c) scattering from trihedral, (d) surface scattering, (e) edge diffraction, (f) tip diffraction, and (g) traveling wave scattering.

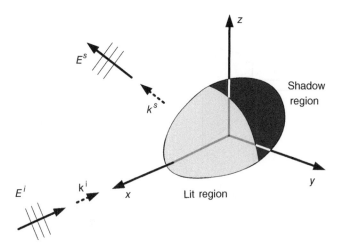

FIGURE 2.2 Electromagnetic scattering from a perfectly conducting object.

images, the scattering from canonical targets (usually man-made objects) provides mostly specular reflections. Therefore, the image consists of a number of localized energy points that are also known as *scattering centers*. Especially at high frequencies, the scattering center model provides various advantages, which will be studied in Chapter 7.

2.2 SCATTERING FROM PECs

In this section, we will derive the far-field EM scattering from a perfectly conducting object. This formulation is closely parallel with the derivations that can be found in References 4 and 5. The geometry of the problem is assumed to be as shown in Figure 2.2 where a PEC object is illuminated by a plane wave with $exp(j\omega t)$ time dependence and incident wave vector of $\mathbf{k^i} = k_o\hat{k}^i = k_x^i\hat{x} + k_y^i\hat{y} + k_z^i\hat{z}$. The incident electric and magnetic wave fields are given by

$$E^i(r) = \hat{u}E_o e^{-j\mathbf{k^i \cdot r}} \qquad (2.1)$$

and

$$H^i(r) = \frac{1}{\omega\mu} k^i \times E^i(r), \qquad (2.2)$$

where E_o and $\hat{u}$ give the magnitude and the polarization direction of the incident electric field. According to the physical optics (PO) theory [6], the current induced on the surface of the object is given by the following well-known formula:

$$J_s(r') = \begin{cases} 2\hat{n}(r') \times H^i(r') & \text{lit region} \\ 0 & \text{shadow region.} \end{cases} \tag{2.3}$$

Here, $\hat{n}(r')$ is the outward unit vector of the illuminated object's surface. The vector r' is defined from the origin to any point on the illuminated surface (S_{lit}) of the object. The scattered electric field at the far-field region along the observation vector of r is given by

$$E^s(r) = -j\omega\mu \iint_{S_{lit}} J_s(r') \left(\frac{e^{-jk_o r}}{4\pi r} e^{jk^s \cdot r'} \right) d^2 r', \tag{2.4}$$

where $k^s = k_o \hat{r}$ is the wavenumber vector in the scattering direction. Putting Equation 2.3 into Equation 2.4 will yield the scattered field equation in terms of the incident electric field as

$$E^s(r) = -\frac{jk_o E_o}{4\pi r} e^{-jk_o r} \iint_{S_{lit}} 2\hat{n}(r') \times \left(\hat{k}^i \times \hat{u} \right) e^{j(k^s - k^i) \cdot r'} d^2 r'. \tag{2.5}$$

This is the scattered electric field in the far field using the physical optics approximation. As will be explored in Chapter 4, this formula will constitute a basis for the derivation of ISAR imaging of objects.

2.3   RADAR CROSS SECTION (RCS)

RCS can be regarded as the measure of the EM energy intercepted and reradiated by an object (or target). The unit of RCS is square-meters (m^2). An RCS parameter is generally used to categorize the object's EM reflectivity or ability to scatter the EM energy for a particular direction and at a particular frequency.

RCS is the main parameter in detection of airplanes, ships, tanks, or, more commonly, military targets. The low observable (or stealth) aircrafts are designed to give very low RCS values so that they cannot be detected by the opponent's radar. These special design features for low-observable platforms include planar body surfaces that reflect the incoming wave to another direction and special radar absorbing material (RAM)-based coating or paint. Passenger airplanes, on the other hand, may have high RCS features due to bare metallic surfaces that reflect almost all of the incoming energy, rounded shape design that scatters the EM wave almost in all directions, and canonical shapes and cavities (such as engine ducts) that scatter the incident field in the backward direction. Furthermore, they are also big in size compared to fighters. Therefore, their RCS values are much higher than specially designed stealth aircrafts.

The concept of RCS plays an important role in radar imaging. In fact, monostatic inverse synthetic aperture radar (ISAR) images of a target are

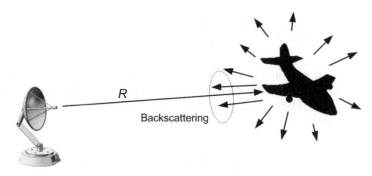

FIGURE 2.3   The EM energy scatters in all directions when it hits a target.

constructed by backscattered field measurements (or RCS measurements). If the backscattered RCS measurements from the target for different frequencies and different look angles are transformed back to where the scattering phenomena originate, what we basically obtained is the "ISAR image" of that target.

2.3.1   Definition of RCS

RCS can be briefly described as the effective echoing area of the target when it is illuminated by a plane wave as illustrated in Figure 2.3. RCS can also be thought of as the measure of an object's ability to scatter EM signals in the direction of the radar receiver. According to the IEEE Standard Definitions of Terms, *RCS* or *the scattering cross section* (*SCS*) or the *echoing area* is defined as the following: "For a given scatterer, upon which a plane wave is incident, that portion of the SCS corresponding to a specified polarization component of the scattered wave" [7].

A more formal definition of the RCS (σ) of an object can be made as the following: "It is the equivalent area intercepting the amount of power that, when scattered isotropically, produces at the radar receiver a power density W^s that is equal to the density scattered by the actual object" [4].

The formula of RCS can be derived easily, as follows: As depicted in Figure 2.3, the object (or the target) is located at R distance from the radar. Incident plane wave from the radar transmitter produces a power density of W^i at the target. If the object (or the target) has an RCS area of σ, the power reflected by that object is then equal to

$$P_r = \sigma \cdot W^i. \tag{2.6}$$

According to the above definition of RCS, this reflected power will reradiate in all directions (isotropically). Therefore, the power density W^s of the reflected wave at the radar receiver is

$$W^s = \frac{P_r}{4\pi R^2}$$

$$= \sigma \cdot \frac{W^i}{4\pi R^2}.$$

(2.7)

Therefore, σ in the above equation can be left alone to give

$$\sigma = 4\pi R^2 \frac{W^s}{W^i}.$$

(2.8)

The formal equation of RCS can be easily obtained as the following:

$$\sigma = \lim_{R \to \infty} \left(4\pi R^2 \frac{W^s}{W^i} \right).$$

(2.9)

In an alternative definition of RCS (σ), it is the measure of the ratio of back-scatter power per unit solid angle (*steradian*) along the radar direction to the power density that is intercepted by the object (or target):

$$\sigma = \frac{4\pi \cdot \{scattered\ power\ toward\ radar\ per\ steradian\}}{\{incident\ power\ density\ at\ the\ target\}}.$$

(2.10)

If W^s is the scattered power density from the target, the amount $\{R^2 W^s\}$ gives the scattered power per steradian at the radar receiver. Assuming that the incident power density is W^i, the RCS of a target can then be rewritten as

$$\sigma = 4\pi R^2 \frac{W^s}{W^i}.$$

(2.11)

As the target is located at the far field of radar, the distance R is brought to the infinity in the above equation so that we reach the formula in (2.9). The RCS can also be expressed using the complex electric and magnetic fields at the target (E^i and H^i) and at the radar receiver (E^s and H^s). That is,

$$\sigma = \lim_{R \to \infty} \left(4\pi R^2 \frac{|E^s|^2}{|E^i|^2} \right)$$

or

$$\sigma = \lim_{R \to \infty} \left(4\pi R^2 \frac{|H^s|^2}{|H^i|^2} \right).$$

(2.12)

When the incident EM wave hits the target, the scattered energy radiates in all directions, as illustrated in Figure 2.3. Only the scattered energy in the direction of radar receiver is included in the RCS calculation. If the radar transmitter and the receiver are collocated, the energy collected by the radar is called the *backscattered* energy. This radar configuration provides *monostatic* setup for RCS measurement. If the radar receiver is placed at a different position than the transmitter, the calculated cross section will provide *bistatic RCS*.

The RCS area of a target does not necessarily bear a direct link with the physical cross-sectional area of that target, but depends on other parameters as well. Besides the aperture of the target seen by the radar (or projected cross section), the EM reflectivity characteristics of the target's surface and the directivity of the radar reflection caused by the target's geometric shape are the key parameters that affect the RCS value. Therefore, RCS can be approximated as the multiplication of (1) projected cross section, S; (2) reflectivity, Γ; and (3) directivity, D as

$$\sigma \approx S \cdot \Gamma \cdot D. \tag{2.13}$$

Here, *projected cross section* is the cross-sectional area of the object along the radar look-angle direction. This gives the projected area of the illuminated region of the object by the incident wave. *Reflectivity* of the target gives the amount (percentage) of the intercepted and reradiated EM energy by the target. *Directivity* is the ratio of scattered energy in the radar direction to the scattered energy from an isotropic scatterer (such that this isotropic scattering is uniform in all directions).

The term RCS is used for a specified polarization component of the reflected energy. If the radar is transmitting in vertical (V) polarization, vertically polarized scattered energy is used for the determination of RCS. Similarly, if the radar is transmitting in horizontal (H) polarization, horizontally polarized scattered energy should be used for the calculation of RCS. The term SCS refers to all types of polarization for transmission and reception. Therefore, SCS presents the all possible polarization types, such as V-V, V-H, H-V, and H-H.

It is essential to point out that the RCS of an object is both angle dependent and frequency dependent. As the look angle toward a target changes, the projected cross section of that target generally changes. Depending on the structure and the material of the target, the reflectivity of the target might also change. Overall, the RCS of the target alters as the look angle varies. Similarly, if the frequency of the EM wave changes, the effective electrical size (or projected cross section) of the target changes as well. Since the EM reflectivity is also a frequency-dependent quantity, the RCS of the target also varies as the frequency of the radar changes. Therefore, an RCS of an object is characterized together with the look angle and the particular frequency of operation. It is also important to note that RCS is independent of the target's distance from the radar.

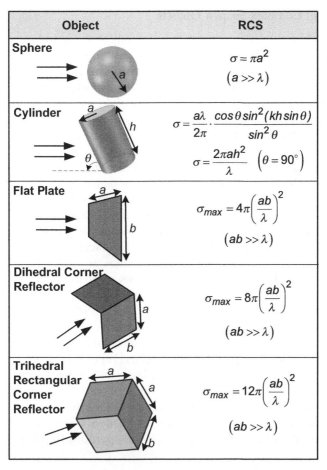

Object	RCS
Sphere	$\sigma = \pi a^2$ $(a \gg \lambda)$
Cylinder	$\sigma = \dfrac{a\lambda}{2\pi} \cdot \dfrac{\cos\theta \sin^2(kh\sin\theta)}{\sin^2\theta}$ $\sigma = \dfrac{2\pi a h^2}{\lambda} \quad (\theta = 90°)$
Flat Plate	$\sigma_{max} = 4\pi\left(\dfrac{ab}{\lambda}\right)^2$ $(ab \gg \lambda)$
Dihedral Corner Reflector	$\sigma_{max} = 8\pi\left(\dfrac{ab}{\lambda}\right)^2$ $(ab \gg \lambda)$
Trihedral Rectangular Corner Reflector	$\sigma_{max} = 12\pi\left(\dfrac{ab}{\lambda}\right)^2$ $(ab \gg \lambda)$

FIGURE 2.4   RCS values for perfectly conducting simple objects (all objects are taken as perfect electrical conductors).

2.3.2   RCS of Simple Shaped Objects

RCS calculation of simply shaped objects can be made via analytical expressions. There has been a lot of interest in formulating the scattered field and the RCS from simple objects such as a sphere, cylinder, ellipsoid, or plate. Since the RCS analyses of such objects are very well documented [8–12], only some of them (those that are important in radar imaging) are listed in Figure 2.4. When usual targets such as airplanes, ships, and tanks are considered, their physical shapes include canonical structures such as corner reflectors, cylinder-like formations, and flat and curved surfaces. Therefore, the knowledge of RCS from basic canonical shapes is important to comprehend the RCS contribution of these substructures on the target to the total RCS.

2.3.3 RCS of Complex Shaped Objects

As given in Figure 2.4, the calculation of canonical shaped objects is analytically formulated. On the other hand, RCS calculation or prediction of complexly shaped objects is usually a difficult task. There exist some numerical approaches by which to estimate the RCS from an arbitrarily shaped object. Some full-wave approaches based on electric field integral equation (EFIE) or magnetic field integral equation (MFIE) techniques [13, 14] are used to calculate the scattering from electrically small targets. The common numerical technique used to implement such approaches is the well-known method of moment (MoM) [15] technique. MoM-based techniques are computationally effective when the electrical size of the target is on the order of, at most, a few wavelengths. At high frequencies when the size of the scatterer is much greater than the wavelength, however, the computation burden of MoM becomes significantly huge such that the computation time and the computation memory requirements are not manageable in simulating the scattering from electrically large and complex targets such as tanks, airplanes, and ships.

To cope with such targets at high frequencies, some hybrid methods [16, 17] that combine different EM techniques in one simulator are used. The most famous one is called the shooting and bouncing ray (SBR) technique that can successfully estimate the scattering from large and complex platforms at high frequencies [18]. The SBR technique efficiently combines the geometric optics (GO) and the physical optics (PO) approaches [19] to get the accurate estimation of the scattering (or the RCS) from large and complex objects at high frequencies and beyond [20, 21]. An example of an RCS calculation from a complex airplane model is shown in Figure 2.5 where the RCS of a 14-m long airplane model is used for the simulation of the monostatic RCS. The calculation is done for the whole azimuth angles at the operating frequency of 2 GHz.

2.4 RADAR RANGE EQUATION

The main goal of a typical radar is to detect the scattered (or backscattered) EM echoes from a target and to extract the information within those EM signatures. The following sequential events happen as the EM wave travels from the transmitter to the receiver:

1 First, the radar signal is generated with the help of the microwave generator (or source).
2 Then the generated signal is transferred to the transmitter by means of transmission lines.
3 The signal is send out via the transmitting antenna.
4 The radar wave travels in the air and reaches the target.

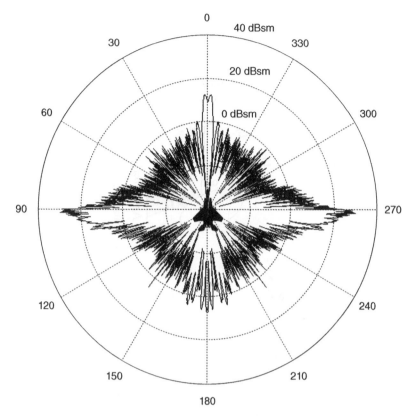

FIGURE 2.5   Simulated RCS (in dBsm) of an aircraft model at 2 GHz as a function of azimuth angles.

5 Only some little portion of the transmitted radar signal is captured by the target and reflected back depending on the RCS value of the target.

6 The reflected wave travels in the air, and only a small fraction of the reflected energy reaches back to the radar receiver.

7 The receiver antenna captures some portion of this energy and passes it to the radar receiver.

8. The radar receiver analyzes this scattered signal to obtain the information about the target that may include its location, size, and velocity.

The *radar range equation* is the mathematical expression that manifests the analysis of what happens to the signal strength while it goes through the above processes.

2.4.1   Bistatic Case

A concept sketch of the above events for the bistatic radar configuration is illustrated in Figure 2.6. First, the microwave generator (usually a *klystron* or

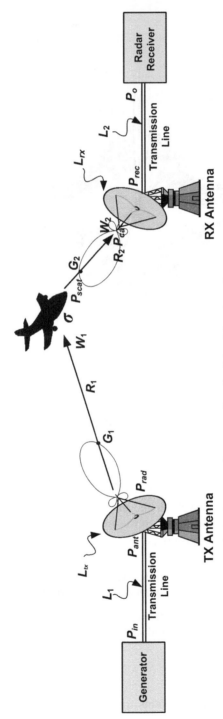

FIGURE 2.6 Geometry for obtaining bistatic radar range equation.

44

a *traveling-wave tube*) produces the radar signal of P_{in} watts. This signal is transferred to a transmitting antenna through a transmission line or a waveguide. Because of the finite conductivity of the lines and other microwave components, some of the power is lost during the transmission. Therefore, the power delivered to the transmitting antenna is given by

$$P_{ant} = \frac{P_{in}}{L_1},$$

(2.14)

where L_1 is the transmit loss mainly due to finite conductivity of the transmission line or dielectric losses. Once the power is delivered to the antenna, the power radiated by the antenna is

$$P_{rad} = \left(1 - |\Gamma_{tx}|^2\right) \cdot P_{ant}$$
$$= \frac{P_{ant}}{L_{tx}}.$$

(2.15)

Here, Γ_{tx} is the reflection coefficient at the terminal of the antenna and is given by

$$\Gamma_{tx} = \frac{Z_a - Z_o}{Z_a + Z_o}.$$

(2.16)

In the above equation, Z_a is the antenna radiation impedance and Z_o is the characteristic impedance of the transmission line connected to the terminals of the antenna. The loss, $1/(1 - |\Gamma_{tx}|^2) \triangleq L_{tx}$, associated by the impedance mismatch at the transmitter, is called the *transmitter loss*. Of course, if the antenna is matched, the radiated power will be equal to the power transmitted to the antenna as

$$P_{rad} = P_{ant}.$$

(2.17)

When the power is radiated by the transmitting antenna at a particular direction with the antenna gain of G_1, power density at the range of R_1 (where the target is located) can be found as

$$W_1 = \frac{G_1 \cdot P_{rad}}{4\pi R_1^2}.$$

(2.18)

The power incident to the target is scattered with an amount of the *equivalent echoing area*, or simply the RCS of the target. Therefore, the scattered power is then equal to

$$P_{scat} = \sigma \cdot W_1$$
$$= \frac{\sigma \cdot G_1 \cdot P_{rad}}{4\pi R_1^2}.$$

(2.19)

Then, the power is reradiated by the target as it reaches the receiver antenna located at R_2 distance away from the target. The power density of the scattered power around the receiver becomes

$$W_2 = \frac{P_{scat}}{4\pi R_2^2}$$
$$= \frac{\sigma \cdot G_1 \cdot P_{rad}}{\left(4\pi R_1 R_2\right)^2}.$$

(2.20)

At this stage, a very small amount of power is available to the receiving antenna. Therefore, it is crucial to catch as much power as possible by using a large aperture antenna. This implies that it is preferable to use largest practical aperture or reflector to collect as much of the incident power as possible. Antennas capture the incident signals with their effective apertures, A_{eff}, but not their actual apertures [15]. In most antennas, effective aperture sizes are smaller than the actual aperture sizes. The power captured by the receiver antenna is then equal to

$$P_{cap} = A_{eff} \cdot W_2$$
$$= \frac{A_{eff} \cdot \sigma \cdot G_1 \cdot P_{rad}}{\left(4\pi R_1 R_2\right)^2}.$$

(2.21)

Antenna effective aperture, A_{eff}, can be written in terms of its gain, G_2 as [15]

$$G_2 = \frac{4\pi \cdot A_{eff}}{\lambda^2}.$$

(2.22)

Replacing effective aperture A_{eff} with the receiver gain G_2 in Equation 2.16 using the relationship in Equation 2.13, one can easily obtain the following equation for the power captured by the receiver as

$$P_{cap} = \frac{\lambda^2 \cdot \sigma \cdot G_1 \cdot G_2 \cdot P_{rad}}{\left(4\pi\right)^3 \cdot \left(R_1 R_2\right)^2}.$$

(2.23)

Some of the captured power is delivered to the transmission line of the receiver if the antenna is not perfectly matched. Then, the received power at the front end of the transmission line is given by

$$P_{rec} = \left(1 - |\Gamma_{rx}|^2\right) \cdot P_{cap}$$
$$= \frac{P_{cap}}{L_{rx}}.$$

(2.24)

Here, Γ_{rx} is the reflection coefficient at the terminal of the receiving antenna and is equal to

$$\Gamma_{rx} = \frac{Z_b - Z_o}{Z_b + Z_o}.$$ (2.25)

In the above equation, Z_b is the receiving antenna's radiation impedance, and Z_o is the characteristic impedance of the transmission line connected to the terminals of the antenna. The loss, $1/(1-|\Gamma_{rx}|^2) \triangleq L_{rx}$, associated by the impedance mismatch at the transmitter, is known as the *receiver loss*. Of course, if the antenna is matched, the received power will be equal to the power captured by the antenna as

$$P_{rec} = P_{cap}.$$ (2.26)

Rewriting Equation 2.18 in terms of the received power, one can easily obtain

$$\begin{aligned} P_{rec} &= \frac{P_{cap}}{L_{rx}} \\ &= \frac{\lambda^2 \cdot \sigma \cdot G_1 \cdot G_2 \cdot P_{rad}}{(4\pi)^3 \cdot (R_1 R_2)^2 \cdot L_{rx}}. \end{aligned}$$ (2.27)

This power is then transmitted to radar receiver by using a transmission line. If there are some electric and/or dielectric losses, L_2, within the line, the power output to the radar receiver will be equal to

$$\begin{aligned} P_{out} &= \frac{P_{rec}}{L_2} \\ &= \frac{\lambda^2 \cdot \sigma \cdot G_1 \cdot G_2 \cdot P_{rad}}{(4\pi)^3 \cdot (R_1 R_2)^2 \cdot L_{rx} \cdot L_2}. \end{aligned}$$ (2.28)

If we insert the input power to the above equation with the help of Equations 2.5 and 2.6, we can get the famous *radar range equation* as below:

$$\frac{P_{out}}{P_{in}} = \frac{\lambda^2 \cdot \sigma \cdot G_1 \cdot G_2}{(4\pi)^3 \cdot (R_1 R_2)^2 \cdot L_{tot}}.$$ (2.29)

In Equation 2.24, L_{tot} is the total loss accounted for all the losses and is given by

$$L_{tot} = L_1 \cdot L_{tx} \cdot L_{rx} \cdot L_2.$$ (2.30)

If both the transmitter and the receiver antenna are perfectly matched and there are no losses inside the transmission lines, then $L_{tot} = 1$; therefore, Equation 2.29 can be simplified to give

$$\frac{P_{out}}{P_{in}} = \frac{\lambda^2 \cdot \sigma \cdot G_1 \cdot G_2}{(4\pi)^3 \cdot (R_1 R_2)^2}.$$ (2.31)

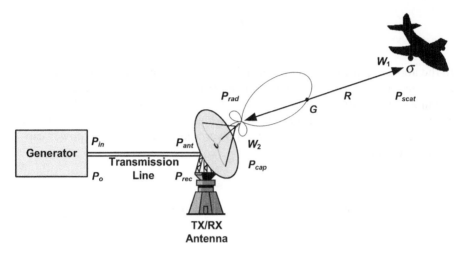

FIGURE 2.7   Geometry for obtaining monostatic radar range equation.

2.4.2   Monostatic Case

In the case of monostatic operation of radar, the same antenna is used for transmitting the radar signal and receiving the backscattered wave from the target (see Fig. 2.7). Therefore, the antenna gains, G_1 and G_2 in Equation 2.27, become identical, say G ($=G_1 = G_2$). Similarly, the target distance from the transmitter and the receiver distance from the target becomes equal, say R ($=R_1 = R_2$). Then the radar range equation can be obtained in its simplified form as shown below:

$$\frac{P_{out}}{P_{in}} = \frac{\lambda^2 \cdot \sigma \cdot G^2}{(4\pi)^3 \cdot R^4}. \tag{2.32}$$

2.5   RANGE OF RADAR DETECTION

While working with radars, another important parameter that should be carefully considered is the detection range of radar, that is, the farthest distance of the target that can be detected over the noise floor of the radar. This distance can be easily calculated starting from the radar range equation. Let us rewrite Equation 2.32 in terms of antenna effective aperture, $A_{eff} = 4\pi \cdot G_2/\lambda^2$, as

$$P_{out} = \frac{P_{in} \cdot A_{eff}^2 \cdot \sigma}{4\pi \cdot \lambda^2 \cdot R^4}. \tag{2.33}$$

The minimum power at the receiver output can only be detected if the received signal is greater than the noise floor, as demonstrated in Figure 2.8. If the

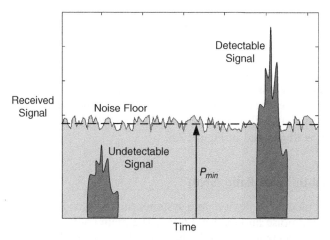

FIGURE 2.8   Minimum receiver power corresponding to maximum range of radar.

power level at the receiver output is lower than the noise floor, this signal cannot be distinguished from the noise mostly produced by the environment and the electronic equipment and therefore can either be considered as noise or clutter.

Considering that the input power to the radar remains unchanged, the output power at the receiver in Equation 2.33 is selected as the minimum detectable signal, P_{min}, for the maximum range distance of R_{max} as

$$(P_{out})_{min} \triangleq P_{min}$$
$$= \left(\frac{P_{in} \cdot A_{eff}^2 \cdot \sigma}{4\pi \cdot \lambda^2 \cdot R^4} \right)_{min} \tag{2.34}$$
$$= \frac{P_{in} \cdot A_{eff}^2 \cdot \sigma}{4\pi \cdot \lambda^2 \cdot R_{max}^4}.$$

Therefore, it is easy to find the maximum range of radar by rearranging Equation 2.34 to leave R_{max} alone as

$$R_{max} = \left(\frac{P_{in}}{P_{min}} \cdot \frac{A_{eff}^2 \cdot \sigma}{4\pi \cdot \lambda^2} \right)^{1/4}. \tag{2.35}$$

The above equation gives the maximum range of an object that can be detectable by the radar. The meaning of "maximum range" is clarified with the following example: Let the input power of a monostatic radar that has an antenna effective area of $A_{eff} = 3$ m^2 be 1 MW. If the sensitivity of this radar is 80 dBmW (10 nW) and it is used to detect a target with an RCS of 10 m^2 at 10 GHz, then the maximum range can be readily calculated by plugging the appropriate numbers into Equation 2.35 to give

$$R_{max} = \left(\frac{10^6}{10^{-8}} \cdot \frac{3^2 \cdot 10}{4\pi \cdot 0.03^2} \right)^{1/4} = 29,867 \text{ m.} \qquad (2.36)$$

If this target is located at the range closer than 29,867 m (or ~30 km), then it will be detected by this radar. However, any object that has a maximum RCS of 10 m² and located beyond 30 km will not be perceived as a target since the received signal level will be lower than the sensitivity level (or the noise floor) of the radar as illustrated in Figure 2.8.

2.5.1 Signal-to-Noise Ratio (SNR)

Similar to all electronic devices and systems, radars must function in the presence of internal noise and external noise. The main source of internal noise is the agitation of electrons caused by heat. The heat inside the electronic equipment can also be caused by environmental sources such as the sun, the earth, and buildings. This type of noise is also known as *thermal noise* [22] in the electrical engineering community.

Let us investigate the SNR of a radar system: Similar to all electronic systems, the *noise power spectral density* of a radar system can be described as the following equation:

$$N_o = k \cdot T_{eff}. \qquad (2.37)$$

Here, $k = 1.381 \times 10^{-23} W/(Hz \times K°)$ is the well-known Boltzman constant, and T_{eff} is the *effective noise temperature* of the radar in degrees Kelvin ($K°$). T_{eff} is not the actual temperature, but is related to the reference temperature via the *noise figure*, F_n, of the radar as

$$T_{eff} = F_n \cdot T_o, \qquad (2.38)$$

where the reference temperature, T_o, is usually referred to as room temperature ($T_o \approx 290 \ K°$). Therefore, *noise power spectral density* of the radar is then being equated to

$$N_o = k \cdot F_n \cdot T_o. \qquad (2.39)$$

To find the value of the *noise power*, P_n, of the radar, it is necessary to multiply N_o with the *effective noise bandwidth*, B_n, of the radar as shown below:

$$\begin{aligned} P_n &= N_o \cdot B_n \\ &= k \cdot F_n \cdot T_o \cdot B_n. \end{aligned} \qquad (2.40)$$

Here, B_n may not be the actual bandwidth of the radar pulse; it may extend to the bandwidth of the other electronic components such as the matched filter at the receiver. Provided that the noise power is determined, it is easy to define the SNR of radar by combining Equations 2.29 and 2.40 as below:

$$
\begin{aligned}
SNR &\triangleq \frac{P_s}{P_n} \\
&= \frac{P_{out}}{P_n} \\
&= \frac{P_{in} \cdot \lambda^2 \cdot \sigma \cdot G_1 \cdot G_2}{(4\pi)^3 \cdot (R_1 R_2)^2 \cdot (k \cdot F_n \cdot T_o \cdot B_n)}.
\end{aligned}
\tag{2.41}
$$

The above equation is derived for the bistatic radar operation. The equation can be simplified to the following for the monostatic radar setup:

$$
SNR = \frac{P_{in} \cdot \lambda^2 \cdot \sigma \cdot G^2}{(4\pi)^3 \cdot R^4 \cdot (k \cdot F_n \cdot T_o \cdot B_n)}.
\tag{2.42}
$$

2.6 RADAR WAVEFORMS

The selection of the radar signal type is mainly decided by the specific role and the application of the radar. Therefore, different waveforms can be utilized for the various radar applications. The most commonly used radar waveforms are

1. Continuous wave (CW)
2. Frequency modulated continuous wave (FMCW),
3. Stepped frequency continuous wave (SFCW),
4. Short pulse, and
5. Chirp (linear frequency modulated [LFM]) pulse.

Next, these waveforms will be investigated while their time and frequency characteristics are demonstrated and their common usages and applications are addressed.

2.6.1 CW

A CW radar system transmits radio wave signals at a particular frequency. If both the radar and the target are stationary, then the frequency of the received CW signal is the same as the transmitted signal. On the other hand, the returned signal's frequency components are shifted from the transmitted frequency if the target is in motion with respect to the radar. This type of shift in the frequency spectrum is called *Doppler frequency shift* and plays an important role in finding the velocity of the target in most radar applications. The concept of Doppler frequency shift is also important for ISAR imaging.

We shall see the use of Doppler frequency shift concept in Range-Doppler ISAR imaging applications in Chapter 6.

The time-domain signal of the CW radar is as simple as

$$s(t) = A \cdot \cos(2\pi f_o t), \qquad (2.43)$$

where f_o is the operating frequency. The frequency spectrum of this CW signal can be readily found by applying the forward Fourier transform operation to Equation 2.34 to get

$$S(f) = \frac{A}{2} \cdot (\delta(f - f_o) + \delta(f + f_o)). \qquad (2.44)$$

An example of a CW radar signal is shown in Figure 2.9. In Figure 2.9a, a purely sinusoidal signal that has a frequency of 1 kHz is drawn. The frequency spectrum of this signal is plotted in Figure 2.9b where two impulses at $f_o = \pm 1$ kHz can be easily seen.

As opposed to pulsed radar systems that use the time delay of the transmitted pulses to find the range of the target, CW radars measure the instantaneous rate of change in the target's range from the radar. This change causes the Doppler shift in the frequency content of the returned EM wave due to the motion of the radar, target, or both. One of the best uses of CW radar is the police radar system that estimates the speed of motor vehicles. A demonstration of CW police radar is shown in Figure 2.10. Assuming that the radar is stationary and transmitting a CW signal with a frequency of f_o, the frequency of the reflected wave from a stationary target is the same as the transmitting frequency f_o. If the target is approaching the radar, the frequency of the reflected wave increases with a shifted amount of f_D which is known as Doppler frequency shift [23] and is given by Fig. 2.10b:

$$f_D = \frac{2v_r}{\lambda_o}. \qquad (2.45)$$

Here, v_r is the radial speed of the moving target, and λ_o is the wavelength corresponding to the frequency of the transmitted wave. In a dual situation, if the target is moving away from the radar, the frequency of the reflected wave is altered such that the Doppler frequency shift produces a negative value. Therefore, the wavelength of the reflected wave increases, and the frequency decreases with an amount of f_D (see Fig. 2.10c).

The use of CW waveform in various radar applications provides the following advantages. First of all, the radars that use CW waveforms are easy to manufacture, thanks to their simple waveform shapes. Second, they can detect any target on the range as far as the power level permits. Therefore, there is no range constraint for detection. Additionally, they can be used in both very low frequency band (e.g., radio altimeters) and very high frequency band (e.g., early warning radars).

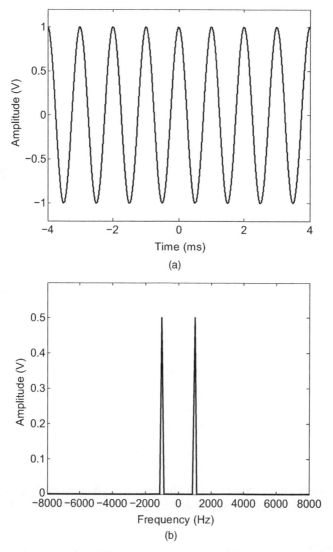

FIGURE 2.9   An example of CW radar waveform in (a) time domain, (b) frequency domain.

CW radars have the following disadvantages. They cannot estimate the range of a possible target. Range is normally measured by the time delay between different pulses created by the radar. In CW radars, however, the waveform is continuous and not pulsed. Furthermore, they can only detect moving targets. Reflected energy from stationary targets is filtered out since their basic operation is based on measuring the Doppler shift in the frequency.

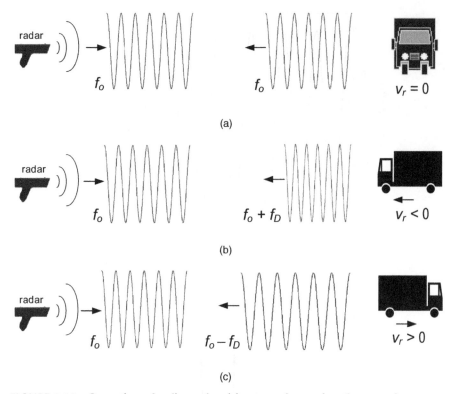

FIGURE 2.10   Operation of police radar: (a) returned wave has the same frequency as the transmitted signal for the stationary target, (b) returned wave's frequency is increased for the approaching target; (c) returned wave's frequency is decreased for the target going away.

Another disadvantage comes from the fact that they maximize the power consumption since they continuously broadcast the outgoing signal.

2.6.2   FMCW

While the CW radar can only estimate the Doppler shift created by the movement of the target with respect to radar, FMCW radar can be used to determine the range of a possible target. The common way to modulate the frequency is done by simply increasing the frequency as the time passes. This type of modulation is also known as *linear frequency modulation* or *chirp modulation*.

The waveform of an LFM continuous wave signal is simply given by

$$s(t) = A \cdot \sin\left(2\pi(f_o \mp \frac{K}{2}t)t \right), \tag{2.46}$$

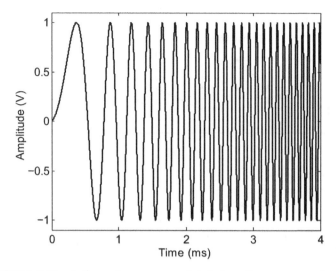

FIGURE 2.11   A linear frequency modulated continuous wave signal.

where A is the signal amplitude, f_o is the starting frequency, and K is the *chirp rate* (or *frequency increase/decrease rate*). In the above equation, the "+" sign indicates an *upchirp* signal and the "−" sign a *downchirp* signal. The instantaneous frequency of this signal can be easily found by taking the time derivative of the phase as

$$f_i(t) = \frac{1}{2\pi} \cdot \frac{\mathrm{d}}{\mathrm{dt}}\left(2\pi(f_o \mp \frac{K}{2}t)t \right)$$
$$= f_o \mp Kt. \tag{2.47}$$

A simple upchirp time-domain signal is illustrated in Figure 2.11. As obvious from the figure, the frequency of the wave increases as time progresses.

In FMCW radar operation, consecutive LFM signals are transmitted by the radar. If the period of serial LFM waves is T, the frequency variation of such waveforms can be represented as in Figure 2.12a. The received signal arrives with a time delay of t_d. This time delay can be determined in the following manner: The difference in the frequency between the transmitted and the received signals, Δf, can be found as below:

$$\Delta f = f_{tx} - f_{rx}$$
$$= (f_o \mp Kt) - (f_o \mp K(t - t_d)) \tag{2.48}$$
$$= \mp K \cdot t_d.$$

Of course, time delay t_d is related to the range, R, of the target by the following equation:

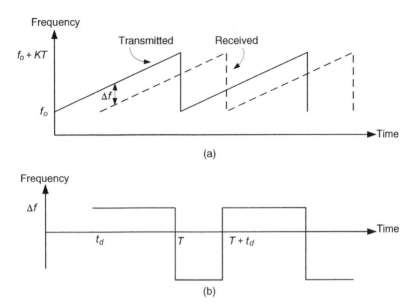

FIGURE 2.12 Operation of LFMCW radar: (a) time-frequency display of the transmitted and received LFMCW signals, (b) the difference in the frequency between the transmitted and the received signals.

$$t_d = \frac{2R}{c}. \tag{2.49}$$

Here, c is the speed of light in the air. Combining Equations 2.48 and 2.49, it is easy to determine the range of the target via

$$R = c\frac{\Delta f}{2K}. \tag{2.50}$$

The block diagram of linear frequency modulated continuous wave (LFMCW) radar is shown in Figure 2.13. The LFMCW generator produces the LFM signal to be broadcasted by the transmitter. The receiver collects the returned wave that is multiplied with the transmitted signal. The output has both the sum and the difference of the transmitted and the received frequencies. As shown in Figure 2.12b, only the positive frequency difference, Δf, is selected. Then the signal is fed to a discriminator that contains a differentiator plus an envelope detector. The output of the discriminator is proportional to the frequency difference, Δf. Once Δf is obtained, the range, R, of the target can be easily obtained via Equation 2.50.

It is also obvious from Figure 2.12 that range ambiguity occurs when $t_d > T$. Therefore, the maximum difference in frequency can be $\Delta f_{max} = KT$, which means that the maximum unambiguous range can be determined as

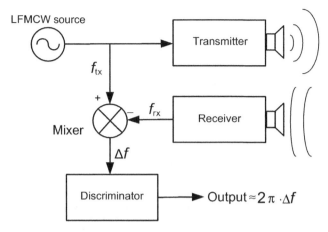

FIGURE 2.13   LFMCW radar block diagram.

$$R_{max} = c\frac{\Delta f_{max}}{2k}$$
$$= c\frac{KT}{2k}$$
$$= c\frac{T}{2}.$$

(2.51)

The above equation suggests that FMCW radar can only be used for short- or mid-range detection of objects. Therefore, it is not suitable for long-range detection.

2.6.3 SFCW

Another popular radar waveform used to determine the range is the SFCW. This signal is formed by emitting a series of single-frequency short continuous subwaves. In generating the SFCW signal, the frequencies between adjacent subwaves are increased by an incremental frequency of Δf as demonstrated in Figure 2.14. For one burst of SFCW signal, a total of N CW signals, each having a discrete frequency of $f_n = f_o + (n-1) \cdot \Delta f$, is sent. Each subwave has a time duration of τ and is of T distance away from the adjacent subwave. The total frequency bandwidth, B, and the frequency increment (or resolution), Δf, can be readily calculated as below:

$$B = (f_{N-1} - f_o) + \Delta f$$
$$= N \cdot \Delta f.$$

(2.52)

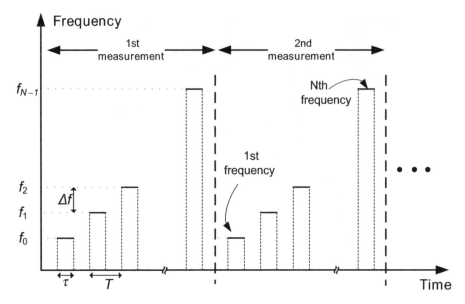

FIGURE 2.14 SFCW signal in time-frequency plane.

$$\Delta f = \frac{B}{N}$$
$$= \frac{(f_{N-1} - f_o) + \Delta f}{N}. \tag{2.53}$$

The SFCW signal can be used to estimate the range of a possible target in the following manner. Suppose that the target is at the range distance of R_o from the radar. With a single measurement of monostatic SFCW radar, the phase of the backscattered wave is proportional to the range as given in the following equation:

$$E^s[f] = A \cdot e^{-j2k \cdot R_o}. \tag{2.54}$$

Here, E^s is the scattered electric field, A is the scattered field amplitude, and k is the wavenumber vector corresponding to the frequency vector of $f = [f_o\ f_1\ f_2 \dots f_{N-1}]$. The number 2 in the phase corresponds to the two-way propagation between radar-to-target and target-to-radar. It is obvious that there is Fourier transform relationship between $(2k)$ and (R). Therefore, it is possible to resolve the range, R_o, by taking the inverse Fourier transform (IFT) of the output of the SFCW radar. The resulted signal is nothing but the *range profile* of the target. The range resolution is determined by the Fourier theory as

$$\Delta r = \frac{2\pi}{2BW_k}$$

$$= \frac{\pi c}{2\pi BW_f} \quad (2.55)$$

$$= \frac{c}{2B},$$

where BW_k and $BW_f \triangleq B$ are the bandwidths in wavenumber and frequency domains, respectively. The maximum range is then determined by multiplying the range resolution by the number of SFCW pulses:

$$R_{max} = N \cdot \Delta r$$

$$= \frac{N \cdot c}{2B}. \quad (2.56)$$

We will demonstrate the operation of SFCW radar with an example. Let us consider a point target which is 50 m away from the radar. Suppose that the SFCW radar's frequencies change from 2 GHz to 22 GHz with the frequency increments of 2 MHz. Using Equations 2.55 and 2.56, one can easily find the range resolution and the maximum range as 0.75 cm and 75 m, respectively. Applying the Matlab routine "Figure2.15.m" to the synthetic backscattered data, the range profile of this point target can be obtained as plotted in Figure 2.15. It is clearly seen from the figure that the point target at the range of 50 m is perfectly pinpointed.

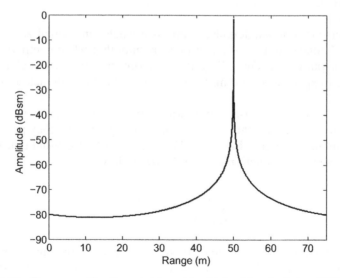

FIGURE 2.15   Range profile of a point target is obtained by the help of SFCW radar processing.

2.6.4 Short Pulse

One of the simplest radar waveforms is the short pulse (or impulse) whose time duration is usually on the order of a few nanoseconds. As calculated in Equation 2.55, the range resolution of a pulsed radar is equal to

$$\Delta r = \frac{c}{2B}, \tag{2.57}$$

where B is the frequency bandwidth of the pulse. According to the Fourier theory, the frequency bandwidth, B, of a pulse is also inversely proportional to its pulse duration as

$$B = \frac{1}{\tau}, \tag{2.58}$$

which means that the range resolution is proportional to its pulse duration as

$$\Delta r = c\frac{\tau}{2}. \tag{2.59}$$

Therefore, to have a good range resolution, the duration of a pulse has to be as small as possible. Common short pulse waveforms are rectangular pulse, single-tone pulse, and single wavelet pulse of different forms. In Figure 2.16a, a rectangular pulse-shape wave is shown, and the spectrum of this signal is plotted in Figure 2.16b. In the frequency domain, a sinc-type pattern is obtained as expected.

Another common single-pulse shape is a single sine signal as plotted in Figure 2.17. Since the time-domain pulse is smoother when compared to the rectangular pulse (see Fig. 2.17a), the spectrum widens, and sidelobe levels decrease as expected according to the Fourier theory as depicted in Figure 2.17b.

Another popular short-duration waveform is called the wavelet signal. Wavelets are much smoother than the sine pulse; therefore, they provide less sidelobes in the frequency domain. In Figure 2.18a, a Mexican-hat type wavelet whose mathematical function is given below is shown:

$$m(t) = \frac{1}{\sqrt{2\pi} \cdot \sigma^3} \left(1 - \left(\frac{t}{\sigma} \right)^2 \right) e^{-\left(\frac{t}{\sqrt{2}\sigma} \right)^2}. \tag{2.60}$$

Since this signal is much smoother than the previous short pulse waveforms that we have presented, the frequency extent of this wavelet is extremely broad. Therefore, it provides an ultrawide band (UWB) spectrum as most of the other short-duration wavelets do as shown in Figure 2.18b.

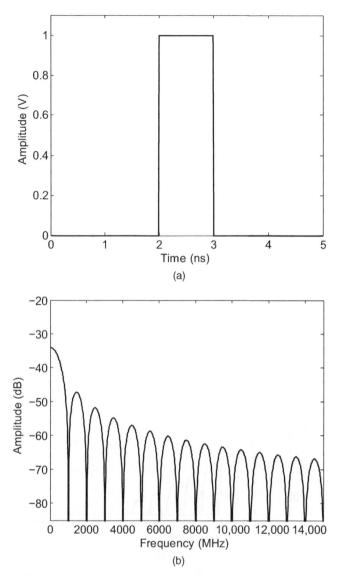

FIGURE 2.16   A short-duration rectangular pulse in (a) time domain, (b) frequency domain.

While these short pulses are good for providing a wide spectrum, they are not practical in terms of providing sufficient energy. This is because of the fact that it is not possible to put great amount of power onto a very small pulse. To circumvent this problem, the pulse is modulated by altering the frequency as time continues to pass. The common practice is to use a chirp waveform to be able to put enough energy onto the pulse, as will be investigated next.

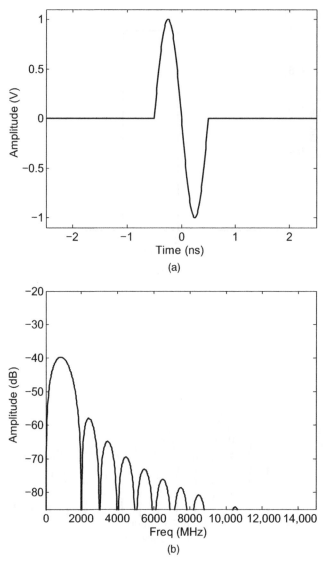

(a)

(b)

FIGURE 2.17 A short-duration single-frequency pulse in (a) time domain, (b) frequency domain.

2.6.5 Chirp (LFM) Pulse

As explained in the previous paragraph, it will not be possible to use a sufficiently wide pulse and achieve a wide bandwidth. If a broadband spectrum is
achieved with an unmodulated, or constant-frequency pulse (as in Fig. 2.19a),
its time duration has to be quite small such that it may not be possible to put
enough energy on it. A solution to this problem is to use a modulated pulse

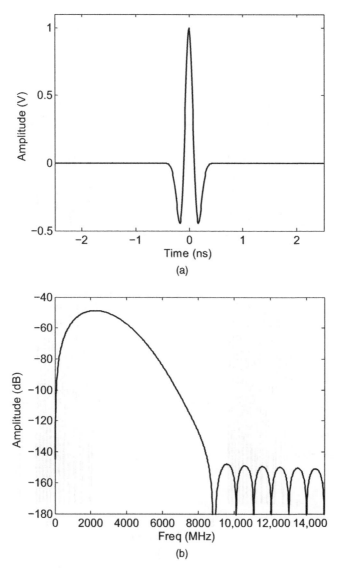

FIGURE 2.18   A short-duration Mexican-hat pulse in (a) time domain, (b) frequency domain.

of sufficient duration such that this modulated waveform provides the required frequency bandwidth for the operation of radar.

The common waveform is the LFM pulse, also known as the chirp pulse, whose waveform is shown in Figure 2.19b. In practice, this waveform is repeated in every T_{PR} intervals for most common radar applications, especially for localization of targets in the range. T_{PR} is called the *pulse repetition interval*

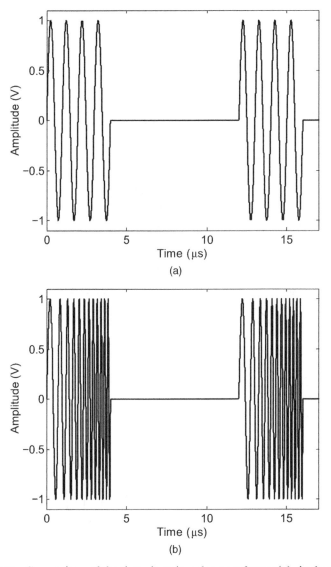

FIGURE 2.19   Comparison of the time-domain pulse waveforms: (a) single-tone pulse, (b) LFM (Chirp) pulse.

(PRI) or *pulse repetition period*. The inverse of this interval gives the *pulse repetition frequency* (PRF), defined as

$$f_{PR} = \frac{1}{T_{PR}}. \tag{2.61}$$

The mathematical expression of the *upward chirp* signal whose frequency is increasing as time passes along the pulse is given as

$$m(t) = \begin{cases} A \cdot \sin\left(2\pi\left(f_o + \dfrac{K}{2}(t - nT_{PR})\right)(t - nT_{PR})\right), & nT_{PR} \le t \le nT_{PR} + \tau \\ 0, & elsewhere \end{cases} \quad (2.62)$$

where n is an integer, τ is the pulse width, and K is the chirp rate. The instantaneous frequency of the pulse is $f_i(t) = f_o + Kt$. It is also possible to form another LFM pulse by decreasing the frequency along the pulse width as shown below:

$$m(t) = \begin{cases} A \cdot \sin\left(2\pi\left(f_o - \dfrac{K}{2}(t - nT_{PR})\right)(t - nT_{PR})\right), & nT_{PR} \le t \le nT_{PR} + \tau \\ 0, & elsewhere \end{cases} \quad (2.63)$$

For the *downward chirp* pulse, the instantaneous frequency is then equal to $f_i(t) = f_o - Kt$.

To demonstrate the broad spectrum of the LFM waveform, the Fourier transform of single-tone and LFM pulse signals in Figure 2.19 is taken and plotted in Figure 2.20. It is clearly seen from this figure that chirp signal provides more frequency bandwidth when compared to constant-frequency pulse.

In radar applications, LFM pulse waveforms are mainly utilized in finding range profiles, and also for synthetic aperture radar (SAR) and ISAR processing as will be discussed in the forthcoming chapters.

2.7 PULSED RADAR

Pulsed radar systems are commonly used especially in SAR and ISAR systems. They transmit and receive a sequence of modulated pulses. Therefore, the same type of pulse is repeated in every T_{PR} interval, or, as it is called, PRI as depicted in Figure 2.21. The range information can be gathered from the two-way trip time (or time delay) between the transmitted and received pulses. Pulsed radar systems have the ability to measure both the range (the radial distance) and the radial velocity of the target.

2.7.1 PRF

As pulses are repeated in T_{PR}, the corresponding PRF of the radar is as given in Equation 2.61. PRF gives the total number of pulses transmitted in every second by the radar. In radar applications, PRF value can be quite critical as it is linked to maximum range of a target, R_{max}, and the maximum Doppler frequency, $f_{D,max}$ (so the maximum target velocity v_{max} of the target), that can

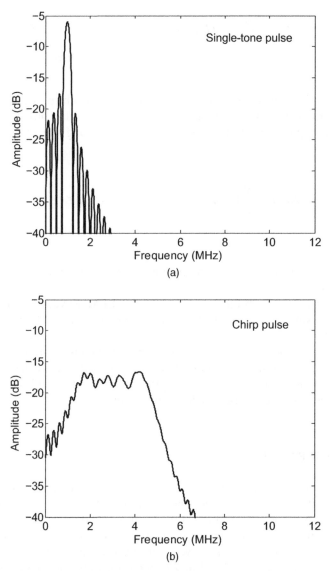

FIGURE 2.20 Comparison of the spectrum of (a) single-tone pulse and (b) LFM pulse. Although both signals use the same time duration, frequency bandwidth of the Chirp waveform is much wider than the single-tone waveform.

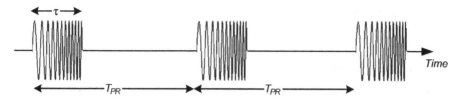

FIGURE 2.21   Pulsed radar systems use a sequence of modulated pulses.

be detectable by the radar. The use of PRF in ISAR range-Doppler processing will be explored in Chapter 6.

2.7.2   Maximum Range and Range Ambiguity

As calculated in Equation 2.59, the range resolution is proportional to the pulse duration as $\Delta r = c \cdot \tau / 2$. Therefore, the smaller the pulse duration, the finer the range resolution we can get. On the other hand, maximum range is determined by time delay between the transmitted and received pulses. Since the pulses are repeated for every T_{PR} seconds, any received pulse that is reflected back from a target at R distant on the range should arrive before the next pulse is transmitted to avoid the ambiguity in the range, that is,

$$T_{PR} \geq \frac{2R}{c}. \tag{2.64}$$

If T_{PR} is fixed, then the range should be less than the following quantity:

$$R \leq c \frac{T_{PR}}{2}. \tag{2.65}$$

Therefore, the maximum range that can be unambiguously detected by the pulsed radar is calculated by the period between the pulses, that is, T_{PR}, as given below:

$$
\begin{aligned}
R_{max} &= c \frac{T_{PR}}{2} \\
&= \frac{c}{2 f_{PR}}.
\end{aligned} \tag{2.66}
$$

This is also called *unambiguous range* since any target within this range is accurately detected by the radar at its true location. However, any target beyond this range will be mislocated in the range as the radar can only display the R_{max} modulus of the target's location along the range axis. To resolve the range ambiguity problem, some radars uses multiple PRFs while transmitting the pulses [23].

2.7.3 Doppler Frequency

In radar theory, the concept of Doppler frequency describes the shift in the center frequency of an incident EM wave due to movement of radar with respect to target. The basic concept of Doppler shift in frequency has been conceptually demonstrated through Figure 2.10 and is defined as

$$f_D = \begin{cases} +\dfrac{2v_r}{\lambda_o}, & \text{for approaching target} \\[2mm] -\dfrac{2v_r}{\lambda_o}, & \text{for moving away target} \end{cases}, \tag{2.67}$$

where v_r is the radial velocity along the radar line of sight (RLOS) direction. Now, we will demonstrate how the shift in the phase (also in the frequency) of the reflected signal from a moving target constitutes. Let us consider an object moving toward the radar with a speed of v_r. The radar produces and sends out pulses with the PRF value of f_{PR}. Every pulse has a time duration (or width) of τ. The illustration of Doppler frequency shift phenomenon is given in Figure 2.22. The leading edge of the first pulse hits the target (see Fig. 2.22I). After a time advance of Δt, the trailing edge of the first pulse hits the target as shown in Figure 2.22II. During this time period, the target traveled a distance of

$$d = v_r \cdot \Delta t. \tag{2.68}$$

Looking at the situation in Figure 2.22II, it is obvious that the pulse distance before the reflection is equal to the distance traveled by the leading (or trailing) edge of the pulse plus the distance traveled by the target as

$$\begin{aligned} p &= x + d \\ c\tau &= c\Delta t + v_r \Delta t. \end{aligned} \tag{2.69}$$

Similarly, the pulse distance after the reflection is equal to the distance traveled by the leading (or trailing) edge of the pulse minus the distance traveled by the target as

$$\begin{aligned} p' &= x - d \\ c\tau' &= c\Delta t - v_r \Delta t. \end{aligned} \tag{2.70}$$

Dividing these last two equations yields

$$\frac{c\tau'}{c\tau} = \frac{c\Delta t - v_r \Delta t}{c\Delta t + v_r \Delta t}. \tag{2.71}$$

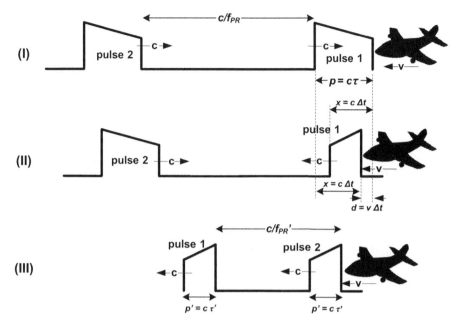

FIGURE 2.22 Illustration of Doppler shift phenomenon: (I) the leading edge of the first pulse in hitting the target at $t = 0$; (II) the trailing edge of the first pulse in hitting the target at $t = \Delta t$; (III) the trailing edge of the second pulse is hitting the target at $t = dt$. During this period, the target traveled a distance of $D = v_r \times dt$.

On the left-hand side of this equation, c terms are canceled, whereas Δt terms are canceled on the right-hand side. Then, the pulse width after the reflection can be written in terms of the original pulse width as

$$\tau' = \tau\left(\frac{c - v_r}{c + v_r}\right). \tag{2.72}$$

The term $(c - v_r)/(c + v_r)$ is known as the *dilation factor* in the radar community. Notice that when the target is stationary ($v_r = 0$), then the pulse duration remains unchanged ($\tau' = \tau$) as expected.

Now, consider the situation in Figure 2.22III. As trailing edge of the second pulse is hitting the target, the target has traveled a distance of

$$D = v_r \cdot dt \tag{2.73}$$

within the time frame of dt. During this period, the leading edge of the first pulse has traveled a distance of

$$x = c \cdot dt. \tag{2.74}$$

On the other hand, the leading edge of the second pulse has to travel a distance of $(c/f_{PR} - D)$ at the instant when it reaches the target. Therefore,

$$\frac{c}{f_{PR}} - D = c \cdot dt$$
$$\frac{c}{f_{PR}} - v_r \cdot dt = c \cdot dt. \tag{2.75}$$

Solving for dt yields

$$dt = \frac{c/f_{PR}}{c + v_r}. \tag{2.76}$$

Putting Equation 2.70 into Equation 2.67, one can get

$$D = \frac{cv_r/f_{PR}}{c + v_r}. \tag{2.77}$$

The new PRF for the reflected pulse is

$$f'_{PR} = \frac{c}{x - D}$$
$$= \frac{c}{c \cdot dt - \dfrac{cv_r/f_{PR}}{c + v_r}} \tag{2.78}$$
$$= \frac{c + v_r}{c \cdot dt + v_r \cdot dt - v_r/f_{PR}}.$$

Using Equation 2.75 inside Equation 2.78, one can get the relationship between the PRFs of incident and reflected waves as

$$f'_{PR} = \frac{c + v_r}{c/f_{PR} - v_r/f_{PR}}$$
$$= f_{PR} \cdot \left(\frac{c + v_r}{c - v_r} \right). \tag{2.79}$$

If the center frequency of the incident and reflected waves are f_0 and f_0', these two frequencies are related to each other with the same factor:

$$f_0' = f_0 \cdot \left(\frac{c + v_r}{c - v_r} \right). \tag{2.80}$$

To find the Doppler shift in the frequency, f_D, we should subtract the center frequency of the incident wave from the center frequency of the reflected wave as

$$
\begin{aligned}
f_D &= f_0' - f_0 \\
&= f_0 \cdot \left(\frac{c + v_r}{c - v_r} \right) - f_0 \\
&= f_0 \frac{2v_r}{c - v_r}.
\end{aligned}
\tag{2.81}
$$

Since it is also obvious that the target velocity is very small compared to the speed of light (i.e., $v_r \ll c$), Equation 2.81 simplifies as

$$
\begin{aligned}
f_D &= f_0 \frac{2v_r}{c} \\
&= \frac{2v_r}{\lambda_0},
\end{aligned}
\tag{2.82}
$$

where λ_0 is the wavelength corresponding to the center frequency of f_0. For the target that is moving away from the radar, the Doppler frequency shift has a negative sign as

$$
\begin{aligned}
f_D &= -f_0 \frac{2v_r}{c} \\
&= -\frac{2v_r}{\lambda_0}.
\end{aligned}
\tag{2.83}
$$

It is clear from these equations that Doppler frequency shift is directly proportional to the velocity of the target. If the velocity increases, the shift in the frequency increases as well. If the target is stationary with respect to radar ($v_r = 0$), then the Doppler frequency shift is zero. The velocity v_r in the equation corresponds to the velocity along the RLOS. If the target is moving along another direction, v_r refers to the velocity projected toward the direction of radar. If the target's velocity is v as illustrated in Figure 2.23, then the Doppler frequency shift in terms of the target's original velocity will be equal to

$$
\begin{aligned}
f_D &= \frac{2v_r}{\lambda_0} \\
&= \frac{2v}{\lambda_0} \cos\theta.
\end{aligned}
\tag{2.84}
$$

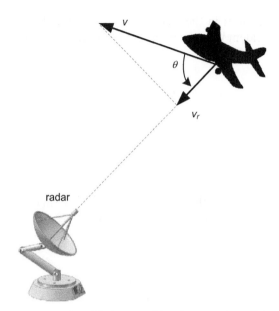

FIGURE 2.23   Doppler shift is caused by the target's radial velocity, v_r.

2.8   MATLAB CODES

Below are the Matlab source codes that were used to generate all of the Matlab-produced Figures in Chapter 2. The codes are also provided in the CD that accompanies this book.

Matlab code 2.1: Matlab file "Figure2-9.m"

```
%-----------------------------------------------------------------
% This code can be used to generate Figure 2.9
%-----------------------------------------------------------------
%---Figure 2.9(a)------------------------------------------------
clear all
close all

fo = 1e3; % set the frequency
t = -4e-3:1e-7:4e-3; % choose time vector
s = cos(2*pi*fo*t); % time domain CW signal
plot(t*1e3,s,'k','LineWidth',2);
set(gca,'FontName', 'Arial', 'FontSize',14,'FontWeight','Bold
');
xlabel('Time [ms]');
ylabel('Amplitude [V]');
axis([-4 4 -1.2 1.2])

%---Figure 2.9(b)------------------------------------------------
N = length(t);
```

```
df = 1/(t(N)-t(1)); % Find frequency resolution
f = -df*(N-1)/2:df:df*(N-1)/2; % set frequency vector

S = fft(s)/N; % frequency domain CW signal
plot(f,fftshift(abs(S)),'k','LineWidth',2);
set(gca,'FontName', 'Arial', 'FontSize',14,'FontWeight','Bold
');
xlabel('Frequency [Hz]');
ylabel('Amplitude [V]');
axis([-.8e4 .8e4 0 .6])
```

Matlab code 2.2: Matlab file "Figure2-11.m"
```
%----------------------------------------------------------------
% This code can be used to generate Figure 2.11
%----------------------------------------------------------------
clear all
close all

fo = 100; % set the base frequency
t = 0:1e-7:4e-3; % choose time vector
k = 3e6; % select chirp rate
s = sin(2*pi*(fo+k*t/2).*t); % time domain FMCW signal

plot(t*1e3,s,'k','LineWidth',2);
set(gca,'FontName', 'Arial', 'FontSize',14,'FontWeight','Bold
');
xlabel('Time [ms]'); ylabel('Amplitude [V]');
axis([0 4 -1.1 1.1])
```

Matlab code 2.3: Matlab file "Figure2-15.m"
```
%----------------------------------------------------------------
% This code can be used to generate Figure 2.15
%----------------------------------------------------------------
clear all
close all

c = .3; % speed of light []
f = 2:.002:22; % choose frequency vector
Ro = 50; % choose range of target [m]
k = 2*pi*f/c; % wavenumber

Es = 1*exp(-j*2*k*Ro); % collected SFCW electric field

df = f(2)-f(1); % frequency resolution
N = length(f); % total stepped frequency points
dr = c/(2*N*df); % range resolution
R = 0:dr:dr*(length(f)-1); %set the range vector

plot(R, 20*log10(abs(ifft(Es))),'k','LineWidth',2)
```

```
set(gca,'FontName', 'Arial', 'FontSize',14,'FontWeight','Bold
');
xlabel('Range [m]'); ylabel('Amplitude [dBsm]');
axis([0 max(R) -90 0])
```

Matlab code 2.4: Matlab file "Figure2-16.m"
```
%----------------------------------------------------------------
% This code can be used to generate Figure 2.16
%----------------------------------------------------------------
clear all
close all

t = 0:0.01e-9:50e-9; % choose time vector
N = length(t);
pulse(201:300)=ones(1, 100); % form rectangular pulse
pulse(N)=0;

% Frequency domain equivalent
dt = t(2)-t(1); % time resolution
df = 1/(N*dt); % frequency resolution
f = 0:df:df*(N-1); % set frequency vector
pulseF = fft(pulse)/N; % frequency domain signal

%---Figure 2.16(a)-----------------------------------------------
plot(t(1:501)*1e9,pulse(1:501),'k','LineWidth',2);
set(gca,'FontName', 'Arial', 'FontSize',14,'FontWeight','Bold
');
xlabel('Time [ns]');
ylabel('Amplitude [V]');
axis([0 5 0 1.1]);

%---Figure 2.16(b)-----------------------------------------------
plot(f(1:750)/1e6,20*log10(abs(pulseF(1:750))),'k','LineWi
dth',2);
set(gca,'FontName', 'Arial', 'FontSize',14,'FontWeight','Bold
');
axis([0 f (750)/1e6 -85 -20]);
xlabel('Frequency [MHz]');
ylabel('Amplitude [dB]');
```

Matlab code 2.5: Matlab file "Figure2-17.m"
```
%----------------------------------------------------------------
% This code can be used to generate Figure 2.17
%----------------------------------------------------------------
clear all
close all

t = -25e-9:0.01e-9:25e-9; % choose time vector
```

```
N = length(t);
sine(2451:2551) = -sin(2*pi*1e9*(t(2451:2551)));% form sine
pulse
sine(N)=0;

% Frequency domain equivalent
dt = t(2)-t(1); % time resolution
df = 1/(N*dt); % frequency resolution
f = 0:df:df*(N-1); % set frequency vector
sineF = fft(sine)/N; % frequency domain signal

%---Figure 2.17(a)-----------------------------------------------
plot(t(2251:2751)*1e9,sine(2251:2751),'k','LineWidth',2);
set(gca,'FontName', 'Arial', 'FontSize',14,'FontWeight','Bold
');
xlabel('Time [ns]');
ylabel('Amplitude [V]');
axis([-2.5 2.5 -1.1 1.1]);

%---Figure 2.17(b)-----------------------------------------------
plot(f(1:750)/1e6,20*log10(abs(sineF(1:750))),'k','LineWi
dth',2);
set(gca,'FontName', 'Arial', 'FontSize',14,'FontWeight','Bold');
axis([0 f (750)/1e6 -85 -20]);
xlabel('Freq [ MHz]');
ylabel('Amplitude [dB]');
```

Matlab code 2.6: Matlab file "Figure2-18.m"
```
%----------------------------------------------------------------
% This code can be used to generate Figure 2.18
%----------------------------------------------------------------
clear all
close all

sigma = 1e-10; % set sigma
t = -25e-9:0.01e-9:25e-9; % choose time vector
N = length(t);
mex(2451:2551)=1/sqrt(2*pi)/sigma^3*(1-t(2451:2551).^2/
sigma^2)...
  .*(exp(-t(2451:2551).^2/2/sigma^2)); % form wavelet
mex = mex/max(mex);
mex(N) = 0;

% Frequency domain equivalent
dt = t(2)-t(1); % time resolution
df = 1/(N*dt); % frequency resolution
f = 0:df:df*(N-1); % set frequency vector
mexF = fft(mex)/N; % frequency domain signal
```

```
%---Figure 2.18(a)-------------------------------------------------
plot(t(2251:2751)*1e9,mex(2251:2751),'k','LineWidth',2);
set(gca,'FontName', 'Arial', 'FontSize',14,'FontWeight','Bold');
xlabel('Time [ns]');
ylabel('Amplitude [V]');
axis([-2.5 2.5 -.5 1.1]);


%---Figure 2.18(a)-------------------------------------------------
plot(f(1:750)/1e6,20*log10(abs(mexF(1:750))),'k','LineWi
dth',2);
set(gca,'FontName', 'Arial', 'FontSize',14,'FontWeight','Bold
');
axis([0 f (750)/1e6 -180 -40]);
xlabel('Freq [ MHz]');
ylabel('Amplitude [dB]');
```

Matlab code 2.7: Matlab file "Figure2-19and20.m"
```
%------------------------------------------------------------------
% This code can be used to generate Figure 2.19 and 2.20
%------------------------------------------------------------------
clear all
close all

fo = 1e6; % choose base frequency
t = 0:1e-9:4e-6;
tt = 0:1e-9:17e-6; % choose time vector
k = 1.0e12; % choose chirp rate
sinep = sin(2*pi*fo*t);
sinep(12001:16001) = sinep; % form CW pulse
sinep (17001) = 0;
m = sin(2*pi*(fo+k*t/2).*t); % form LFM pulse
m(12001:16001) = m;
m (17001) = 0;


%---Figure 2.19(a)-------------------------------------------------
plot(tt*1e6,sinep,'k','LineWidth',2);
set(gca,'FontName', 'Arial', 'FontSize',14,'FontWeight','Bold
');
xlabel('Time [\mus]');
ylabel('Amplitude [V]');
axis([0 17 -1.1 1.1])


%---Figure 2.19(b)-------------------------------------------------
plot(tt*1e6,m,'k','LineWidth',2);
set(gca,'FontName', 'Arial', 'FontSize',14,'FontWeight','Bold
');
xlabel('Time [\mus]');
ylabel('Amplitude [V]');
axis([0 17 -1.1 1.1])
```

```
% Frequency domain equivalent
df = 1/170e-6; % frequency resolution
f = 0:df:df*170000; % set frequency vector
sinep = sin(2*pi*fo*t);
sinep (170001) = 0;
m = sin(2*pi*(fo+k*t/2).*t);
m (170001) = 0;


fsinep = fft(sinep)/length(t); % spectrum of CW pulse
fm = fft(m)/length(t);% spectrum of LFM pulse

%---Figure 2.20(a)--------------------------------------------------
plot(f(1:2000)/1e6,20*log10(abs(fsinep(1:2000))),'k','LineWi
dth',2);
set(gca,'FontName', 'Arial', 'FontSize',14,'FontWeight','Bold
');
xlabel('Frequency [MHz]');
ylabel('Amplitude [dB]');
text(8,-10,'Single tone pulse')
axis([0 12 -40 -5])

%---Figure 2.20(b)--------------------------------------------------
plot(f(1:2000)/1e6,20*log10(abs(fm(1:2000))),'k','LineWi
dth',2);
set(gca,'FontName', 'Arial', 'FontSize',14,'FontWeight','Bold
');
xlabel('Frequency [MHz]');
ylabel('Amplitude [dB]');
text(8,-10,'Chirp pulse')
axis([0 12 -40 -5])
```

REFERENCES

1 J. C. Stover. *Optical scattering: Measurement and analysis*, 2nd ed. SPIE Press, Washington, USA, 1995.

2 Y. Zhong-cai, S. Jia-Ming, and W. Jia-Chun. Validity of effective-medium theory in Mie scattering calculation of hollow dielectric sphere. 7th International Symposium on Antennas Propagation & EM Theory (ISAPE '06), pp. 1–4, 2006.

3 U. Brummund and B. Mesnier. A comparative study of planar Mie and Rayleigh scattering for supersonic flowfield diagnostics. 18th Intern. Congress Instrumentation in Aerospace Simulation Facilities (ICIASF 99), 42/1–4210, 1999.

4 T. H. Chu and D. B. Lin. Microwave diversity imaging of perfectly conducting objects in the near field region. *IEEE Trans Microwave Theory Tech* 39 (1991) 480–487.

5 R. Bhalla and H. Ling. ISAR image formation using bistatic data computed from the shooting and bouncing ray technique. *J Electromagn Waves Appl* 7(9) (1993) 1271–1287.

6   C. A. Balanis. *Advanced engineering electromagnetics*. Wiley, New York, 1989.

7   Antenna Standards Committee of the IEEE Antennas and Propagation Society, IEEE Standard Definitions of Terms, IEEE Sed 145-1993, The Institute of Electrical and Electronics Engineers, NewYork, 28.

8   R. J. Sullivan. *Microwave radar imaging and advanced concepts*. Artech House, Norwood, MA, 2000.

9   G. T. Ruck, D. Barrick, W. Stuart, and C. Krichbaum. *Radar cross section handbook*, vol. 1. Plenum Press, New York, 1970.

10 J. W. Crispin and K. M. Siegel. *Methods for radar cross-section analysis*. Academic, New York, 1970.

11 M. I. Skolnik. *Radar handbook*, 2nd ed. McGraw-Hill, New York, 1990.

12 K. F. Warnick and W. C. Chew. Convergence of moment-method solutions of the electric field integral equation for a 2-D open cavity. *Microw Opt Tech Lett* 23(4) (1999) 212–218.

13 Ö. Ergül and L. Gürel. Improved testing of the magnetic-field integral equation. *IEEE Microw Wireless Comp Lett* 15(10) (2005) 615–617.

14 C. A. Balanis. *Antenna theory, analysis and design*. Harper & Row, New York, 1982.

15 E. Ekelman and G. Thiele. A hybrid technique for combining the moment method treatment of wire antennas with the GTD for curved surfaces. *IEEE Trans Antennas Propag* AP-28 (1980) 831.

16 T. J. Kim and G. Thiele. A hybrid diffraction technique—General theory and applications. *IEEE Trans Antennas Propag* AP-30 (1982) 888–898.

17 H. Ling, K. Chou, and S. Lee. Shooting and bouncing rays: Calculating the RCS of an arbitrarily shaped cavity. *IEEE Trans Antennas Propag* AP-37 (1989) 194–205.

18 R. Bhalla and H. Ling. A fast algorithm for signature prediction and image formation using the shooting and bouncing ray technique. *IEEE Trans Antennas Propagat* 43 (1995) 727–731.

19 E. F. Knott, J. F. Shaeffer, and M. T. Tuley. *Radar cross section*, 2nd ed. Artech House, Norwood, MA, 1993.

20 F. Weinmann. Ray tracing with PO/PTD for RCS modeling of large complex objects. *IEEE Trans Antennas Propag* AP-54 (2006) 1797–1806.

21 H. Ling, R. Chou, and S. Lee. Shooting and bouncing rays: Calculating RCS of an arbitrary cavity. *IEEE Trans Antennas Propag Intern Symp* 24 (1986) 293–296.

22 J. Johnson. Thermal agitation of electricity in conductors. *Phys Rev* 32 (1928) 97.

23 B. R. Mahafza. *Radar systems analysis and design using MATLAB*, 2nd ed. Chapman & Hall/CRC, Boca Raton, FL, 2000.

CHAPTER THREE

Synthetic Aperture Radar

Synthetic aperture radar (SAR) is a high-resolution airborne and spaceborne remote sensing technique for imaging remote targets on a terrain or more generally on a scene. In 1951, Carl Wiley realized that if the echo signal is collected when the radar is moving along a straight path, the Doppler spectrum of the received signal can be used to synthesize a much longer aperture so that very close targets in the along-track dimension can be resolved [1]. In 1953, the first measured SAR image was formed when a C-46 aircraft was used to map a section of Key West, Florida [2, 3]. The first on-board satellite SAR system was developed by the National Aeronautics and Space Administration (NASA) researchers and put on Seasat in 1978. This remarkable satellite provided a lot of data for oceanographic applications. After Seasat, several satellites carrying SAR systems have been launched by different countries. Russian Almaz (1987), European ERS-1 (1991) and ERS-2 (1995), and Canadian Radarsat (1995) were among some of them. The first space-shuttle mission that has a SAR module was SIR-A (shuttle imaging radar). After SIR-A was launched aboard the space shuttle Columbia in 1981, other spaceborne SAR missions were followed. SIR-B (1984) and spaceborne imaging radar-C/X-band synthetic aperture radar SIR-C/X-SAR (1994) acquired SAR images in multiple frequencies and polarizations for more advanced applications such as interferometric and polarimetric mapping of terrains.

Although SAR has been primarily utilized for surveillance applications such as detection of opponents' territories, buildings, airplanes, and tanks, it has also found many real-world applications ranging from geophysics to archeology. For the last couple of decades, several air and space vehicles have been mapping Earth's surface to better understand and interpret terrains and

Inverse Synthetic Aperture Radar Imaging with MATLAB Algorithms, First Edition.
Caner Özdemir.
© 2012 John Wiley & Sons, Inc. Published 2012 by John Wiley & Sons, Inc.

associated geological events. The use of SAR in predicting volcano eruptions, coseismic displacement fields, and glacier motions are some of the various applications in different sciences. In geology, SAR applications are mainly focused on topography and topographic changes and assessment of hazards such as potential for floods, volcanic eruptions, and earthquakes. In ecology, SAR systems are being utilized for land cover classification, inundation mapping, and biomass measurements [4]. SAR has also been used in environmental science for forest classification, deforest monitoring, hazard monitoring, oil spill, and the detection of squatters in cities. SAR usage in hydrology is primarily focused on detection of soil moisture and snow water equivalence. In oceanography, SAR systems can map ocean currents, winds, ocean surface features, sea ice thickness, and coastal processes. In agriculture, its main usage is for crop monitoring. The SAR concept is also used for ice sheet and glacier research, including ice velocity determination, seasonal melt monitoring, and the study of icebergs. SAR-based systems are utilized for detecting and imaging subsurface objects ranging from mines to archeological substances.

3.1   SAR MODES

The modes of SAR operation can be divided into three according to the radar antenna's scanning operation. As illustrated in Figure 3.1a, when the radar collects the electromagnetic (EM) reflectivity of the region alongside which it travels, observing a strip of a terrain parallel to the flight path, this mode is called *side-looking SAR* or *strip-map SAR*. When the radar tracks and focuses its illumination to a fixed, particular area of interest as shown in Figure 3.1b, this mode is named *spotlight SAR*.

Another mode of SAR operation is called *scan SAR*, which is especially used when the radar is flying at high altitude and to obtain a swath wider than the ambiguous range [5]. This enhancement in swath costs degradation in range resolution. For this mode, the illumination area is divided into several segments, and each segment is assigned to the observation of a different swath. As the radar platform moves, radar illuminates one segment for a time period and then switches to illuminate another one. This switching is accomplished in a methodology such that the desired swath width is covered and no empty segment is left as the platform progresses on its track as demonstrated in Figure 3.1c.

3.2   SAR SYSTEM DESIGN

A generic SAR system block diagram is shown in Figure 3.2. Let us take a closer look at the subsystems that constitute a SAR system. All the timing and control signals are produced by the processor control unit. First, the SAR signal linear frequency modulated (LFM) pulse or the stepped frequency

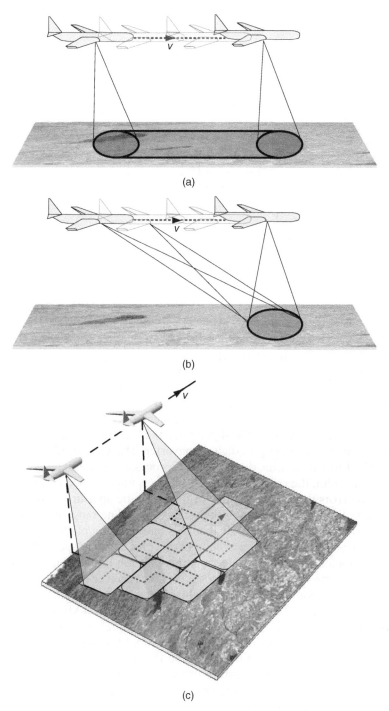

FIGURE 3.1   Modes of SAR: (a) side-looking SAR, (b) spotlight SAR, (c) scan SAR.

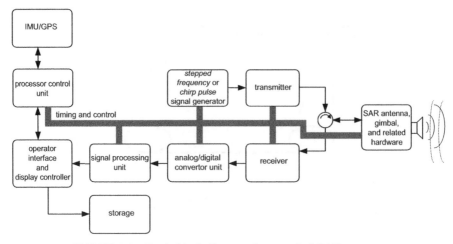

FIGURE 3.2 Basic block diagram for a typical SAR system.

waveform) is generated by the waveform generator and passed to the transmitter. Most SAR systems use either a single antenna or two closely situated
antennas for transmission and reception such that the system works generally
at the monostatic configuration. A SAR antenna, gimbal, and antenna beam-
former can be used to form and direct the main beam along the direction of
the scene or target. After the transmitted SAR signal is reflected back from
the scene or target, the received signal is collected by the SAR antenna and
passed to the receiver. Then, the signal after the receiver's output is sampled
and digitized by the analog-to-digital converter. The unfocused SAR image
can then be formed after the digital signal is processed by the signal processing
unit. At this point, the constructed SAR image is unprocessed or raw, that is,
it contains errors due to instabilities of the synthetic aperture caused by the
platform motion and also due to other effects, including range migration and
range walk. It is obvious that SAR theory is based on the platform's uniform
motion along a straight line or a circular path with a constant speed. Any other
motion characteristics that deviate from this uniform motion such as yawing,
rolling, or pitching will cause error (or shift) in the received phase that contains
the information about the locations of the scatterers on the scene. To be able
to correct these effects caused by such instabilities, most SAR systems carry
inertial measurement unit (IMU) and global positioning system (GPS) sensors
to record the mission history of the flight. The raw SAR image can be displayed
on the operator's scene. The raw image and the data provided by IMU/GPS
sensors can also be stored for future processing or downlinked to a remote
station for further signal and image analysis. Post-processing of raw SAR data
to successfully correct such errors results in what is called the focused SAR
image.

3.3   RESOLUTIONS IN SAR

SAR has garnered its fame because it can provide fine resolutions both in range and cross-range dimensions. The term *range (slant-range)* corresponds to the line-of-sight distance from the radar to the target to be imaged. The term *cross range (transverse range, azimuth* or *along track)* is used for the dimension that is perpendicular to range or parallel to the radar's along-track axis. High resolution in range is obtained in SAR operations by using a wide bandwidth transmitted signal, generally the frequency modulated or the so-called *chirp waveform.* Good resolution in the cross-range dimension is achieved by coherently processing the target's EM scattering measured at different aspects while the radar is generally moving along a straight path.

With this construct, SAR can provide images that have comparable resolutions to those of optic imaging systems. In fact, SAR can do even better since it can operate day and night, as well as in cloudy and rainy weather conditions. Furthermore, unlike photographic images that contain only the amplitude information of the target's reflectivity of light, SAR provides both the amplitude and the phase information of the scattered EM field from the scene. This leads to formation of interferometric SAR (IFSAR) images that may present the information about the third dimension, height, so that three-dimensional (3D) SAR images of the scene can also be formed.

The term "synthetic aperture" is commonly used in the literature because the idea behind SAR is to synthesize the effect of large aperture radar by means of a collection of small-size real aperture radars (RARs; see Fig. 3.3a). This is accomplished by moving the small aperture radar along with the imaginary aperture axis to emulate a much longer aperture as depicted in Figure 3.3b. Typically, the radar is put on an airborne or a spaceborne platform, and the ground's EM reflectivity is measured at different time instants while the radar platform is moving through its along-track path (see Fig. 3.4). Coherent processing of EM echoes from the illuminated area at different frequencies and apertures makes it possible to form the two-dimensional (2D) image of the terrain or the scene.

The range processing of a SAR system is identically the same as it is with conventional radar. Therefore, the range resolution, Δr, in SAR is the same as in the case of conventional radar as

$$\Delta r = \frac{c}{2B}. \tag{3.1}$$

On the other hand, the frequency bandwidth B of the pulsed waveform in SAR operation is selected wide enough to achieve finer resolutions. According to the Fourier theory, the time duration of each pulse is then required to be very short. On the other hand, it is very hard to put enough energy into a very short pulse. Since very small portion of the transmitted power scatters back to the radar, it will be almost impossible to sense the received signal above

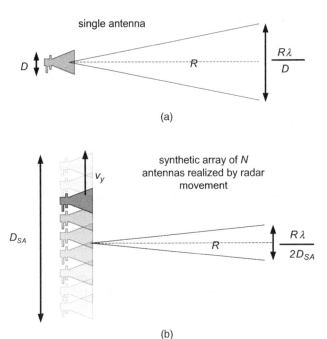

FIGURE 3.3 (a) A real aperture of a single antenna, (b) a synthetic aperture of N antennas.

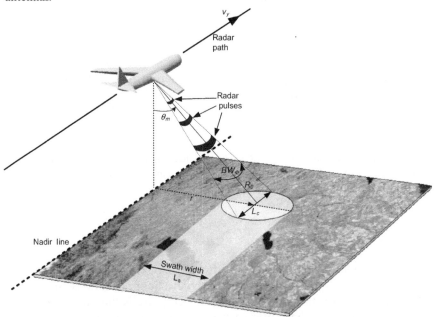

FIGURE 3.4 Basic SAR operation: Radar platform is moving to synthesize the effect of a long, real aperture radar.

the noise floor for such a short pulse. To circumvent this problem, an LFM or chirp waveform that has an instantaneous frequency of

$$f_i(t) = f_o + B \cdot \frac{t}{T_o}; \quad -\frac{T_o}{2} \le t \le \frac{T_o}{2} \tag{3.2}$$

is utilized. Here, f_o is the starting frequency, T_o is the pulse duration, and B is the total bandwidth. LFM signal has the suited property of providing required bandwidth with longer pulse duration as explained in detail in Chapter 2, Section 2.6.5. For example, 1 μs duration chirp pulse with 1 GHz bandwidth provides a range resolution of 15 cm which is quite good for typical SAR applications.

A good cross-range resolution in the SAR system is achieved by forming a synthetic line antenna of length D_{SA} by moving the radar antenna along the straight path (see Fig. 3.3b). If a single antenna is used, the cross-range resolution is expected to be

$$\Delta y = \frac{R\lambda}{2D}, \tag{3.3}$$

where λ is the operational wavelength and D is the aperture of the antenna. However, if a synthetic array of antennas is formed by moving a single antenna along a synthetic length of D_{SA}, we should theoretically expect a cross-range resolution of

$$\Delta y = \frac{R\lambda}{D_{SA}}. \tag{3.4}$$

Notice that the effective synthetic aperture size is twice that of a real array [6]; in reality, therefore, Δy becomes equal to

$$\Delta y = \frac{R\lambda}{2D_{SA}}. \tag{3.5}$$

The detailed derivation (Eq. 3.5) will be given in Section 3.7.2. If the SAR platform, for instance, collects the scattered EM wave at the center frequency of 10 GHz from a target of 20 km away and for a synthetic aperture of 2 km, then the cross-range resolution is 15 cm, which is much better than in the RAR case.

3.4 SAR IMAGE FORMATION: RANGE AND AZIMUTH COMPRESSION

SAR imagery is, in fact, based on successive signal processing algorithms called *range compression* and *azimuth compression*. The usual raw SAR data are in

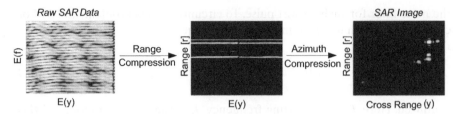

FIGURE 3.5 SAR image is formed by applying range compression and azimuth compression algorithms independently.

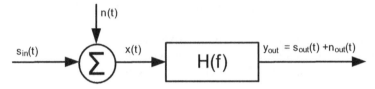

FIGURE 3.6 The matched filter receiver.

the form of a 2D multifrequency, multiaspect (or multispatial) scattered field data, as shown in Figure 3.5. The range compression and azimuth compression are usually applied independently to obtain the ultimate SAR image (see Fig. 3.5).

3.5   RANGE COMPRESSION

As mentioned earlier, the LFM or chirp waveform is usually utilized to compress the range signal. The word *compression* is used because the frequency content of the chirp signal is changing rapidly such that the full bandwidth B corresponds to a time duration of $T_p = 1/B$, which is much shorter than the actual pulse duration, T_o. Because of this property, the LFM pulse is also called the *stretch waveform* [7,8] in radar literature. Range compression is commonly achieved by applying a matched filter and the pulse compression procedure afterward to maximize the signal-to-noise ratio (SNR) and obtain the range compressed waveform.

3.5.1   Matched Filter

The matched filter is the optimal linear filter commonly used in radar signal processing to maximize the SNR of the reflected wave in the presence of noise, as demonstrated in Figure 3.6. Since the transmitted signal is known, the reflected signal with noise is tested with the same type of signal to maximize the signal energy within the reflected wave over the additive noise. One common procedure that utilizes the matched filter operation is called the *pulse compression*.

If the matched filter receiver has the frequency response, $H(f)$, then the impulse response of the filter is $h(t) = \mathcal{F}^{-1}\{H(f)\}$. The receiver is assumed to collect the signal $s_{in}(t)$ together with an additive white noise, $n(t)$. Although it is not necessary, the Gaussian white noise (GWN) is commonly assumed.

The autocorrelation function of $n(t)$ is

$$\mathcal{R}_n(t) = \int_{-\infty}^{\infty} n(\tau)n^*(t-\tau)d\tau$$
$$= \frac{N_o}{2}\delta(t).$$

(3.6)

Here, ()* denotes the complex conjugate operation. Therefore, the power spectral density (PSD) of the noise can be found by Fourier transforming the autocorrelation function as

$$PSD_n(f) = F\{\mathcal{R}_n(t)\}$$
$$= \frac{N_o}{2}.$$

(3.7)

Notice that the PSD of noise is frequency independent, that is, constant for all frequencies. The receiver filter has the output that is the convolution of the input signals, $s_{in}(t)$ and $n(t)$, with the filter impulse response of $h(t)$. Therefore, the output signal, $s_{out}(t)$, and the output noise, $n_{out}(t)$, are equal to

$$s_{out}(t) = h(t) * s_{in}(t)$$
$$n_{out}(t) = h(t) * n(t).$$

(3.8)

Taking the Fourier transform (FT) of Equation 3.8, we get

$$S_{out}(f) = H(f) \cdot S_{in}(f)$$
$$N_{out}(f) = H(f) \cdot N(f).$$

(3.9)

Let us define the SNR value at the output of the filter as the ratio of the signal power to the noise power at the output of the filter:

$$SNR_{out} = \frac{|S_{out}(t)|^2}{|n_{out}(t)|^2}.$$

(3.10)

Notice that the PSD of noise at the output can be written in terms of PSD of noise at the input and the filter characteristic as

$$PSD_{n_{out}}(f) = |H(f)|^2 \cdot PSD_n(f)$$
$$= |H(f)|^2 \cdot \frac{N_o}{2}.$$

(3.11)

Therefore, the output noise power, $P_{n_{out}}$, can now be calculated as given below:

$$P_{n_{out}} \triangleq |n_{out}(t)|^2$$
$$= \int_{-\infty}^{\infty} PSD_{n_{out}}(f) df \qquad (3.12)$$
$$= \frac{N_o}{2} \cdot \int_{-\infty}^{\infty} |H(f)|^2 df.$$

Similarly, power of the output signal can be calculated as

$$P_{S_{out}} \triangleq |s_{out}(t)|^2$$
$$= |\mathcal{F}^{-1}\{S_{out}(f)\}|^2 \qquad (3.13)$$
$$= \left| \int_{-\infty}^{\infty} H(f) \cdot S_{in}(f) \cdot e^{j2\pi ft} df \right|^2.$$

If the filter provides a time delay, t_i, and the signal is measured at that time instant, Equation 3.13 can be rewritten as

$$P_{S_{out}} = \left| \int_{-\infty}^{\infty} H(f) \cdot S_{in}(f) \cdot e^{j2\pi ft_i} df \right|^2. \qquad (3.14)$$

According to Schwarz inequality [8], Equation 3.14 can be readily decomposed into

$$P_{S_{out}} \leq \int_{-\infty}^{\infty} |H(f)|^2 df \cdot \int_{-\infty}^{\infty} |S_{in}(f) \cdot e^{j2\pi ft_i}|^2 df$$
$$= \int_{-\infty}^{\infty} |H(f)|^2 df \cdot \int_{-\infty}^{\infty} |S_{in}(f)|^2 df \qquad (3.15)$$
$$= \int_{-\infty}^{\infty} |H(f)|^2 df \cdot P_{S_{in}},$$

where $P_{S_{in}}$ is the signal's power at the input of the filter. Then, substituting Equations 3.15 and 3.13 into Equation 3.10, we get

$$SNR_{out} \triangleq \frac{P_{S_{out}}}{P_{n_{out}}}$$
$$\leq \frac{P_{S_{in}} \cdot \int_{-\infty}^{\infty} |H(f)|^2 df}{\frac{N_o}{2} \cdot \int_{-\infty}^{\infty} |H(f)|^2 df} \qquad (3.16)$$
$$= \frac{2P_{S_{in}}}{N_o}.$$

Therefore, the SNR value at the output is always less than or equal to $2P_{s_{in}}/N_o$. According to the Schwarz inequality [8], the inequality in Equation 3.15 becomes equal if

$$\left(S_{in}(f)\cdot e^{j2\pi f t_i}\right)^* = k \cdot H(f) \tag{3.17}$$

or

$$\begin{aligned} H(f) &= \frac{1}{k}\cdot S_{in}^*(f)\cdot e^{-j2\pi f t_i} \\ &= C\cdot S_{in}^*(f)\cdot e^{-j2\pi f t_i}, \end{aligned} \tag{3.18}$$

where k and C are some constants and can be taken as 1 for simplicity. When Equation 3.18 is satisfied, then the output SNR reaches its maximum value of $2P_{s_{in}}/N_o$. The filter in Equation 3.18 that maximizes the SNR at the filter output and that is matched to the spectrum of the input signal as $|H(f)| = |S_{in}(f)|$ is called the *matched filter*. Knowing the frequency response of the filter, the impulse response or the time-domain characteristics of the matched filter can be found by inverse Fourier transformation (Eq. 3.18) as

$$\begin{aligned} h(t) &= \mathcal{F}^{-1}\{H(f)\} \\ &= \int_{-\infty}^{\infty} S_{in}^*(f)\cdot e^{-j2\pi f t_i}\cdot e^{j2\pi f t}dt \\ &= \int_{-\infty}^{\infty} S_{in}^*(f)\cdot e^{j2\pi f(t-t_i)}dt \\ &= \left[\int_{-\infty}^{\infty} S_{in}(f)\cdot e^{j2\pi f(t_i-t)}dt\right]^* \\ &= s_{in}^*(t_i-t). \end{aligned} \tag{3.19}$$

3.5.1.1 Computing Matched Filter Output via Fourier Processing

The output of the matched filter can be easily computed via fast Fourier transform (FFT) processing. The filter output, $y_{out}(t)$, can be written in terms of the received signal and the impulse function of the filter as the convolution below:

$$y_{out}(t) = x(t)*h(t). \tag{3.20}$$

Taking the FT of Equation 3.16, we get the following equation

$$Y(f) = X(f)\cdot H(f). \tag{3.21}$$

Therefore, Equation 3.20 can be rewritten in a more appropriate way as given below:

$$y_{out}(t) = IFT\{FT\{x(t)\}\cdot H(f)\}. \tag{3.22}$$

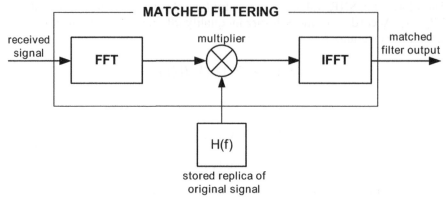

FIGURE 3.7   The matched filter realization via FFT processing.

The block diagram for realizing the matched filter operation is shown in Figure 3.7. If the received signal is sampled and digitized, its FFT is multiplied with the matched filter transfer function (or the stored replica of the original transmitted signal that is given in Equation 3.18). The result of the multiplication undergoes an inverse fast Fourier transform (IFFT) operation to obtain the final signal at the output of the matched filter.

3.5.1.2   Example for Matched Filtering   An example is provided to demonstrate the use of matched filtering operation. As given in Figure 3.8a, a rectangular pulse waveform is assumed to be transmitted by the radar. This pulse travels 4 μs in time to reach the receiver (see Fig. 3.8b). Additive white Gaussian noise (AWGN), whose time-domain characteristics are shown in Figure 3.8c, is assumed to be present in the propagation medium. The receiver collects the shifted version of the original waveform distorted with AWGN as depicted in Figure 3.8d. After the received signal is collected, the matched filtering operation is employed by using Equation 3.22. Therefore, the resultant filter output in Figure 3.8e is obtained since the matched filter operation has maximized the SNR. It is clearly seen from this figure that the noise is highly suppressed. The shape of the output signal is in the form of a triangular pulse since the matched filter output is nothing but the convolution of the input signal with the filter's impulse response which has the same shape as the input signal, that is, rectangular pulse. As expected, the output signal makes its maximum at the time instant of 4 μs.

3.5.2   Ambiguity Function

In pulsed radar signal processing, the *ambiguity function* is used to acquire insight about the usage of different waveforms for various radar applications. Ambiguity function is a 2D function of time delay and Doppler frequency. In fact, the ambiguity function can be regarded as the output of the matched filter

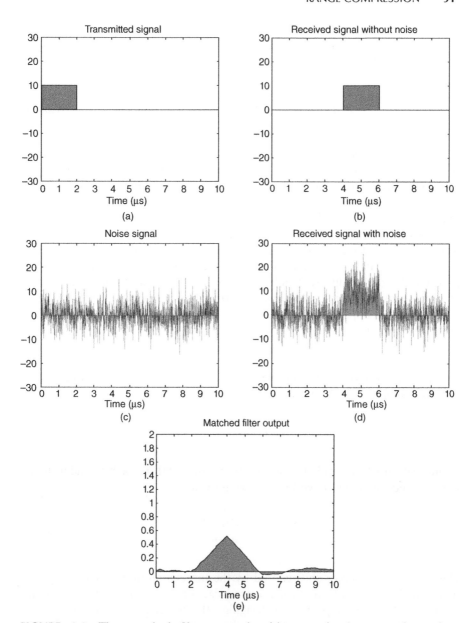

FIGURE 3.8 The matched filter example: (a) transmitted rectangular pulse, (b) received pulse signal without noise, (c) normal distributed noise signal, (d) received signal with an additive noise, and (e) the output of the matched filter.

that is used to compress the radar pulse as will be explained in Section 3.5.1.3. The formula of the ambiguity function is given by the correlation of the pulse with its delayed version both in time and frequency as given below:

$$\chi_g(t, f_D) = \left| \int_{-\infty}^{\infty} g(\tau) \cdot g^*(\tau - t) \cdot e^{-j2\pi f_D \tau} d\tau \right|. \tag{3.23}$$

Usually, the time shift, t, represents the reflected waveform and the frequency shift, f_D, represents the Doppler frequency shift due to a moving target or moving radar or both. If the target is stationary with respect to radar (i.e., $f_D = 0$), the ambiguity function reduces to *autocorrelation function* of the signal $g(t)$. As the formula in Equation 3.23 suggests, the ambiguity function describes the intrusion caused by the range and/or Doppler shift of a target when compared to a reference (or pilot) target of the same type located at $(t = 0, f_D = 0)$. Therefore, $\chi(t, f_D)$ can be conveniently used by radar designers to test and evaluate different radar waveforms by comparing their range and Doppler frequency shift resolutions such that the appropriate waveform for a particular radar application can be judged. To visualize this examination, a 3D plot (called *radar ambiguity diagram*) of the ambiguity function versus time delay and Doppler frequency shift is used.

3.5.2.1 *Relation to Matched Filter*   Suppose a monotone transmitted signal that oscillates at a single frequency as

$$s_{tx}(t) \sim e^{-j2\pi f_c t} \tag{3.24}$$

If the scatterer has a radial translational motion with respect to radar, then the received signal will have the following form:

$$\begin{aligned} s_{rx}(t) &\sim s_{tx}(t) \cdot e^{-j2\pi f_D t} \\ &\sim e^{-j2\pi (f_c + f_D) t}. \end{aligned} \tag{3.25}$$

Here, f_D is the Doppler shift caused by the target's motion. When using the matched filter, the pattern of the received signal is an unknown phenomenon. Therefore, the best approach in designing the matched filter is to use the replica of the transmitted signal. When the target is moving, the matched filter output can be written as

$$\begin{aligned} s_{out}(t) &= h(t) * s_{rx}(t) \\ &= h(t) * s_{tx}(t) \cdot e^{-j2\pi f_D t} \\ &= \int_{-\infty}^{\infty} h(t - \tau) \cdot s_{tx}(\tau) \cdot e^{-j2\pi f_D \tau} d\tau. \end{aligned} \tag{3.26}$$

As mentioned above, $h(t)$ can be conventionally selected to be equal to $s_{tx}^*(-t)$. Then,

$$h(t - \tau) = s_{tx}^*(-(t - \tau))$$
$$= s_{tx}^*(\tau - t).$$

(3.27)

Therefore, the matched filter output becomes equal to

$$s_{out}(t) = \int_{-\infty}^{\infty} s_{tx}(\tau) \cdot s_{tx}^*(\tau - t) \cdot e^{-j2\pi f_D \tau} d\tau$$
$$\triangleq \chi_{s_{tx}}(t, f_D).$$

(3.28)

Comparing this result with the definition of the ambiguity function in Equation 3.23, the output of the matched filter is the same as the ambiguity function of the transmitted waveform, s_{tx}.

Another implementation of the ambiguity function is that it is nothing but the *cross-correlation* of $s_{tx}(t)$ and $s_{tx}(t) \cdot e^{-j2\pi f_D t}$. Therefore, the plot of ambiguity function with respect to the time variable, t, and the Doppler frequency variable, f_D, presents a 2D map of ambiguity of the received waveform to the transmitted waveform in terms of time shift and Doppler shift. Time shift is normally due to the fact that the target is located at a finite distance in range and Doppler shift is generally caused by the target's motion. Therefore, radar designers utilize the radar ambiguity diagram to test different radar waveforms in terms of uncertainties or the resolutions both in time and the Doppler shift. Next, we will demonstrate radar ambiguity diagrams for some basic radar waveforms.

3.5.2.2 *Ideal Ambiguity Function* The ideal ambiguity function must provide perfect resolution such that any neighboring targets can be resolved regardless of how they close they are to each other. In the radar ambiguity diagram, the ideal ambiguity function is the 2D impulse (or Dirac delta) function located at $(t = 0, f_D = 0)$ as shown in Figure 3.9. Therefore, the ideal ambiguity function can be mathematically represented as

$$\chi_{ideal}(t, f_D) = \delta(t, f_D).$$

(3.29)

The term "ideal" is used because this function represents no ambiguities at all by providing zero time delay and zero Doppler shift. The ideal ambiguity function cannot be generated practically. This is because there is no real signal that can generate the ambiguity function of $\chi_{ideal}(t, f_D) = \delta(t, f_D)$ after applying the correlation integral of Equation 3.23.

3.5.2.3 *Rectangular-Pulse Ambiguity Function* One of the simplest radar signals is the rectangular pulse, whose mathematical expression can be written as

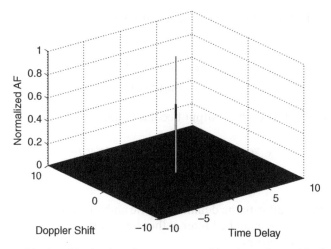

FIGURE 3.9 Ideal ambiguity function represented by the radar ambiguity diagram.

$$g(t) = \begin{cases} A & 0 \leq t \leq T \\ 0 & elsewhere. \end{cases} \tag{3.30}$$

Inserting Equation 3.30 into Equation 3.23, the ambiguity function for the rectangular pulse can be calculated analytically as follows:

$$\chi(t, f_D) = \begin{cases} A^2(T - |t|) \cdot sinc(f_D(T - |t|)) & |t| \leq T \\ 0 & elsewhere. \end{cases} \tag{3.31}$$

Here, $sinc(\cdot)$ represents the *sinus cardinalis* function whose mathematical expression is simply

$$sinc(t) = \frac{\sin(\pi t)}{\pi t}. \tag{3.32}$$

A numerical example is illustrated in Figure 3.10 where a normalized plot of ambiguity function for a rectangular pulse of width $T = 1$ µs and amplitude of $A = 1$ is shown.

3.5.2.4 LFM-Pulse Ambiguity Function As will be shown later, LFM, or chirp pulse is one of the most suitable waveforms to have a range resolution. Therefore, we will investigate the ambiguity caused by this type of pulse by plotting the corresponding ambiguity function using a radar ambiguity diagram. The mathematical expression for an upward LFM pulse that has a width of T is given as

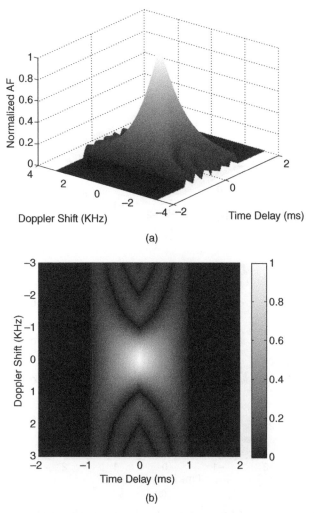

FIGURE 3.10 Normalized rectangular pulse ambiguity function (a) in the radar ambiguity diagram, (b) on the 2D time delay-Doppler shift plane.

$$g(t) = \begin{cases} A \cdot e^{-j\pi Kt^2} & 0 \le t \le T \\ 0 & elsewhere. \end{cases} \tag{3.33}$$

Inserting Equation 3.29 into Equation 3.23, the ambiguity function for the LFM pulse can be calculated analytically as follows:

$$\chi(t, f_D) = \begin{cases} A^2 (T - |t|) \cdot sinc((Kt - f_D) \cdot (T - |t|)) & |t| \le T \\ 0 & elsewhere. \end{cases} \tag{3.34}$$

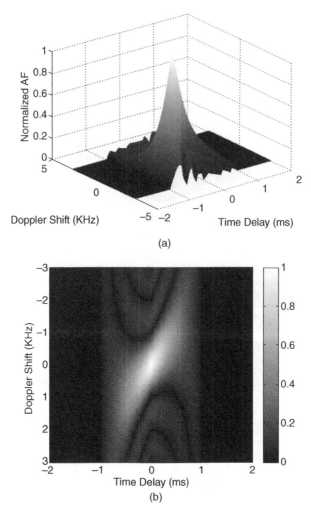

FIGURE 3.11 Normalized LFM pulse ambiguity function (a) in the radar ambiguity diagram, (b) on the 2D time delay-Doppler shift plane.

A numerical example is illustrated in Figure 3.11 where the normalized plot of the ambiguity function for the rectangular pulse of width $T = 1$ µs and amplitude of $A = 1$ is shown. The chirp rate parameter, K, is chosen as $2 \cdot 10^6$. As shown from the figure, as the time delay changes, Doppler shift also changes due to the nature of the LFM function.

3.6 PULSE COMPRESSION

Where imaging radar is concerned, using LFM waveform and applying *pulse compression* procedure is the common practice to obtain the range compressed waveform and achieve a good range resolution. The use of LFM waveform provides sufficient pulse duration to be able to boost enough energy for a good SNR value at the receiver. For a monotone pulse waveform of duration T_1, the range resolution is

$$\Delta r = \frac{c}{2} T_1. \tag{3.35}$$

Therefore, the shorter the pulse duration, the sharper the range resolution. On the other hand, it is very hard to put enough energy into a very short pulse due to the fact that the energy of a signal, $s(t)$, is proportional to its duration, T_1, as

$$E = \int_o^{T_1} |s(t)|^2 \, dt. \tag{3.36}$$

If the noise power density is N_o, the SNR value at the matched filter output becomes $2E/N_o$. Then, if a monotone pulse waveform is used, the peak power over the short period of T_1 should be boosted up significantly, which may not be possible using microwave tubes. To offer a solution to this problem, Oliver [9] has suggested using LFM (or chirp) waveform such that the transmitted pulse—which provides the same bandwidth of monotone pulse—is stretched over a longer period of time. Hence, it will be possible to elevate the signal energy by increasing the duration time of T_1 without amplifying the peak power. Pulse compression scheme is concluded by applying the matched filter processing on reception to "compress" the received signal to a sufficiently short duration signal in time so that a fine resolution in range is achieved.

3.6.1   Detailed Processing of Pulse Compression

As the transmitted signal, LFM (chirp) pulse waveform given below is used

$$S_{tx} = \begin{cases} e^{j2\pi\left(f_c t + \frac{K \cdot t^2}{2}\right)} & |t| \le T_1 / 2 \\ 0 & elsewhere, \end{cases} \tag{3.37}$$

where K is the chirp rate and it has a positive sign for the upchirp case or a negative sign for the downchirp situation. T_1 is the pulse duration. The magnitude of the pulse is normalized to 1 for simplicity. The instantaneous frequency of the transmitted pulse can be calculated as the time derivative of its phase as

$$\begin{aligned} f_i &= \frac{1}{2\pi} \frac{d}{dt} \left\{ 2\pi \left(f_c t + \frac{K \cdot t^2}{2} \right) \right\} \\ &= f_c + K \cdot t. \end{aligned} \tag{3.38}$$

If the upchirp signal is chosen, the frequency of the transmitted signal is increasing from $(f_c - K \cdot T_1/2)$ to $(f_c + K \cdot T_1/2)$. The time-frequency plot of this upchirp waveform is illustrated in Figure 3.12a. For the downchirp wave-form, the frequency is dropping from $(f_c + K \cdot T_1/2)$ to $(f_c - K \cdot T_1/2)$ as its time-frequency plot is demonstrated in Figure 3.12b. Therefore, the bandwidth of the LFM pulse (either upchirp or downchirp) is

$$B = K \cdot T_1. \tag{3.39}$$

If the time-bandwidth product, D, is defined as the multiplication of the sig-nal's duration in time with the total bandwidth of the signal, we can calculate the time-bandwidth product of the chirp signal as

$$\begin{aligned} D &= B \cdot T_1 \\ &= K \cdot T_1^2. \end{aligned} \tag{3.40}$$

This value is also called the *dispersion factor* and *compression ratio* as will be demonstrated later in this chapter.

A typical upchirp signal has the waveform characteristics given in Equation 3.33. If there exists a point target at the range of R_i, then the backscattered signal at the receiver will have a time delay of $t_i = 2R_i/c$. Therefore, the received signal will have the following phase term:

$$s(t) \sim e^{j2\pi\left(f_c(t-t_i)+\frac{K}{2}\cdot(t-t_i)^2\right)}. \tag{3.41}$$

Here, the target is assumed to be stationary, and therefore, there is no Doppler shift at the frequency content of the received signal. Otherwise, the frequency should be replaced with $f_c' = f_c + \Delta f$ where Δf is the Doppler shift due to the motion of the target with respect to radar. As demonstrated in Equation 3.19, the matched filter's impulse response is related to the original signal as

$$\begin{aligned} h(t) &= s_{in}^*(-t) \\ &= e^{j2\pi\left(f_c t-\frac{K\cdot t^2}{2}\right)}, |t| \le T_1/2. \end{aligned} \tag{3.42}$$

Then, the matched filter output signal, $s_{out}(t)$, can be written as the convolution between the received signal and the matched filter's impulse response as

$$\begin{aligned} s_{out}(t) &= \int_{-\infty}^{\infty} s(t-\tau)h(\tau)d\tau \\ &= \int_{-T_1/2}^{T_1/2} e^{j2\pi\left(f_c(t-\tau-t_i)+\frac{K}{2}\cdot(t-\tau-t_i)^2\right)} e^{j2\pi\left(f_c\tau-\frac{K\cdot\tau^2}{2}\right)}d\tau \\ &= e^{j2\pi\left(f_c(t-t_i)+\frac{K}{2}\cdot(t-t_i)^2\right)} \cdot \int_{-T_1/2}^{T_1/2} e^{-j2\pi[K(t-t_i)\tau]}d\tau \\ &= e^{j2\pi\left(f_c(t-t_i)+\frac{K}{2}\cdot(t-t_i)^2\right)} \cdot \frac{\sin(K\pi T_1(t-t_i))}{K\pi(t-t_i)} \\ &= T_1 \cdot e^{j2\pi\left(f_c(t-t_i)+\frac{K}{2}\cdot(t-t_i)^2\right)} \cdot \text{sinc}[KT_1(t-t_i)]. \end{aligned} \tag{3.43}$$

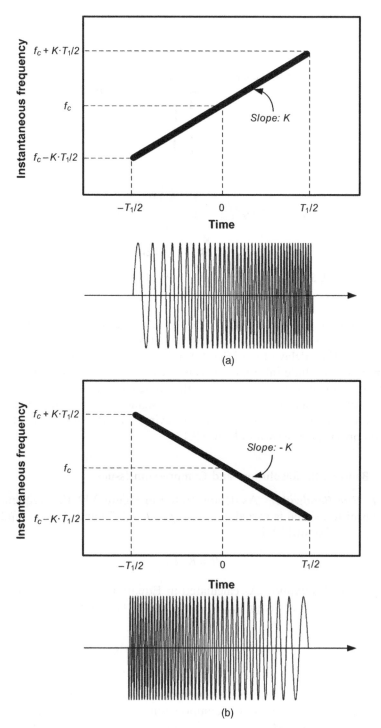

FIGURE 3.12   Time-frequency plot for an (a) upchirp signal, (b) downchirp signal.

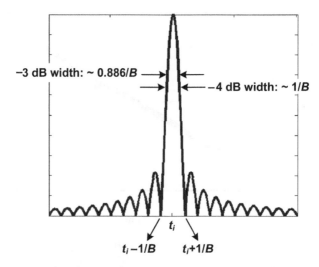

FIGURE 3.13   Output signal after the pulse compression process.

If the output signal in Equation 3.43 is carefully examined, one can easily notice that the first phase term signal is the delayed version of the transmitted signal with a time delay of $t_i = 2R_i/c$. Since it is only a phase signal, it does not present any amplitude information. This phase term is usually suppressed by applying an envelope detector after the matched filtering operation. The second term in Equation 3.43 is an amplitude signal that has a maximum value at the time instant of t_i and has many decaying sidelobes in time as the sinc function provides as shown in Figure 3.13.

3.6.2   Bandwidth, Resolution, and Compression Issues

3.6.2.1   The Bandwidth   As demonstrated in Figure 3.12, the frequency of the transmitted signal is varied between $(f_c - K \cdot T_1/2)$ and $(f_c + K \cdot T_1/2)$, so the frequency bandwidth is

$$B = K \cdot T_1. \tag{3.44}$$

3.6.2.2   The Resolution   As calculated in Equation 3.43, the magnitude of the output signal after the pulse compression has the *sinc*-type characteristics of

$$\begin{aligned} |s_{out}(t)| &= T_1 \cdot |\text{sinc}[KT_1(t - t_i)]| \\ &= T_1 \cdot |\text{sinc}[B(t - t_i)]|. \end{aligned} \tag{3.45}$$

A plot of Equation 3.45 is provided in Figure 3.13 to comment on the resolution of the output signal and the compression ratio. As seen from the figure, the first nulls occur at B away from the center of the sinc at t_i. That is to say, if there is a second point scatterer that shows up in time at $\Delta t = 1/B$ in time, it can be resolved from the first point scatterer and be detected. Therefore, the resolution in time is $\Delta t = 1/B$. Then the resolution in range can be found via the following calculation:

$$\Delta r = \Delta t \cdot \frac{c}{2}$$
$$= \frac{c}{2B},$$

(3.46)

which is identical to the result that was previously calculated and presented in many different places of this book.

3.6.2.3   The Compression   The term *compression* is used for comparing the time duration of the input and output signals that are available during the pulse compression operation. The ratio of the transmitted pulse width to the compressed signal's main lobe width is described as the *pulse compression ratio*.

Initially, the transmitted pulse has a width of T_1. The output signal has the sinc-type characteristic that has an infinite extent in time, theoretically. The -3 dB width of the main lobe of the output time signal is approximately $0.886/B$. The effective width of the main lobe is usually taken as the -4 dB width which is about $1/B$. Therefore, the compression ratio, CR, can be calculated easily as

$$CR = \frac{\text{width of input signal}}{\text{width of output signal}}$$
$$= \frac{T_1}{1/B}$$
$$= \frac{T_1}{1/(K \cdot T_1)}$$
$$= K \cdot T_1^2$$
$$\triangleq D,$$

(3.47)

which is also found to be equal to the *time-bandwidth product*, D. Therefore, to increase the pulse compression ratio, the time-bandwidth product should be increased (or vice versa).

3.6.3   Pulse Compression Example

An example is provided to demonstrate the use of matched filtering and pulse compression in SAR/ISAR imaging. Figure 3.14a shows a chirp waveform representing the transmitted signal from the radar. The normal distributed random noise (Fig. 3.14b) is added such that the received signal is collected as shown in Figure 3.14c. A time delay of 1 µs is assumed to have occurred between the transmit and receive times of the radar.

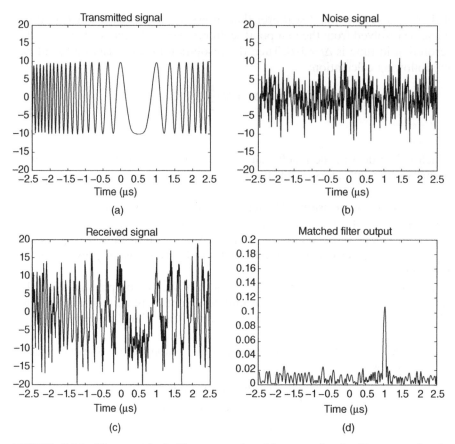

FIGURE 3.14 The matched filter example: (a) transmitted chirp-type signal, (b) normal distributed noise signal, (c) received signal (transmitted + noise), and (d) the output of the matched filter.

As seen from Figure 3.14c, the received waveform is highly distorted due to noise. The output signal after matched filtering is plotted in Figure 3.14d where the received echo is shorter in time due to the feature of the chirp signal. Also, the SNR is magnified such that the output signal is observed at its correct location of 1 μs in time.

3.7   AZIMUTH COMPRESSION

3.7.1   Processing in Azimuth

The compression along the azimuth (or along the cross range) is achieved by utilizing a long synthetic antenna formed by the radar motion. The geometry for SAR azimuth processing is drawn in Figure 3.15. Let there be a point scat-

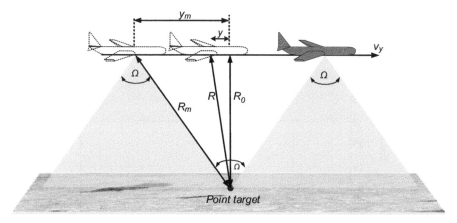

FIGURE 3.15   Geometry for synthetic aperture radar (SAR).

terer at R_o away from the center of the synthetic aperture while the radar platform is moving with a constant speed of v_y. When the radar travels a distance of $y = v_y \cdot t$ along the aperture, the scatterer's distance from the radar becomes $R = (R_o^2 + y^2)^{1/2}$. For practical applications, of course, y is very small when compared to R_o. Therefore, the following approximation to the second order can be made:

$$
\begin{aligned}
R &= R_o \left(1 + \frac{y^2}{R_o^2} \right)^{1/2} \\
&\approx R_o + \frac{y^2}{2R_o}.
\end{aligned}
$$

(3.48)

This approximation provides some error in the phase of the received signal. To understand how this approximation is acceptable, let us consider a SAR operation situation where $R_o = 8$ km, central operational frequency is 6 GHz, and the -3 dB beamwidth of the SAR antenna is $\Omega = 5°$. This means that radar starts to observe the point target when $y_m = R_o \cdot \tan(\Omega/2) \cong 350$ m away from the target's projection on the y-axis. Therefore, the point target on the scene is visible to the radar for a synthetic aperture length of $Y_m = 2 * y_m = 700$ m. In Figure 3.16a, the actual radial distance and estimated radial distance are plotted together to be compared with each other for the selected SAR parameters. As seen from the figure, the formula in Equation 3.48 pretty well estimates the radial distance of the scatterer from the radar. The associated error in the radial distance per wavelength is plotted in Figure 3.16b for the effective synthetic aperture length of ~700 m. The phase error in λ is almost zero for most of this effective synthetic aperture and reaches only small fractions of λ at the edges of the effective synthetic aperture. When the -3 dB beamwidth

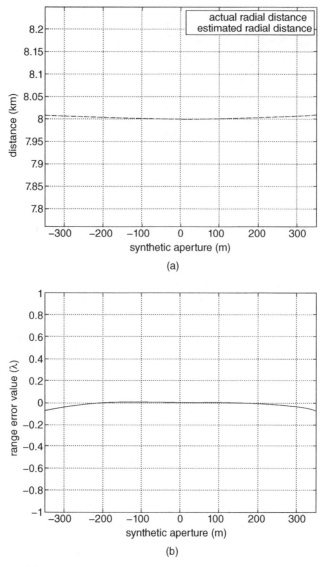

FIGURE 3.16 (a) Actual and estimated radial distance of a scatterer at 8 km away by an antenna whose beamwidth is 5°; (b) corresponding error in range estimation per wavelength.

of the SAR antenna is not small and R_o is much bigger, the effective synthetic aperture length can reach a bigger value such that a correction scheme may be needed to control the error associated with the approximation in Equation 3.48.

The EM wave that possesses two-way propagation between the radar and the scatterer has a phase term of

$$\Phi(y) \sim \exp(-j2kR)$$

$$= \exp\left(-j\left(\frac{4\pi}{\lambda}\right)\left(R_o + \frac{y^2}{2R_o}\right)\right)$$

$$= \exp\left(-j\left(\frac{4\pi}{\lambda}\right)R_o\right) \cdot \exp\left(-j\frac{2\pi y^2}{\lambda R_o}\right) \qquad (3.49)$$

$$= constant \cdot \exp\left(-j\frac{2\pi y^2}{\lambda R_o}\right).$$

By neglecting the phase term that has no time dependency, we can write the phase term of the received signal as

$$\Phi(y) \sim \exp\left(-j\frac{2\pi v_y^2 \cdot t^2}{\lambda R_o}\right)$$

$$\Phi(t) \sim \exp(-j\gamma t^2), \qquad (3.50)$$

where $\gamma = 2\pi v_y^2/(\lambda R_o)$. Then, the Doppler shift in frequency can be found by taking a time derivative of the phase:

$$f_D(t) = \frac{1}{2\pi}\frac{\partial}{\partial t}\left(-\frac{2\pi v_y^2 \cdot t^2}{\lambda R_o}\right)$$

$$= -\frac{2v_y^2 \cdot t}{\lambda R_o}. \qquad (3.51)$$

The processing in cross-range dimensions can be made by integrating the whole aperture return over the integration time, T_s. This method is called *unfocused SAR* since uniform illumination is assumed during T_s. Then, the output signal of the azimuth SAR processing is obtained via the following averaging integral:

$$s_{out}^{unfocused}(y) = \frac{1}{T_s}\int_{-T_s/2}^{T_s/2} \exp\left(-j\frac{2\pi}{\lambda R_o} \cdot (v_y t - y)^2\right) dt. \qquad (3.52)$$

In fact, Equation 3.52 is in the form of *Fresnel* integral [10, 11]. Therefore, the cross-range compressed data can be obtained numerically by solving the Fresnel integral in Equation 3.52 in unfocused-SAR imaging.

   Noticing that the signal in Equation 3.50 is in a similar form as the chirp signal, the processing in azimuth can also be carried out in a similar manner as range. In unfocused SAR, the path lengths from different aperture points differ. This leads to a resolution mismatch as the radar moves along its track. In photography, a lens is used to focus the rays from object to image plane. In SAR, this is achieved by adjusting the path lengths from all aperture points to be the same by applying a matched filter. This process is called the *focused-*

SAR operation. So, if the matched filter processing is applied to the azimuth SAR data in a similar manner, then the output of the filter can be found via

$$s_{out}^{focused}(t) = \frac{1}{T_s}\int_{-T_s/2}^{T_s/2}\exp\left(-j\gamma(t+\tau)^2\right)\cdot\exp\left(j\gamma\tau^2\right)d\tau$$
$$= sinc(\gamma T_s t)\cdot\exp\left(-j\gamma t^2\right) \quad (3.53)$$
$$= sinc\left(\frac{2v_y^2 T_s}{\lambda R_o}t\right)\cdot\exp\left(-j\frac{2v_y^2}{\lambda R_o}t^2\right).$$

The second term in the above equation is just the phase term and does not have any effect on the amplitude. Therefore, only the first term represents the envelope of the azimuth compressed SAR data.

3.7.2 Azimuth Resolution

The first nulls bandwidth of the sinc function in Equation 3.53 is equal to $\Delta t_y = \lambda R_o/(2v_y^2 T_s)$. Therefore, the cross-range resolution, Δy, is equal to

$$\Delta y = v_y\cdot\Delta t_y$$
$$= v_y\cdot\frac{\lambda R_o}{2v_y^2 T_s} \quad (3.54)$$
$$= \frac{\lambda R_o}{2v_y T_s}.$$

Noticing that the term $(v_y T_s)$ equals D_{SA} which is the length of the synthetic aperture, Δy becomes equal to

$$\Delta y = \frac{\lambda R_o}{2D_{SA}}. \quad (3.55)$$

Synthetic aperture length, D_{SA}, can also be written in terms of the longest dimension, L, of the radar antenna as

$$D_{SA} = \frac{\lambda R_o}{L}. \quad (3.56)$$

Inserting Equation 3.56 into Equation 3.55, we get

$$\Delta y = \frac{\lambda R_o}{2(\lambda R_o / L)}$$
$$= \frac{L}{2}. \quad (3.57)$$

Therefore, the cross-range resolution is independent of target distance R_o and the wavelength λ and depends only on the aperture of the real antenna mounted on the platform for the focused SAR, which is very interesting. This result offers that the smaller the size of the SAR antenna, the better the cross-range resolution. Although this result seems a little surprising, it is actually

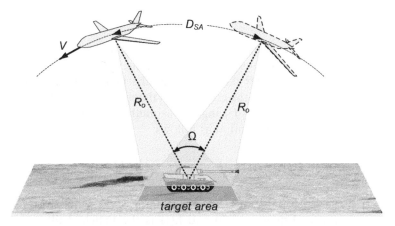

FIGURE 3.17   Spotlight SAR with a circular flight path.

not. According to the antenna theory, the width of the main beam of the antenna increases as the aperture (or size) of the antenna decreases. Therefore, the illuminated synthetic aperture line becomes larger for shorter antenna. This means that any point along the synthetic aperture axis is observed by the radar for a longer period of time. As a result, finer cross-range resolution can be achieved.

3.7.3   Relation to ISAR

As demonstrated in Figure 3.17, when the focused-SAR operation is considered, the synthetic aperture path is an arc formed by the angular bandwidth of Ω. The length of the synthetic aperture is related to the angular bandwidth via

$$D_{SA} = R_o \cdot \Omega. \qquad (3.58)$$

Then, the cross-range or azimuth resolution can be found via

$$\begin{aligned} \Delta y &= \frac{\lambda R_o}{2 D_{SA}} \\ &= \frac{\lambda R_o}{2(R_o \cdot \Omega)} \\ &= \frac{\lambda}{2\Omega} \\ &= \frac{c}{2 f_c \cdot \Omega}, \end{aligned} \qquad (3.59)$$

where f_c is the frequency of operation. This result is also identical for the case of the ISAR operation to be demonstrated in Chapter 4. Therefore, the SAR operation with circular flight path is analogous to the ISAR operation.

3.8   SAR IMAGING

Generally, processing steps in range and cross-range domains are computationally intensive. This is because the correlation integral has to be calculated for every pixel of the 2D SAR data. However, the image formation in SAR can be accelerated by utilizing the convolution theorem that makes it possible to use the FT. With this in mind, range or azimuth compressed data can be written in the form of a convolution as follows:

$$I(t) = \int_{-\infty}^{\infty} s(\tau) \cdot h(t-\tau) d\tau = s(t) * h(t), \tag{3.60}$$

where $s(t)$ is the chirp pulse in range or azimuth SAR signal whose phase variation is given in Equation 3.50 and $h(t)$ is the matched filter impulse response. As the convolution in time corresponds to multiplication in frequency according to the Fourier theory, the convolution in Equation 3.60 can easily be speeded up by utilizing the FT concept, as shown below:

$$I(t) = \mathcal{F}\{\mathcal{F}^{-1}\{s(t)\} \cdot \mathcal{F}^{-1}\{h(t)\}\}. \tag{3.61}$$

Therefore, it is clear that the shape of the resulting image response is determined by the FT of the pulse envelope. When it is a rectangular pulse, the image response comes out to be a *sinc* function. This response is also known in radar imaging as *point spread function* (PSF). Since the length of the collected SAR data has to be finite, then the limits in the Fourier integral become also finite. Therefore, the spreading effect is always unavoidable in SAR imagery. In some cases, a *sinc* type PSF can be problematic when a weak scatterer happens to be located very close to a strong scatterer. Since the first sidelobe level of the *sinc* function is about 13 dB lower than the main beam, the weak scatterer may not be detected due to high sidelobe level of the sinc spreading. Therefore, it is usually preferable to use smooth pulse envelopes such as Hanning, Hamming, or Kaiser-type windowing whose FTs offer much weaker sidelobe levels. Figure 3.18 shows the compressed signal for a Hanning weighted chirp signal. It is clear that the sidelobes are pretty well suppressed compared to *sinc* response. Therefore, such weightings provide better *peak-to-sidelobe ratio* (PSLR) at the price of increased main-lobe beamwidth, that is, worse resolution.

3.9   EXAMPLE OF SAR IMAGERY

An example of measured SAR imagery is given in Figure 3.19. This image shows the space-shuttle SAR image of San Francisco, California. The raw SAR data were collected by the SIR-C/X-SAR when it flew aboard the space-

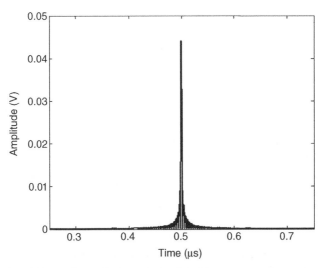

FIGURE 3.18 Hanning windowed range compressed data.

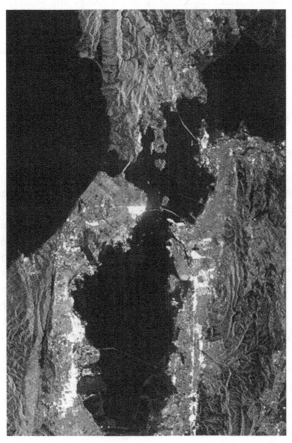

FIGURE 3.19 SIR-C/X-SAR image of San Francisco, California. (Courtesy NASA/ JPL-Caltech.)

shuttle Endeavour in 1994. The SAR system on SIR-C/X was able to collect L, C, and X band raw SAR data over 225 km above the earth surface for multiple frequencies and polarizations in aiming to better understand the global environment and how it is changing. The SIR-C/X-SAR system was designed to provide surface resolutions of 30 m both in range and cross-range dimensions.

The particular image in Figure 3.19 demonstrates the most general features of a typical SAR image. For instance, rough surfaces such as mountain terrains show up in the SAR image as the mixture of bright and dark spots since they experience scattering in all directions. On the other hand, smooth surfaces and regions of calm water such as sea or lake surfaces appear to be black. These surfaces behave like mirrors and scatter EM energy according to the Snell's law; therefore, almost no energy scatters back in the direction of the radar. Hills and other large-scale targets tend to appear bright on the side where illuminated and dark on the other side where there is no EM illumination (shadowing). Human-made objects such as buildings and vehicles behave like corner reflectors and they appear as brighter spots. When a very strong point scatterer (e.g., specular point) exists, it shows up as a bright cross in the image due to strong sidelobes of the PSR in range and cross-range dimensions. Most of these features can be observed in the SAR image in Figure 3.19.

3.10 PROBLEMS IN SAR IMAGING

3.10.1 Range Migration

During SAR operation, the radar platform first approaches and moves away from any scatterer that is entering its main antenna beam. Therefore, the range distance of any scatterer depicts a parabolic curve called *range curvature* while it is illuminated by the radar beam. This ambiguity in range distance may result in shifting of scatters to the nearby range cells while integrating the raw SAR data over the synthetic aperture time interval. This phenomenon is called *range migration*. For spaceborne SAR, this curvature can be significant since the range extent and the integration time is longer compared to the airborne SAR. On the other hand, the antenna's footprint on the ground is much shorter for the airborne SAR. Therefore, range-delay shift of the scatterer may be less than the range resolution, which may cause no migration in the range.

Another cause of range migration occurs from the phenomenon called *range walk*. When the integration time is long, the rotation of earth causes shifting of the position of the scatterer with respect to radar. The same effect happens when the ground target is moving. All these causes may result in migration of the scatters between the image cells. To have a focused image, *range migration correction* procedure in the Doppler frequency domain is required [12, 13].

Both the range curvature and the range walk possess a parabolic track within the real beam of the radar. If the range error resulted by these effects is δR, the total range can be written as

$$R + \delta R = \left(R^2 + \left(\frac{v_y \cdot T_s}{2} \right)^2 \right)^{1/2}. \tag{3.62}$$

Here, v_p is the velocity of the scatterer, and T_s is the integration time in SAR azimuth processing. In fact, the integration time can also be written in terms of target's velocity v_y and single-look angular extent Ω as $T_s = R \cdot \Omega / v_y$. After substituting this into Equation 3.62 and taking the binomial expansion of the range error, one can get

$$\begin{aligned} \delta R &= \left(R^2 + \left(\frac{R \cdot \Omega}{2} \right)^2 \right)^{1/2} - R \\ &= R \left(1 + \left(\frac{\Omega}{2} \right)^2 \right)^{1/2} - R \\ &= R \left(1 + \frac{1}{2} \left(\frac{\Omega}{2} \right)^2 + H.O.T. \right) - R \\ &\cong R \frac{\Omega^2}{8}. \end{aligned} \tag{3.63}$$

Then, with Δr being the range resolution, the migration along the range cells can be found by

$$N \approx \frac{\delta R}{\Delta r} = R \frac{\Omega^2}{8 \Delta r}. \tag{3.64}$$

If this number turns out to be more than 1, the range migration correction has to be done accordingly. If not, no correction is needed.

3.10.2 Motion Errors

The basic theory of SAR relies on the assumption that the scene or the target is stationary. If a scatterer in the scene is moving, the Doppler shift posed by the scatterer's line-of-sight velocity sets the "incorrect" distance information about the position of the scatterer to the phase of the received EM wave. When the scatterer is moving so fast, it occupies several pixels in the image during the integration interval of SAR. Therefore, the image of the scatterer is blurred like a comet in the sky. This is analogous to optical photography when the object is moving so fast that it occupies a greater amount of space until the lens is closed. When the scatterer is in slow motion, the SAR image may not be blurred. However, the location of the scatterer will still not be correct due to Doppler shift. To minimize these motion errors, *motion compensation* techniques have to be applied. The details of these techniques will be given in Chapter 8.

3.10.3 Speckle Noise

The size of one SAR resolution cell (i.e., a pixel) can vary from a few centimeters to tens of meters depending on the size of the synthetic aperture and the bandwidth in frequency. Generally, the Earth's surface is uneven within one resolution cell. Therefore, one pixel of SAR may not resolve the detailed features of the Earth's surface. Since the EM energy diffracted from such uneven patches will have various phase values, the net effect on the image pixel can be constructive or destructive. The resulting noise-like behavior is known as *ground clutter* or *speckle noise*. One way to reduce speckle noise in SAR imaging is to use multilook processing [14]. Another method is to apply statistical filtering of speckle noise by using proper probability distribution models such as Rayleigh distribution [15, 16].

3.11 ADVANCED TOPICS IN SAR

3.11.1 SAR Interferometry

This is one of the most unique applications of SAR imagery. When compared with optical imaging, which has only the amplitude information of the reflected light, SAR images contain both the amplitude and the phase information of the backscattered EM energy of the scene. Therefore, if the multilook images of the scene from different elevations are obtained, the comparison of the phases may lead to resolution of the third dimension (i.e., height) to have 3D interferometric images. The first demonstrated IFSAR was presented by Graham when he configured a cross-track interferometer in 1974 [17]. Goldstein used an along-track SAR interferometry configuration to measure radial velocity of moving targets in 1987 [18]. In 1988, Gabriel obtained an elevation map of a terrain with the data collected by SIR-B by means of a single antenna with repeat pass [19].

In IFSAR, a second image of the same scene is used to extract the third dimension. This can be achieved with either an airborne/spaceborne vehicle carrying two-radar system or a single antenna vehicle with repeat passes over the same scene. For the latter situation, suppose that vehicle is moving in the y-direction with a speed of v_y and an altitude of h from a reference scatterer point on the scene. Therefore, the phase of the round-trip EM wave that is reflected from the reference point is

$$\vartheta_1 = \exp\left(-jk(x^2 + y^2 + (z-h)^2)^{1/2}\right). \tag{3.65}$$

On the next pass, the vehicle keeps the same height and speed, but its track is shifted by a distance of d in the x-direction. Therefore, the phase of the back-scattered EM wave from the same reference point becomes

$$\vartheta_2 = \exp\left(-jk((x-d)^2 + y^2 + (z-h)^2)^{1/2}\right). \qquad (3.66)$$

Multiplying the first phase with the complex conjugate of the second phase and expanding the difference to the first order of d yields

$$\vartheta_1 \cdot \vartheta_2^* = \exp\left(j\frac{4\pi xd}{\lambda R}\right), \qquad (3.67)$$

where $R = (x^2 + y^2 + (z-h)^2)^{1/2}$. In this equation, R, y, and x are already known after applying the range compression, azimuth compression, and Doppler processing. If the separation d between the two passes is known, the only unknown parameter, z (i.e., height), can be easily retrieved from Equation 3.67. This is the basic approach in IFSAR processing.

An example of IFSAR is shown in Figure 3.20. This image of Mount Etna, Italy, was obtained by the SIR-C/X-SAR in 1994. The image covers an area of 51.2 km by 22.6 km. Different elevation regions surrounding the volcanic Mount Etna are illustrated with different color values.

3.11.2 SAR Polarimetry

For a conventional SAR, both the receiver and the transmitter are designed to have only one polarization, that is, horizontal (H) or vertical (V). However,

FIGURE 3.20   (SIR-C/X-SAR) Interferometric image of Mount Etna, Italy. (Courtesy NASA/JPL-Caltech.)

to extract the full scattering feature of a scene, all polarization signatures, *HH*, *HV*, *VH*, and *VV*, have to be collected. When identical polarizations are used for transmit and receive, the SAR system is said to be *copolarized*. When the radar collects the received field at orthogonal polarization to that of the transmitted field, then the SAR system is *cross polarized*.

The images obtained by these two polarization cases may differ due to the different scattering features of the terrain. For smooth surfaces such as oceans, lakes, and deserts, copolarized SAR images differ significantly from cross-polarized ones. However, both copolarized and cross-polarized SAR setups produce similar images for rough surfaces like mountains and forests. Man-made objects are always found to appear in the image for any type of polarization since most of them involve corner reflector type features. Therefore, it may be possible to identify the features and material types of terrain structure, such as rock type, with the help of SAR polarimetry. Good polarimetric SAR images were presented by Held [20] and Sullivan [21].

3.12 MATLAB CODES

Below are the Matlab source codes that were used to generate all of the Matlab-produced figures in Chapter 3. The codes are also provided in the CD that accompanies this book.

Matlab code 3.1: Matlab file "Figure3-8.m"

```
%------------------------------------------------------------------
% This code can be used to generate Figure 3.8
%------------------------------------------------------------------
clear all
close all

%--- transmitted signal ------------------------------------------
fc = 8e8; % initial frequency
To = 10e-6; %pulse duration
N = 200; %sample points
td = 4e-6; % delay

t = 0:To/(5*N-1):To; %time vector
tt = t*1e6; %time vector in micro seconds

s = 10*ones(1,N);
s(5*N) =0; % transmitted signal replica
sr = s;% replica
M=round(td/To*(5*N));% shift amount
ss=circshift(sr.',M);ss=ss.';

%---Figure 3.8(a) ------------------------------------------------
h1=figure;
h = area(tt,sr);
```

```
set(h,'FaceColor',[.5 .5 .5])
set(gca,'FontName', 'Arial', 'FontSize',12,'FontWeight',
'Bold');
title('transmitted signal');
xlabel(' Time [\mus]')
axis([min(tt) max(tt) -30 30]);
%---Figure 3.8(b)-----------------------------------------------
h1=figure;
h = area(tt,ss)
set(h,'FaceColor',[.5 .5 .5])
set(gca,'FontName', 'Arial', 'FontSize',12,'FontWeight',
'Bold');
title('received signal without noise');
xlabel(' Time [\mus]')
axis([min(tt) max(tt) -30 30]);
%---Figure 3.8(c)-----------------------------------------------
%--- Noise Signal ----------------------------------
n=5*randn(1,5*N);
% Plot noise signal
h1=figure;
h=area(tt,n)
set(h,'FaceColor',[.5 .5 .5]);
set(gca,'FontName', 'Arial', 'FontSize',12,'FontWeight',
'Bold');
xlabel(' Time [\mus]'),title('noise signal ');
axis([min(tt) max(tt) -30 30]);
%---Figure 3.8(d)-----------------------------------------------
%--- Received Signal ----------------------------------
x=ss+n;
h1=figure;
h=area(tt,x)
set(h,'FaceColor',[.5 .5 .5]);
set(gca,'FontName', 'Arial', 'FontSize',12,'FontWeight',
'Bold');
xlabel(' Time [\mus]'),title('received signal with noise');
axis([min(tt) max(tt) -30 30]);
%---Figure 3.8(e)-----------------------------------------------
%--- Matched Filtering ----------------------------------
X=fft(x)/N;
S = conj(fft(sr)/N);
H=S;
Y=X.*H;
y=ifft(Y);

% Plot matched filter output
h1=figure;
h=area(tt,real(y));
set(h,'FaceColor',[.5 .5 .5]);
set(gca,'FontName', 'Arial', 'FontSize',12,'FontWeight',
'Bold');xlabel(' Time [\mus]')
```

```
axis([min(tt) max(tt) -.1 2]); title('matched filter output
');
```

Matlab code 3.2: Matlab file "Figure3-9.m"
```
%------------------------------------------------------------
% This code can be used to generate Figure 3.9
%------------------------------------------------------------
clear all
close all

tau = -10:.1:10; fd=tau;L=length(fd);
dummy = ones(L,L);
ideal = fftshift(ifft2(dummy));
mesh(tau,fd,abs(ideal));
colormap(gray)
set(gca,'FontName', 'Arial', 'FontSize',14,'FontWeight','Bold');
xlabel ('Time Delay')
ylabel ('Doppler Shift')
zlabel ('Normalized AF')
```

Matlab code 3.3: Matlab file "Figure3-10.m"
```
%------------------------------------------------------------
% This code can be used to generate Figure 3.10
%------------------------------------------------------------
clear all
close all

T = 1e-3;
A = 1;
tau = -2e-3:1e-5:2e-3;
fd = -3e3:10:3e3;fd=fd.';
X1 = sinc(fd*(T-abs(tau)));
TT = (T-abs(tau));
p = find(TT<0);
TT(p) = 0;
X2 = A*A*ones(length(fd),1)*TT;
X = X1.*X2;
X = X/max(max(abs(X)));

%---Figure 3.10(a)------------------------------------------
mesh(tau*1e3,fd*1e-3,abs(X));
colormap(gray)
set(gca,'FontName', 'Arial', 'FontSize',14,'FontWeight','Bold');
xlabel ('Time Delay [ms]')
ylabel ('Doppler Shift [KHz]')
zlabel ('Normalized AF')

%---Figure 3.10(b)------------------------------------------
imagesc(tau*1e3,fd*1e-3,abs(X));
```

```
colormap(gray)
colorbar
set(gca,'FontName', 'Arial', 'FontSize',14,'FontWeight','Bold');
xlabel ('Time Delay [ms]')
ylabel ('Doppler Shift [KHz]')
```

Matlab code 3.4: Matlab file "Figure3-11.m"
```
%---------------------------------------------------------------
% This code can be used to generate Figure 3.11
%---------------------------------------------------------------
clear all
close all

T = 1e-3;
A = 1;
k = 2e6;
tau = -2e-3:1e-5:2e-3;
fd = -3e3:10:3e3;fd=fd.';
TT = (T-abs(tau));
dummy=k*(ones(length(fd),1)*tau)-fd*ones(1,length(tau));
X1 = sinc(dummy.*(ones(length(fd),1)*TT));
p = find(TT<0);
TT(p) = 0;
X2 = A*A*ones(length(fd),1)*TT;
X = X1.*X2;
X = X/max(max(abs(X)));

%---Figure 3.11(a)----------------------------------------------
mesh(tau*1e3,fd*1e-3,abs(X));
colormap(gray)
set(gca,'FontName', 'Arial', 'FontSize',14,'FontWeight','Bold');
xlabel ('Time Delay [ms]')
ylabel ('Doppler Shift [KHz]')
zlabel ('Normalized AF')

%---Figure 3.11(b)----------------------------------------------
imagesc(tau*1e3,fd*1e-3,abs(X(length(X):-1:1,:)));
colormap(gray);
colorbar
set(gca,'FontName', 'Arial', 'FontSize',14,'FontWeight','Bold');
xlabel ('Time Delay [ms]')
ylabel ('Doppler Shift [KHz]')
```

Matlab code 3.5: Matlab file "Figure3-14.m"
```
%---------------------------------------------------------------
% This code can be used to generate Figure 3.14
%---------------------------------------------------------------
clear all
close all
```

```
%--- transmitted signal ------------------------------------
fc = 8e8; % initial frequency
BWf = 10e6; % frequency bandwidth
To = 5e-6; %pulse duration
Beta = BWf/To;
N = 400; %sample points
td = 1e-6; % delay

t = -To/2:To/(N-1):To/2; %time vector
tt = t*1e6; %time vector in micro seconds
f =fc:BWf/(N-1):(fc+BWf);% frequency vector

s = 10*cos(2*pi*(fc*(t-td)+Beta*((t-td).^2)));% transmitted
signal
sr = 10*cos(2*pi*(fc*t+Beta*(t.^2)));% replica

%---Figure 3.14(a)---------------------------------------------
h = figure;plot(tt,s, 'k','LineWidth',2)
set(gca,'FontName', 'Arial', 'FontSize',12,'FontWeight',
'Bold');
title('transmitted signal');
xlabel(' Time [\mus]')
axis([min(tt) max(tt) -20 20 ]);
%---Figure 3.14(b)---------------------------------------------
%--- Noise Signal ---------------------------------------------
n = 5*randn(1,N);
% Plot noise signal
h = figure;plot(tt,n, 'k','LineWidth',2)
set(gca,'FontName', 'Arial', 'FontSize',12,'FontWeight',
'Bold');
xlabel(' Time [\mus]'),title('noise signal ');
axis([min(tt) max(tt) -20 20 ]);
%---Figure 3.14(c)---------------------------------------------
%--- Received Signal ------------------------------------------
x = s+n;
% Plot received signal
h = figure;plot(tt,x, 'k','LineWidth',2)
set(gca,'FontName', 'Arial', 'FontSize',12,'FontWeight',
'Bold');
xlabel(' Time [\mus]'),title('received signal');
axis([min(tt) max(tt) -20 20 ]);
%---Figure 3.14(d)---------------------------------------------
%--- Matched Filtering ----------------------------------------
X = fft(x)/N;
S = conj(fft(sr)/N);
H = S;
Y = X.*H;
y = fftshift(ifft(Y));
```

```
%----Plot matched filter output---------------------------------
h = figure;plot(tt,abs(y), 'k','LineWidth',2)
set(gca,'FontName', 'Arial', 'FontSize',12,'FontWeight',
'Bold');
xlabel(' Time [\mus]')
axis([min(tt) max(tt) 0 .2]);title('Matched filter output ');
```

Matlab code 3.6: Matlab file "Figure3-16.m"
```
%---------------------------------------------------------------
% This code can be used to generate Figure 3.16
%---------------------------------------------------------------
clear all
close all

Phi_3dB = 5 * pi/180 ; % 3dB beamwidth of the antenna : 5
degrees
R0 = 8e3; % radial distance of the scatterer
f = 6e9; % frequency
lam = 3e8/f; % wavelength

X_max = R0 *tan(Phi_3dB/2); % maximum cross-range extend
x = -X_max:2*X_max/99:X_max; % cross-range vector

R = R0 *(1+x.^2/R0^2).^(0.5); % real range distance
R_est = R0+x.^2/2/R0; % estimated range distance

%---Figure 3.16(a)-----------------------------------------------
h = figure;
plot(x,R/1e3,'k-','LineWidth',1);
hold
plot(x,R_est/1e3,'k.','LineWidth',4);
hold;
grid on
set(gca,'FontName', 'Arial', 'FontSize',14,'FontWeight','Bold
');
legend('actual radial distance','estimated radial distance')
xlabel('synthetic Aperture [m]')
ylabel('distance [km]')
axis([min(x) max(x) R0/1e3-.25 R0/1e3+.25])

%---Figure 3.16(b)-----------------------------------------------
h = figure;plot(x,(R-R_est)/lam,'k','LineWidth',2);
grid on
set(gca,'FontName', 'Arial', 'FontSize',14,'FontWeight','Bold
');
xlabel('synthetic Aperture [m]')
ylabel('range error value [\lambda]')
axis([min(x) max(x) -1 1])
```

REFERENCES

1 C. Wiley. *Pulsed Doppler Radar Method and Means*, US Patent 3,196,436, 1954.

2 C. W. Sherwin, J. P. Ruina, and R. D. Rawcliffe. Some early developments in synthetic aperture radar systems. *IRE Trans Mil Electron* MIL-6(2) (1962) 111–115.

3 L. J. Cutrona, et al. Optical data processing and filtering systems. *IRE Trans Inf Theory* IT-6 (1960) 386–400.

4 http://southport.jpl.nasa.gov/nrc/chapter7.html (accessed 09.11.2011).

5 P. Lacomme, J.-P. Hardange, J.-C. Marchais, and E. Normant. *Air and spaceborne radar systems: An introduction.* William Andrew Publishing/Noyes LLC, Norwich, NY, 2001.

6 B. R. Mahafza. *Radar systems analysis and design using MATLAB*, 2nd ed. Chapman & Hall/CRC, Boca Raton, FL, 2000.

7 W. J. Caputi, Jr. Stretch: A time-transformation technique. *IEEE Trans* AES- 7 (1971) 269–278.

8 V. I. Bityutskov. Bunyakovskii inequality, in M. Hazewinkel, ed. *Encyclopaedia of mathematics.* Springer, D, 2001.

9 B. M. Oliver. *Not with a bang, but a chirp.* Bell Telephone Labs, Techn. Memo, MM-51-150-10, case 33089, March 8, 1951.

10 D. Wehner. *High resolution radar.* Artech House, Norwood, MA, 1987.

11 N. Levanon. *Radar principles.* Wiley-Interscience, New York, 1988.

12 J. M. Lopez-Sanchez and J. Fortuny-Guasch. 3-D radar imaging using range migration techniques. *IEEE Trans Antennas Propagat* 48(5) (2000) 728–737.

13 J. Fortuny-Guasch and J. M. Lopez- Sanchez. Extension of the 3-D range migration algorithm to cylindrical and spherical scanning geometries. *IEEE Trans Antennas Propagat* 49(10) (2001) 1434–1444.

14 R. J. Sullivan. *Microwave radar imaging and advanced concepts.* Artech House, Norwood, MA, 2000.

15 C. Oliver and S. Quegan. *Understanding synthetic aperture radar images.* Artech House, Boston, MA, 1998.

16 F. N. S. Medeiros, N. D. A. Mascarenhas, and L. F. Costa. Evaluation of speckle noise MAP filtering algorithms applied to SAR images. *Int J Remote Sens* 24 (2003) 5197–5218.

17 L. C. Graham. Synthetic interferometer radar for topographic mapping. *Proc IEEE* 62(2) (1974) 763–768.

18 R. M. Goldstein and H. A. Zebker. Interferometric radar measurements of ocean surface currents. *Nature* 328(20) (1987) 707–709.

19 A. K. Gabriel and R. M. Goldstein. Crossed orbit interferometry: Theory and experimental results from SIR-B. *Int J Remote Sens* 9(5) (1988) 857–872.

20 D. N. Held, W. E. Brown, and T. W. Miller. Preliminary results from the NASA/JPL multifrequency multipolarization SAR. Proceedings of the 1988 IEEE National Radar Conference, pp. 7–8, 1988.

21 R. J. Sullivan, et al. Polarimetric X/L/C band SAR. Proceedings of the 1988 IEEE National Radar Conference, pp. 9–14, 1988.

CHAPTER FOUR

Inverse Synthetic Aperture Radar Imaging and Its Basic Concepts

Inverse synthetic aperture radar (ISAR) is a powerful signal processing technique for imaging moving targets in range-Doppler (or range and cross-range) domains. As *range* (or slant range) is defined as the axis parallel to the direction of propagation from radar toward the target, *cross range* is defined as the perpendicular axis to the range direction. An ISAR image has the ability to successfully display the dominant scattering regions (hot points), that is, *scattering centers* on the target. ISAR processing is normally used for the identification and classification of targets. The classic two-dimensional (2D) ISAR image is constructed by collecting the scattered field for different look angles and Doppler histories. Although ISAR processing is similar to synthetic aperture radar (SAR) processing, ISAR imaging procedure has some conceptual differences when compared to the SAR imagery.

4.1 SAR VERSUS ISAR

SAR generally refers to the case where the radar platform is moving while the target stays stationary (see Chapter 3, Figs. 3.1 and 3.3). The required spatial (or angular) diversity is accomplished by the movement of radar around the target or terrain. On the other hand, the term ISAR is used for scenarios when the radar is stationary and the targets are in motion, such as with airplanes, ships, and tanks, as illustrated in Figure 4.1. As similar to the SAR

Inverse Synthetic Aperture Radar Imaging with MATLAB Algorithms, First Edition.
Caner Özdemir.
© 2012 John Wiley & Sons, Inc. Published 2012 by John Wiley & Sons, Inc.

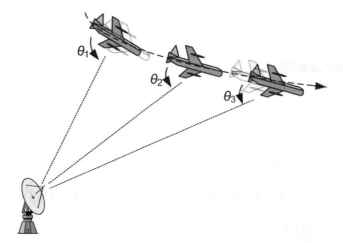

FIGURE 4.1 Inverse synthetic aperture radar (ISAR) geometry.

operation, the required range resolution is achieved by using finite frequency bandwidth of the transmitted signal for the ISAR case as well. As depicted in Figure 4.1, stationary radar collects the scattering data from the target for different look angles by utilizing the target movement. While the target is moving, the look angle of target is assumed to be changing with respect to radar line of sight (RLOS) axis to have a succesful ISAR image. This angular diversity in the ISAR data set is used to resolve different points along the cross-range axis. The details of these concepts will be thoroughly explained in the forthcoming subsections. In terms of the collected echo data set, ISAR geometry, in fact, can also be thought of as similar to spotlight SAR geometry with circular flight paths as illustrated in Figure 4.2.

A more detailed comparison can be made through Figure 4.3 where the ISAR problem and the analogous spotlight SAR problem are illustrated. As can be seen from Figure 4.3a, the radar moving along a circular path collects the backscattered field data from the stationary target for an angular width of Ω for the spotlight SAR geometry. In Figure 4.3b, on the other hand, the stationary radar collects the backscattered field data from a rotating target. The same set of reflectivity data can be obtained if the target rotates the same angular width of Ω provided that the radar in either case is tracked to the target and has the same frequency bandwidth.

In most SAR scenarios, of course, the radar moves along a straight path rather than a circular path, as depicted in Figure 4.4. Therefore, there will be a path length difference, dR, when compared to the ideal case shown in Figure 4.3a. When the integration angle is small and the target is at a sufficient range distance, R, away from the radar, the path length difference of the received signal will be relatively small as well. Assuming that the path length difference dR is smaller than the wavelength, the phase of the received signal for the

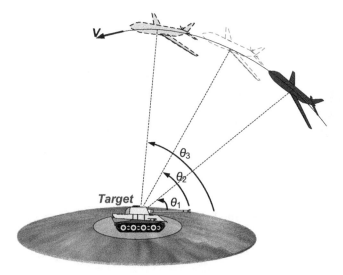

FIGURE 4.2   Spotlight SAR geometry with circular flight path is analogous to ISAR geometry.

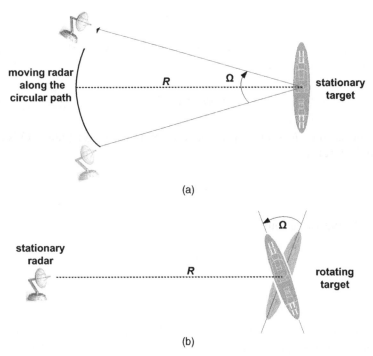

FIGURE 4.3   SAR-to-ISAR transition: (a) spotlight SAR with circular flight path, (b) ISAR.

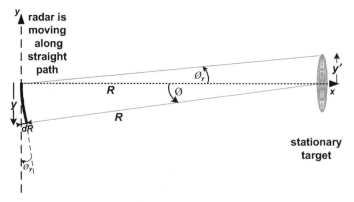

FIGURE 4.4   Spotlight SAR with straight flight path.

straight path operation will have the following extra delay term when compared to the circular path operation:

$$\varphi(y) = -2k \cdot (dR)$$
$$= -2k \cdot y \cdot \sin \emptyset_y. \tag{4.1}$$

The factor 2 stands for the two-way propagation between the radar and the target. As seen from Figure 4.4,

$$\sin \emptyset_y \cong y'/R. \tag{4.2}$$

Furthermore, since R is already assumed to be much larger than y, $\emptyset$ is quite small and therefore

$$y \cong \emptyset \cdot R. \tag{4.3}$$

So, substituting Equations 4.2 and 4.3 into Equation 4.1, we get

$$\varphi(\emptyset) \cong -2k \cdot \emptyset \cdot y'$$
$$= -2\pi \cdot \left(\frac{2\emptyset}{\lambda} \right) \cdot y'. \tag{4.4}$$

From this equation, one can easily realize that the Fourier relation between the aspect variable $\emptyset$ and the target's cross-range variable y' is evident. Therefore, the scattering point along the target's cross-range axis can be resolved by the following cross-range resolution:

$$\delta y' = \frac{1}{BW(2\emptyset / \lambda)}$$

$$= \frac{\lambda}{2\Omega}$$

$$= \frac{c}{2f\Omega}, \tag{4.5}$$

where Ω is the total viewing angle (or the angular width) of the aspect variable, $\emptyset$. This clearly states that if the target's reflectivity is collected over a larger aspect angle (or over a longer synthetic aperture), a finer resolution in the cross-range dimension can be attained.

In real-world practices, of course, the target may also have translational motion components such that the range R is changing while the target is rotating. This situation causes shifting of scatterers in the range from profile to profile and may result in image blurring, which will be studied in Chapters 6 and 9.

4.2   THE RELATION OF SCATTERED FIELD TO THE IMAGE FUNCTION IN ISAR

We will now demonstrate how the scattered field or the reflectivity function from a target can be related to the image function of this target. We first start with the formula that was presented in Chapter 2, Section 2.1 at which the scattered electric field from any perfectly conducting object is shown to be equal to

$$\vec{E}^s(\vec{r}) = -\frac{jk_o E_o}{4\pi r} e^{-jk_o r} \iint_{S_{lit}} 2\hat{n}(\vec{r}') \times (\hat{k}^i \times \hat{u}) e^{j(\vec{k}^s - \vec{k}^i)\cdot\vec{r}'} d^2\vec{r}'. \tag{4.6}$$

Here, k^i and k^s are the incident and scattered wavenumber vectors, $\hat{n}(r')$ is the outward surface unitary normal vector, $\hat{k}^i$ is the unit vector in incident wave direction, E_o and $\hat{u}$ are the magnitude and the polarization unit vector of the incident wave, and S_{lit} is the illuminated part of the object's surface. Now, let us assume that the receiving antenna has a particular polarization such that it collects the scattered field in the $\hat{v}$ direction. Then, we can rewrite Equation 4.6 as given below:

$$\hat{v} \cdot \vec{E}^s(\vec{r}) = -\frac{jk_o E_o}{4\pi r} e^{-jk_o r} \iiint_{-\infty}^{\infty} O(\vec{r}') e^{j(\vec{k}^s - \vec{k}^i)\cdot\vec{r}'} d^3\vec{r}', \tag{4.7}$$

where $O(\vec{r}')$ can be treated as the *scalar object shape function* (OSF) [1, 2] that is specified as

$$O(\vec{r}') = \hat{v} \cdot \left[2\hat{n}(\vec{r}') \times (\hat{k}^i \times \hat{u}) \right] \cdot \delta(S(\vec{r}')). \tag{4.8}$$

In the above equation, the argument of the impulse function is defined in the following way:

$$S(\vec{r}') = \begin{cases} \neq 0 & \vec{r}' \in S_{lit} \\ 0 & \vec{r}' \in S_{shadow} \end{cases}. \tag{4.9}$$

Also, notice that the surface integral in Equation 4.6 is replaced by a volume integral over the entire three-dimensional (3D) space as given in Equation 4.6. If we define the 3D Fourier transform (FT) of the OSF $O(r')$ as

$$\tilde{O}(\vec{k}) = \iiint_{-\infty}^{\infty} O(\vec{r}') e^{j\vec{k}\cdot\vec{r}'} d^3\vec{r}', \tag{4.10}$$

then the scattered electric field in the $\hat{v}$ direction can be rewritten as

$$\hat{v} \cdot \vec{E}^s(\vec{r}) = \left(-\frac{jk_o E_o}{4\pi r} e^{-jk_o r} \right) \tilde{O}(\vec{k}^s - \vec{k}^i). \tag{4.11}$$

The term $\{\hat{v} \cdot \vec{E}^s(\vec{r})\}$ gives the scattered field along the $\hat{v}$-direction according to the projection slice theorem. This result clearly shows that the scattered electric field from a target is directly proportional to 3D FT of its OSF. An ISAR image can, in fact, be regarded as the display of this OSF onto the 2D plane or in the 3D box.

It is also worthwhile to mention that OSF varies with respect to the look angle and the frequency of operation. As we shall demonstrate in Sections 4.5 through 4.7, ISAR image is directly related to the 2D or 3D inverse Fourier transform (IFT) of the scattered electric field as similar to OSF.

4.3   ONE-DIMENSIONAL (1D) RANGE PROFILE

ISAR image can be regarded as the display of range and cross-range profiles of the target in the 2D range/cross-range plane. Before understanding the meaning of the ISAR image, therefore, it is fundamentally important to appreciate the meaning of the range profile and the cross-range profile.

A range profile is the returned waveform shape from a target that has been illuminated by the radar with sufficient frequency bandwidth. If the illuminating wave is a time-domain pulse, then the reflected signal collected by the receiver will have 1D characteristics, typically field intensity (or radar cross-section area) versus time (or range) as illustrated in Figure 4.5. If the illuminating signal is the stepped frequency waveform, then the IFT of the received signal characterizes the 1D range profile of the target.

The physical meaning of the range profile is clarified through the case in Figure 4.5 where a range profile of an airplane is illustrated: As the incident

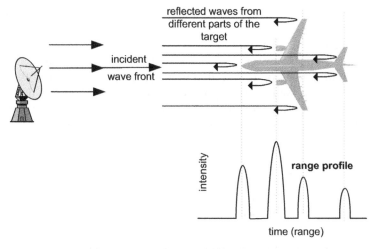

FIGURE 4.5   Range profile of a target.

waveforms pass along with the target, some of the energy will reflect back toward the radar from so-called scattering centers on the target. If these scattering centers are located at different range distances from the radar, they will return at different time instants to the radar receiver so that they can be distinguished in the corresponding 1D range profile. As shown in Figure 4.5, the sources of backscattering points may lie on the cockpit, motor duck, wings, tails, or some other points of the plane. Of course, it will not be possible to resolve the scattering centers that are in the same range by utilizing the range profile concept as they show up in the same range bin (or time location). The range profile is also named in the radar literature as the radar signature since the returned waveform shape is unique for a specific target as different targets provide different range profiles.

Instead of displaying the range profile versus time, it is more physically meaningful to present it versus distance, or range. Then, the range axis can easily be scaled via

$$r = c \cdot t, \tag{4.12}$$

where r is range, t is time, and c is the speed of light.

Let us examine how we can obtain the range profile by processing the frequency-diverse returned (or backscattered) wave. Let us assume that there exist N point scatterers along the down range (assumed along the x-axis) each located at a different x_i location. Then, the backscattered electric field at the far-field can be approximated as

$$\begin{aligned} E^s(f) &\cong \sum_{i=1}^{N} A_i \cdot e^{-j2k \cdot x_i} \\ &= \sum_{i=1}^{N} A_i \cdot e^{-j2\pi \left(\frac{2f}{c} \right) x_i}, \end{aligned} \tag{4.13}$$

where A_i is the backscattered field amplitude for the point scatterer at x_i and $k = 2\pi f/c$ is the corresponding wave number for the frequency f. The number "2" in the exponential accounts for the two-way propagation between the radar and the point scatterers. Assuming that the backscattered field is collected at the far field along the $-x$ direction and the phase center of the scene is taken as $x = 0$, the sign of the exponential should be the same as the sign of the points "x_i"s. With this construct, the range profile can be constructed by inverse Fourier transforming this frequency diverse field with respect to $(2f/c)$ as given below:

$$E^s(x) = F^{-1}\{E^s(f)\}$$
$$= \int_{-\infty}^{\infty} \left[\sum_{i=1}^{N} A_i \cdot e^{-j2\pi\left(\frac{2f}{c}\right)x_i} \right] e^{j2\pi\left(\frac{2f}{c}\right)x} d\left(\frac{2f}{c}\right). \qquad (4.14)$$

In the above equation, both the summing and the integration operators are linear and therefore can be interchanged as

$$E^s(x) = \sum_{i=1}^{N} A_i \cdot \int_{-\infty}^{\infty} e^{j2\pi\left(\frac{2f}{c}\right)(x-x_i)} d\left(\frac{2f}{c}\right). \qquad (4.15)$$

Then, the integral in Equation 4.15 perfectly vanishes to impulse (or Dirac delta) function $\delta(\cdot)$ as given below:

$$E^s(x) = \sum_{i=1}^{N} A_i \cdot \delta(x - x_i). \qquad (4.16)$$

Here, $E^s(x)$ represents the range profile as a function of range, x. Therefore, the point scatterers located at different x_i locations are perfectly pinpointed in the range axis with their associated backscattered field amplitudes of "A_i"s. Of course, the result in Equation 4.16 is valid for the infinite bandwidth. In real applications, however, the backscattered field data can only be collected within a finite bandwidth of frequencies, say, ranging from f_L to f_H. Then, the limits of the integral in Equation 4.15 should be changed to give

$$E^s(x) = \sum_{i=1}^{N} A_i \cdot \int_{f_L}^{f_H} e^{j2\pi\left(\frac{2f}{c}\right)(x-x_i)} d\left(\frac{2f}{c}\right). \qquad (4.17)$$

One can proceed with $E^s(x)$ by taking the definite integral in Equation 4.17 as

$$E^s(x) = \sum_{i=1}^{N} A_i \cdot \frac{1}{j2\pi} \left(e^{j2\pi\left(\frac{2f_H}{c}\right)(x-x_i)} - e^{j2\pi\left(\frac{2f_L}{c}\right)(x-x_i)} \right). \qquad (4.18)$$

Defining the center frequency as $f_c = (f_L + f_H)/2$ and the frequency bandwidth as $B = f_H - f_L$, the result in Equation 4.18 can be simplified to yield

$$E^s(x) = \sum_{i=1}^{N} A_i \cdot e^{j2k_c(x-x_i)} \left(\frac{e^{j2\pi\left(\frac{B}{c}\right)(x-x_i)} - e^{-j2\pi\left(\frac{B}{c}\right)(x-x_i)}}{j2\pi} \right). \qquad (4.19)$$

Here, $k_c = 2\pi f_c/c$ is the wave number corresponding to the center frequency. The above result can be simplified as

$$E^s(x) = \left(\frac{2B}{c} \right) \sum_{i=1}^{N} A_i \cdot e^{j2k_c(x-x_i)} \cdot sinc\left(\frac{2B}{c}(x-x_i) \right). \qquad (4.20)$$

In the above equation, $sinc(\cdot)$ is the sinc (or sinus cardinalis) function already defined in Equation 3.36. The exponential in Equation 4.20 is just the phase term and has the unitary amplitude. The second term, the sinc, is the amplitude term that specifies the shape function of the point scatterer at x_i. Therefore, the scattering centers on the range are centered at the true locations of "x_i"s with their corresponding field amplitudes, "A_i"s. According to the Fourier theory, sinc defocusing around the scattering centers in the range profile pattern is unavoidable due to finite bandwidth of the radar signal. This defocusing is also known as *point spread function* (PSF) or *point spread response* (PSR) in the radar literature and will be investigated later in various parts of this book.

A very common way of constructing the range profile of a target is accomplished by illuminating it via the *stepped frequency continuous wave* (SFCW) signal. In the SFCW setup, the radar transmits a continuous wave signal modulated at N different stepped frequencies of $f_1, f_2, \ldots, f_N$ and collects the scattered field intensities, $E^s[f]$, for these N discrete frequency values. Then, the time-domain range profile can easily be constructed by applying the inverse fast Fourier transform (IFFT) operation as

$$E^s[t] = \mathcal{F}^{-1}\{E_s[f]\}. \qquad (4.21)$$

Afterward, the time axis can readily be transformed to range axis by the simple relation of $x = c \cdot t$ to get $E^s[x]$. If the frequency bandwidth is B, then the resolution in range, Δr, is

$$\Delta x = \frac{c}{2B}. \qquad (4.22)$$

Each sample, distributed by Δx in the range, is called a range bin or a range cell. The total viewed range (or range extend), X_{max}, is then equal to

$$\begin{aligned} X_{max} &= N \cdot \Delta x \\ &= \frac{N \cdot c}{2B}. \end{aligned} \qquad (4.23)$$

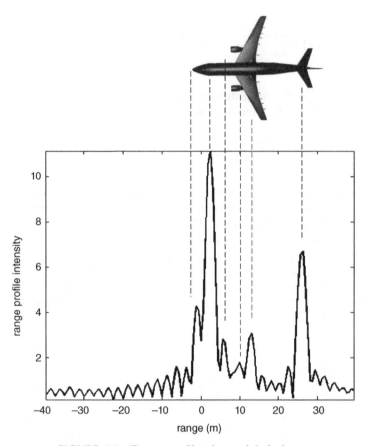

FIGURE 4.6   Range profile of a model airplane.

This is the "window frame" in the range or the *range extent* that can be viewed by the range profile. Therefore, X_{max} should be greater than the length of the target in the range to avoid any ambiguity that may cause aliasing.

An example of the range profile concept is demonstrated in Figure 4.6 where the range profile of a commercial airplane is obtained. The electromagnetic simulation of the backscattered electric field is carried out by a physical optics (PO) and shooting and bouncing ray (SBR) [3] code that can estimate the scattering from complex targets at high frequencies. The backscattered electric field is collected at the far field from the nose of the airplane between 3.97 GHz and 4.03 GHz for a total of 32 discrete frequencies. The corresponding range profile is obtained as shown in Figure 4.6 by taking the 1D IFFT of the frequency diverse backscattered electric field. As observed from the figure, the major scattering occurs from the nose of the airplane, the engine ducts, the wings, and the tail.

The concept of range profile plays an important role in radar imaging. It can be used as a standard tool for extracting the scattering centers and also for determining the length of a target. It may provide essential information for the classification of the objects in the applications of *automatic target recognition* (ATR).

4.4 1D CROSS-RANGE PROFILE

While a *range profile* can be obtained by processing the frequency-diverse radar return from a target, in a similar manner, a *cross-range profile* can be formed by collecting the radar returns from a target at different look angles, as illustrated in Figure 4.7. The aspect width of the look angles is used to resolve the required cross-range points to form the 1D cross-range profile, while the range profile is obtained by treating the backscattered field at single look angle, but different frequencies; the cross-range profile is analogously acquired by processing the backscattered field at one frequency, but different look angles.

Let us assume that there exist P point scatterers located at different (x_i, y_i) points. Our goal is to obtain the cross-range profile so that we can resolve the y_i locations of these scatterer points. The backscattered electric field at the far field for different look angles can be approximated as

$$E^s(\emptyset) = \sum_{i=1}^{P} A_i \cdot e^{-j2\bar{k}\cdot\bar{n}}, \qquad (4.24)$$

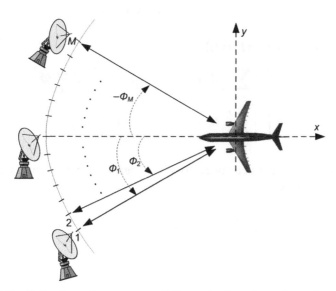

FIGURE 4.7 Collecting radar returns at different look angles to form the cross-range profile of a target.

where A_i is the backscattered field amplitude for the each point scatterer and $\vec{r_i}$ is the vector from the origin to the point scatterer at (x_i, y_i). The $\vec{k} \cdot \vec{r_i}$ argument in Equation 4.24 can be written as follows:

$$
\begin{aligned}
\vec{k} \cdot \vec{r_i} &= (k_x \hat{x} + k_y \hat{y}) \cdot (x_i \hat{x} + y_i \hat{y}) \\
&= k_x \cdot x_i + k_y \cdot y_i \\
&= k\cos\emptyset \cdot x_i + k\sin\emptyset \cdot y_i.
\end{aligned}
\tag{4.25}
$$

Therefore, the backscattered field is equal to

$$
E^s(\emptyset) = \sum_{i=1}^{P} A_i \cdot e^{-j2k\cos\emptyset \cdot x_i} \cdot e^{-j2k\sin\emptyset \cdot y_i}.
\tag{4.26}
$$

For small values of look angles, $\cos\emptyset$ is approximated to 1 and $\sin\emptyset$ is approximated to $\emptyset$. Therefore, this equation reduces to

$$
\begin{aligned}
E^s(\emptyset) &= \sum_{i=1}^{P} B_i \cdot e^{-j2k\emptyset \cdot y_i} \\
&= \sum_{i=1}^{P} B_i \cdot e^{-j2\pi\left(\frac{2f}{c}\right)\emptyset \cdot y_i}.
\end{aligned}
\tag{4.27}
$$

Here, B_i is a constant and equal to $A_i \cdot \exp(-j2k \cdot x_i)$. In the above equation, there exists a Fourier relationship between $(2f/c)\emptyset$ and y_i. Therefore, taking the 1D IFT of Equation 4.27 with respect to $(2f/c)\emptyset$, it is possible to resolve "y_i"s in the cross range using the following equation:

$$
\begin{aligned}
E^s(y) &= \mathcal{F}^{-1}\{E_s(\emptyset)\} \\
&= \int_{-\infty}^{\infty} \left[\sum_{i=1}^{P} B_i \cdot e^{-j2\pi\left(\frac{2f}{c}\right)\emptyset \cdot y_i} \right] e^{j2\pi\left(\frac{2f}{c}\right)\emptyset \cdot y} d\left(\frac{2f}{c}\emptyset\right) \\
&= \int_{-\infty}^{\infty} \left[\sum_{i=1}^{P} B_i \cdot e^{j2\pi\left(\frac{2f}{c}\right)\emptyset \cdot (y-y_i)} \right] d\left(\frac{2f}{c}\emptyset\right)
\end{aligned}
\tag{4.28}
$$

In the above equation, linear operators of integration and summation can be interchanged to give

$$
\begin{aligned}
E^s(y) &= \sum_{i=1}^{P} B_i \cdot \int_{-\infty}^{\infty} e^{j2\pi\left(\frac{2f}{c}\right)\emptyset \cdot (y-y_i)} d\left(\frac{2f}{c}\emptyset\right) \\
&= \sum_{i=1}^{P} B_i \cdot \delta(y - y_i).
\end{aligned}
\tag{4.29}
$$

Here, $E^s(y)$ represents the cross-range profile function as a function of y. Therefore, the point scatterers located at different cross-range locations y_i are perfectly pinpointed in the cross-range axis with corresponding backscattered field amplitudes. The result in Equation 4.29 would be valid if the backscattered field is collected at an infinite number of look angles. Of course, this is impossible in real-life applications. For practical implementation of cross-range profiling, therefore, the limits in the above integration should be changed to finite values of "Ø" as given below:

$$E_s(y) = \sum_{i=1}^{M} B_i \cdot \int_{-\Omega/2}^{\Omega/2} e^{j2\pi\left(\frac{2f}{c}\right)\emptyset \cdot (y-y_i)} d\left(\frac{2f}{c}\emptyset\right). \qquad (4.30)$$

Here, Ω gives total angular width in collecting the backscattered field. The definite integral in Equation 4.26 can be calculated as

$$
\begin{aligned}
E_s(y) &= \sum_{i=1}^{P} B_i \cdot \frac{1}{j2\pi}\left(e^{j2\pi\left(\frac{f}{c}\right)\Omega\cdot(y-y_i)} - e^{-j2\pi\left(\frac{f}{c}\right)\Omega\cdot(y-y_i)}\right) \\
&= \left(\frac{2f}{c}\Omega\right)\cdot\sum_{i=1}^{P} B_i \cdot \left[\frac{\sin\left(2\pi\left(\frac{f}{c}\right)\Omega\cdot(y-y_i)\right)}{2\pi\left(\frac{f}{c}\right)\Omega}\right] \qquad (4.31) \\
&= \left(\frac{2f}{c}\Omega\right)\cdot\sum_{i=1}^{P} B_i \cdot sinc\left[\frac{2f}{c}\cdot\Omega(y-y_i)\right].
\end{aligned}
$$

In the above equation, the impulse function in Equation 4.29 distorts to sinc function due to the finite width of angles as expected.

An example of 1D cross-range profile imaging is illustrated in Figure 4.8 where the cross-range profile of the same airplane is obtained. The scattered electric field is collected at the far-field around the nose direction of the airplane at the single frequency of 4 GHz. The look angle is varied between −1.04° and 1.01° in the azimuth plane for a total of 64 discrete frequencies. The corresponding cross-range profile is acquired as depicted in Figure 4.8 by taking the 1D fast Fourier transform (FFT) of the aspect diverse backscattered electric field. This cross-range profile maps the scattering points from the nose of the airplane, the engine ducts, and the wings.

4.5   2D ISAR IMAGE FORMATION (SMALL BANDWIDTH, SMALL ANGLE)

For bistatic ISAR configuration, the transmitter and the receiver are positioned at different locations in space. If the radar operates as both the transmitter and the receiver, this scenario is called *monostatic* ISAR, which is the common practice in most real-world applications. In this section, we will present a simplified ISAR imaging theory for the monostatic case.

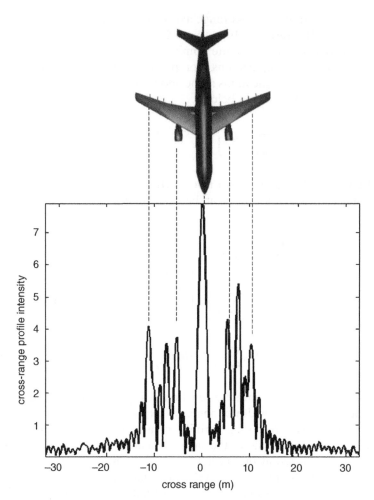

FIGURE 4.8   Cross-range profile of a model airplane.

Since the 2D ISAR image is nothing but the display of range profile in one axis and the cross-range profile in the other axis, the scattered field should be collected for various frequencies and aspects (i.e., look angles) to be able to generate the 2D ISAR image as depicted in Figure 4.9a. In this figure, $\hat{k}$ vector is assumed to lie on the 2D $k_x - k_y$ plane. Collected data set is generated in the spatial-frequency domains, namely, k_x and k_y. If the backscattered electric field data are gathered within the finite bandwidth of frequencies, B, and within a finite width of angles, Ω, then the 2D data occupy a nonuniform grid in the $k_x - k_y$ space (see Fig. 4.9b). However, if both B and Ω are sufficiently small, the data grid in $k_x - k_y$ space approaches to equally spaced linear grid. This situation makes it possible to make use of fast inverse Fourier transform in forming the ISAR image as will be explained in detail below.

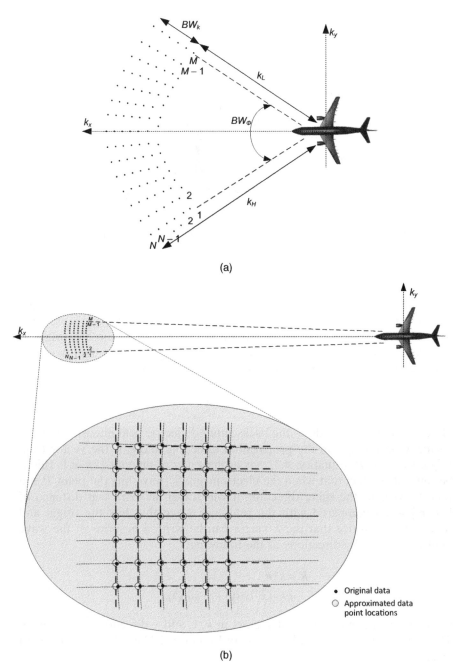

(a)

(b)

FIGURE 4.9 (a) Collection of ISAR raw data in Fourier space for the monostatic case (2D case), (b) ISAR data collection for small-bandwidth and small-angle case.

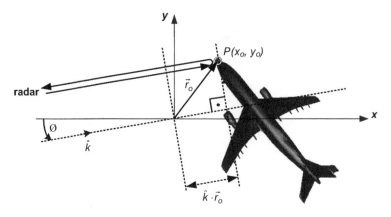

FIGURE 4.10   Geometry for monostatic ISAR imaging (2D case).

The algorithm for 2D ISAR imaging is provided for the monostatic case. Let us start with the algorithm such that a point scatterer $P(x_o, y_o)$ is assumed to be situated on the target as illustrated in Figure 4.10. Taking the origin as the phase center of the geometry, the far-field backscattered field from the point scatterer at an azimuth angle $\varnothing$ can be approximated as

$$E^s(k, \varnothing) \cong A_o \cdot e^{-j2\vec{k}\cdot\vec{r}_0}. \tag{4.32}$$

Here, A_o is the amplitude of the backscattered electric field intensity, $\vec{k}$ is the vector wave number in the propagation direction, and $\vec{r}_0$ is the vector from origin to point P. Equation 4.32 has a phase lag of amount $(2\vec{k}\cdot\vec{r}_0)$. This is because of the fact that when the electromagnetic wave hits the point P and reflects back in the same direction, it will travel an extra trip distance of $(2\vec{k}\cdot\vec{r}_0)$ when compared with the reference wave that hits the origin and reflects back. Notice that the $\vec{k}$ vector can be written in terms of the wave numbers in x and y directions, as the following:

$$\begin{aligned}
\vec{k} &= k \cdot \hat{k} \\
&= k \cdot (\hat{x} \cdot cos\varnothing + \hat{y} \cdot sin\varnothing),
\end{aligned} \tag{4.33}$$

where $\hat{k}$, $\hat{x}$, and $\hat{y}$ are the unit vectors in k, x, and y directions, respectively. Then, the argument in the phase term of Equation 4.32 can be reorganized to yield

$$\begin{aligned}
\vec{k} \cdot \vec{r}_o &= k \cdot (\hat{x} \cdot cos\varnothing + \hat{y} \cdot sin\varnothing) \cdot (\hat{x} \cdot x_o + \hat{y} \cdot y_o) \\
&= kcos\varnothing \cdot x_o + ksin\varnothing \cdot y_o \\
&= k_x \cdot x_o + k_y \cdot y.
\end{aligned} \tag{4.34}$$

Therefore, we can rewrite Equation 4.32 as

$$E^s(k,\emptyset) = A_o \cdot e^{-j2k\cos\emptyset\cdot x_o} \cdot e^{-j2k\sin\emptyset\cdot y_o}. \tag{4.35}$$

This equation offers two separate phase terms as a function of both the spatial frequency variable, k, and the look-angle variable, $\emptyset$. If these phase terms are carefully examined, the Fourier relationships between $(2k\cos\emptyset)$–and–x, and $(2k\sin\emptyset)$–and–y can be easily noticed. Therefore, the ISAR image can be generated in range and cross-range domains by the convenience of the 2D IFT.

When the practical ISAR imaging is concerned, the standard procedure is to use a small frequency bandwidth of B and a small aspect width, Ω, while collecting the echoed data set. This is called *small-bandwidth small-angle* ISAR imaging. In this standard procedure of ISAR, the frequency bandwidth, B, is small compared to center frequency of operation, f_c. In practice, the bandwidths that are less than one-tenth of the center frequency are considered to be sufficiently small. Then the wave number in the second phase term of Equation 4.35 can be approximated as

$$\begin{aligned} k &\cong k_c \\ &= 2\pi f_c / c, \end{aligned} \tag{4.36}$$

where k_c is the wave number corresponding to the center frequency, f_c. In a similar manner, if the look-angle width Ω is small, the following approximations hold true:

$$\begin{aligned} \cos\emptyset &\cong 1 \\ \sin\emptyset &\cong \emptyset. \end{aligned} \tag{4.37}$$

In practice, angular widths that are at most $5°$ to $6°$ are generally considered as small. Then the scattered electric field from point P can be approximated to

$$E^s(k,\emptyset) = A_o \cdot e^{-j2k\cdot x_o} \cdot e^{-j2k_c\emptyset\cdot y_o}. \tag{4.38}$$

To be able to use the advantages of FT, we reorganize Equation 4.35 as

$$E^s(k,\emptyset) = A_o \cdot e^{-j2\pi\left(\frac{2f}{c}\right)x_o} \cdot e^{-j2\pi\left(\frac{k_c\emptyset}{\pi}\right)y_o}. \tag{4.39}$$

Then the ISAR image in the x–y plane can be obtained by taking the 2D IFT of Equation 4.39 as

$$\mathcal{F}_2^{-1}\{E^s(k,\varnothing)\} = A_o \cdot \mathcal{F}_1^{-1}\left\{e^{-j2\pi\left(\frac{2f}{c}\right)x_o}\right\} \cdot \mathcal{F}_1^{-1}\left\{e^{-j2\pi\left(\frac{k_c\varnothing}{\pi}\right)y_o}\right\}$$

$$E^s(x,y) = A_o \cdot \left[\int_{-\infty}^{\infty} e^{-j2\pi\left(\frac{2f}{c}\right)x_o} \cdot e^{j2\pi\left(\frac{2f}{c}\right)x} d\left(\frac{2f}{c}\right)\right]$$

$$\cdot \left[\int_{-\infty}^{\infty} e^{-j2\pi\left(\frac{k_c\varnothing}{\pi}\right)y_o} \cdot e^{j2\pi\left(\frac{k_c\varnothing}{\pi}\right)y} d\left(\frac{k_c\varnothing}{\pi}\right)\right] \qquad (4.40)$$

$$= A_o \cdot \delta(x-x_o, y-y_o)$$

$$\triangleq ISAR(x,y).$$

Here $\delta(x,y)$ represents the 2D impulse function on the x–y plane. As is obvious from Equation 4.40, the point P manifests itself in the ISAR image as a 2D impulse function located at (x_o, y_o) with the correct electromagnetic reflectivity coefficient of A_o.

The backscattered electric field from a target can be approximated as the sum of scattering from a finite number of single point scatterers, called scattering centers, on the target as shown below:

$$E^s(k,\varnothing) \cong \sum_{i=1}^{P} A_i \cdot e^{-j2\bar{k}\cdot\bar{n}}. \qquad (4.41)$$

Here, the backscattered electric field from a target is approximated as the sum of backscattered field from M different scattering centers on the target. While A_i represents the complex backscattered field amplitude for the ith scattering center, $\vec{r}_i = x_i \cdot \hat{x} + y_i \cdot \hat{y}$ is called the displacement vector from origin to the location of the ith scattering center. Then, the ISAR image of the target can be found by taking the 2D inverse Fourier integral of the 2D backscattered field data as

$$ISAR(x,y) = \iint_{-\infty}^{\infty} \{E^s(k,\varnothing)\} \cdot e^{j2\pi\left(\frac{2f}{c}\right)x} e^{j2\pi\left(\frac{k_c\varnothing}{\pi}\right)y} d\left(\frac{2f}{c}\right) d\left(\frac{k_c\varnothing}{\pi}\right) \qquad (4.42)$$

Assuming that the backscattered signal can be represented by a total of scattering centers, the small-bandwidth small-angle ISAR image can then be approximated as

$$ISAR(x,y) \cong \iint_{-\infty}^{\infty} \sum_{i=1}^{M} A_i \cdot e^{-j2\bar{k}\cdot\bar{n}} \cdot e^{j2\pi\left(\frac{2f}{c}\right)x} e^{j2\pi\left(\frac{k_c\varnothing}{\pi}\right)y} d\left(\frac{2f}{c}\right) d\left(\frac{k_c\varnothing}{\pi}\right)$$

$$= \sum_{i=1}^{M} A_i \cdot \iint_{-\infty}^{\infty} e^{-j2\bar{k}\cdot\bar{n}} \cdot e^{j2\pi\left(\frac{2f}{c}\right)x} e^{j2\pi\left(\frac{k_c\varnothing}{\pi}\right)y} d\left(\frac{2f}{c}\right) d\left(\frac{k_c\varnothing}{\pi}\right)$$

$$= \sum_{i=1}^{M} A_i \cdot \iint_{-\infty}^{\infty} e^{j2\pi\left(\frac{2f}{c}\right)(x-x_i)} e^{j2\pi\left(\frac{k_c\varnothing}{\pi}\right)(y-y_i)} d\left(\frac{2f}{c}\right) d\left(\frac{k_c\varnothing}{\pi}\right) \qquad (4.43)$$

$$= \sum_{i=1}^{M} A_i \cdot \delta(x-x_i, y-y_i).$$

Therefore, the resultant ISAR image is composed of nothing but the sum of M scattering centers with their electromagnetic reflectivity coefficients. Of course, the limits of the integral in Equation 4.43 have to be finite in practice due to the fact that the field data can be collected within a finite bandwidth and finite aspect width. Therefore, the practical ISAR image response distorts from the *impulse* function to the sinc function as will be demonstrated in Chapter 5.

4.5.1 Range and Cross-Range Resolutions

Range and the cross-range resolutions in ISAR determine the quality of the resultant image. Therefore, these parameters should be taken into account while applying the ISAR imaging procedure. When the 2D backscattered field data are collected and numerically stored, the Fourier integral in Equation 4.36 is calculated numerically with the help of discrete Fourier transform (DFT).

4.5.1.1 Range Resolution As is obvious from the first phase term in Equation 4.39, there is a direct Fourier relationship between the frequency variable f and the range distance variable x. Representing the frequency bandwidth as B, the Fourier theory imposes the following for the range resolution of ISAR:

$$\Delta x = \frac{1}{BW\left(\dfrac{2f}{c}\right)}$$
$$= \frac{c/2}{BW_f} \tag{4.44}$$
$$= \frac{c}{2B}.$$

Therefore, higher frequency bandwidth offers better resolution in the range direction. To achieve a range resolution value of 15 cm, for example, the backscattered electric field data should be collected for a frequency bandwidth of 1 GHz.

4.5.1.2 Cross-Range Resolution In a similar manner, the cross-range resolution Δy can be calculated as the following: By looking at the second phase term in Equation 4.39, the Fourier relationship between the aspect variable $\emptyset$ and the cross-range distance variable y. If the backscattered data are collected within the finite bandwidth of Ω in azimuth (or elevation) angles, the cross-range resolution of ISAR is equal to

$$\Delta y = \frac{1}{BW\left(\frac{k_c \varnothing}{\pi}\right)}$$

$$= \frac{\pi / k_c}{BW_\varnothing}$$

$$= \frac{\lambda_c}{2\Omega} \qquad\qquad (4.45)$$

$$= \frac{c}{2 f_t \Omega}$$

where λ_c corresponds to the wavelength for the center frequency, f_c. Equation 4.45 suggests the higher the angle width, the better the resolution in the cross-range direction. To acquire a resolution of 15 cm in the cross-range direction, for instance, the backscattered field data should be collected within an angular width of 5.73° for the center frequency operation of 10 GHz.

4.5.2   Range and Cross-Range Extends

Once the range and cross-range resolutions are determined, selection of the number of sampling points determines the spatial extends in these domains, that is, how much the image window extends in range and cross-range directions in the ISAR image. If the frequency bandwidth is sampled by N_x times and the angular width is sampled by N_y times, the corresponding image domain extends are given as

$$X_{max} = N_x \cdot \Delta x$$

$$= \frac{N_x \cdot c}{2B}$$

$$Y_{max} = N_y \cdot \Delta y \qquad\qquad (4.46)$$

$$= \frac{N_y \cdot \lambda_c}{2\Omega}.$$

For the same example above, if the 2D frequency-aspect data are collected over 256 sampling points in each domain, the image size becomes 38.4 m by 38.4 m, which is more than enough to image a fighter aircraft.

4.5.3   Imaging Multi-Bounces in ISAR

ISAR imaging is based on a single-bounce assumption of the scattered waves. On the other hand, there is no doubt that there may be some multiple-bounce mechanisms as the electromagnetic wave hits the target and bounces around it. Since the conventional ISAR imaging procedure is based on a single-bounce situation, these multi-bounces will not be correctly mapped in the ISAR image to the actual scattering locations of the target. In fact, higher order scattering

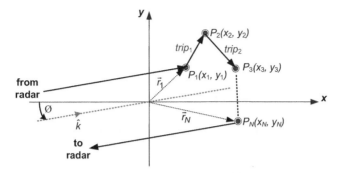

FIGURE 4.11   Geometry for imaging multibounce mechanisms in ISAR.

mechanisms will simply be delayed in the down range and dislocated in the cross range, as will be shown next.

To illustrate how multi-bounces are mapped in the ISAR image, let us consider a total of N-bounce mechanism as illustrated in Figure 4.11. Assuming that the phase center of the geometry is the origin, the corresponding scattered electric field can be written in the following form:

$$E^s\left(k_x, k_y\right) = A \cdot e^{-j\left(\vec{k}\cdot\vec{r}_1 + k \cdot \sum_{n=1}^{N-1} trip_n + \vec{k}\cdot\vec{r}_N\right)}, \tag{4.47}$$

where A is the complex scattered field intensity after N reflections, $trip_n = |\vec{r}_{n+1} - \vec{r}_n|$. We can also refer to $\left(\sum_{n=1}^{N-1} trip_n\right)$ as $(trip_{tot})$ which is the total trip between the first and the last hit points. Note that displacement vectors $\vec{r}_1$ and $\vec{r}_N$ equal to

$$\begin{aligned} \vec{r}_1 &= x_1 \cdot \hat{x} + y_1 \cdot \hat{y} \\ \vec{r}_N &= x_N \cdot \hat{x} + y_N \cdot \hat{y}. \end{aligned} \tag{4.48}$$

Also the wavenumber vector can be written in terms of axis variables and the incident angle variable $\emptyset$ as

$$\vec{k} = k\cos\emptyset \cdot \hat{x} + k\sin\emptyset \cdot \hat{y}. \tag{4.49}$$

Substituting Equations 4.48 and 4.49 into Equation 4.47, we can rewrite the scattered electric field with a more organized phase term, as shown below:

$$E^s\left(k_x, k_y\right) = A \cdot e^{-jk\left((x_1 + x_N)\cos\emptyset + (y_1 + y_N)\sin\emptyset + trip_{tot}\right)}. \tag{4.50}$$

Under small-bandwidth and small-angle ISAR case, the following approximations can be made:

$$k(x_1 + x_N)cos\emptyset \cong k(x_1 + x_N)$$
$$k(y_1 + y_N)sin\emptyset \cong k_c(y_1 + y_N)\emptyset. \tag{4.51}$$

Therefore, the final approximated scattered electric field becomes

$$E^s(k_x, k_y) \cong A \cdot e^{-jk(x_1 + x_N + trip_{tot})} \cdot e^{-jk_c(y_1 + y_N)\emptyset}. \tag{4.52}$$

Now, we apply the 2D IFT procedure to this scattered field to get the ISAR image. Then,

$$E^s(x, y) = F_2^{-1}\{E^s(k, \emptyset)\}$$
$$= \iint_{-\infty}^{\infty} \{E^s(k_x, k_y)\}e^{jk_x x} \cdot e^{jk_y y} dk_x\, dk_y. \tag{4.53}$$

For the small-bandwidth and small-angle ISAR, $k_x = 2k$ and $k_y = 2kc\emptyset$ as previously shown. Therefore, the final ISAR image for an N-point multibounce mechanism will be obtained via

$$E^s(x, y) = A \cdot \iint_{-\infty}^{\infty} e^{jk_x\left(x - \frac{(x_1 + x_N + trip_{tot})}{2}\right)} \cdot e^{jk_y\left(y - \frac{(y_1 + y_N)}{2}\right)} dk_x\, dk_y$$
$$= A \cdot \delta\left(x - \frac{(x_1 + x_N + trip_{tot})}{2}, y - \frac{(y_1 + y_N)}{2}\right). \tag{4.54}$$

This result shows that the image of a multiple-bounce scattering is delayed in the range and dislocated in the cross range as illustrated in Figure 4.12. As easily observed from Equation 4.54, the location of a multibounce scattering in the ISAR image occurs at the following down-range and cross-range points:

$$x' = \frac{(x_1 + x_N + trip_{tot})}{2}$$
$$y' = \frac{(y_1 + y_N)}{2}. \tag{4.55}$$

Therefore, multiple-bounce features in the ISAR image, if properly interpreted, do carry useful information for understanding the physical phenomenon behind the scattering of the electromagnetic wave from the target.

When there is only single bounce, of course, $x_N = x_1$, $y_N = y_1$ and $trip_{tot} = 0$. Therefore, the result in Equation 4.54 readily reduces to the result of a single bounce as

$$E^s(x, y) = A \cdot \delta(x - x_1, y - y_1). \tag{4.56}$$

An interesting case of multi-bounce occurs from a *90° dihedral corner reflector* as demonstrated in Figure 4.13. All the multi-bounces from this geometry have the same travel distance (or time) that the fictitious point scatterer at the

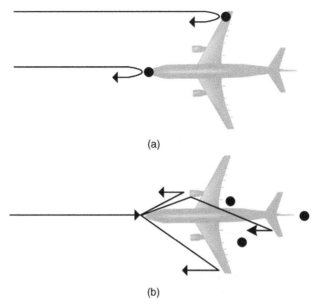

(a)

(b)

FIGURE 4.12 (a) Single-bounce mechanisms are correctly mapped in ISAR. (b) Multibounce mechanisms are delayed in range and dislocated in cross range as their images may show up out of the target.

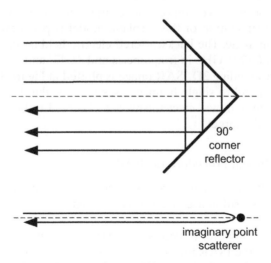

FIGURE 4.13 All double bounces from a 90° corner reflector have the same travel distance as that of a single bounce from an imaginary point scatterer that is supposed to be present at the corner of the plates.

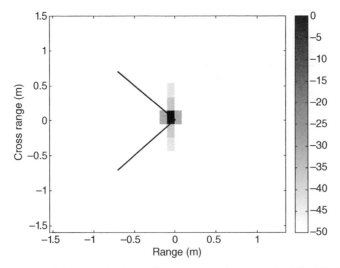

FIGURE 4.14 ISAR image of a 1 m × 1 m corner reflector at 10 GHz. The images of all multi-bounces show up at the corner.

corner of both plates would do. A simulation of a perfectly conducting 90° corner reflector of two identical plates of 1 m × 1 m is carried out around 10 GHz. The simulated scattered field is carried out along the direction of symmetry line of the reflector for an angular bandwidth of 9.2° and the frequency bandwidth of 0.8 GHz. The resulting ISAR image is shown in Figure 4.14, where all the multi-bounces coincide at the corner in the image.

An example of an ISAR image that contains multi-bounces is shown in Figure 4.15. The simulation of this airplane model is performed around 45° from the nose-on angle. The backscattered electric field is collected between 5.8154 GHz and 6.1731 GHz in frequencies and 41.47° and 48.42° in azimuth angles. The corresponding 2D ISAR image is plotted in Figure 4.15. While hot spots that correspond to single-bounce cases occur within the outline of the airplane, some multibounce mechanisms are observed to be located outside of the outline of the target.

4.5.4 Sample Design Procedure for ISAR

The basic algorithm for designing an ISAR image is illustrated in Figure 4.16. The steps of the algorithm are given briefly, in order:

Step 1: The key point for a successful ISAR image is to start the procedure by selecting the ISAR image size, that is, range and cross-range window extends. If the range window extend is X_{max} and the cross-range window extend is Y_{max}, then the size of the ISAR image, X_{max} by Y_{max}, should be selected to cover the actual size of the target to be imaged. It is

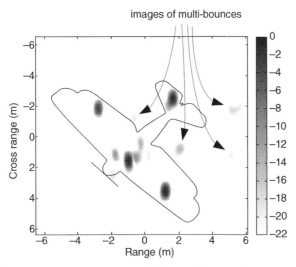

FIGURE 4.15 ISAR image of a plane model from 45° from the nose on. Some multi-bounces show up themselves outside of the plane's outline.

important to note that the size of the target changes in the ISAR image according to the look angle of the radar.

Step 2: The other key selection is the range and cross-range resolutions, Δx and Δy, respectively. These numbers are so critical that they define how many pixels will lie on the target. Therefore, these resolutions are directly linked to the quality of the ISAR image. After the resolution in the ISAR image is decided, the sampling points in range, N_x, and the sampling points in cross range, N_y, can be calculated using the formulas below:

$$N_x = \frac{X_{max}}{\Delta x}$$
$$N_y = \frac{Y_{max}}{\Delta y}. \tag{4.57}$$

If the target's range size is 15 m and the cross-range size is 12 m (which are nominal figures for a fighter aircraft) and the resolutions in both domains are selected as 15 cm, then the target's range will be displayed with 100 range pixels (or bins) while the target's cross range will be displayed by 80 cross-range pixels (or bins).

Step 3: Once the ISAR size is determined, the resolutions in frequency, Δf, and aspect, $\Delta\emptyset$, can be determined by utilizing the Fourier relationships between frequency-and-range and angle-and-cross-range in Equation 4.39 as demonstrated below:

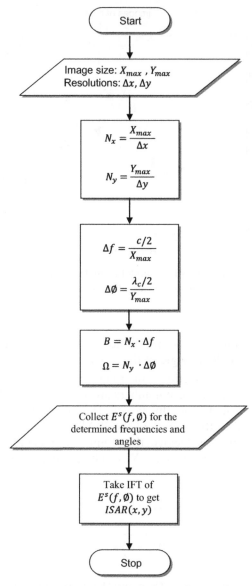

FIGURE 4.16   Flowchart for the basic ISAR imaging algorithm.

$$\Delta f = \frac{c/2}{X_{max}}$$

$$\Delta\emptyset = \frac{\lambda_c/2}{Y_{max}}. \tag{4.58}$$

Then, the frequency bandwidth, B, and the angular width, Ω, will be equal to

$$B = N_x \cdot \Delta f = \frac{N_x \cdot c}{2 \cdot X_{max}}$$

$$\Omega = N_y \cdot \Delta\emptyset = \frac{N_y \cdot \lambda_c}{2 \cdot Y_{max}}. \tag{4.59}$$

Step 4: If the frequencies will be centered around f_c and the radar look angles will be centered around $\emptyset_c$, then the backscattered electric field should be collected for the following multiple frequencies and angles:

$$f = \left[\left(f_c - \frac{N_x\Delta f}{2} \right) \left(f_c - \left(\frac{N_x}{2} - 1 \right)\Delta f \right) \dots (f_c) \dots \left(f_c + \left(\frac{N_x}{2} - 1 \right)\Delta f \right) \right]_{1\times N_x}$$

$$\emptyset = \left[\left(\emptyset_c - \frac{N_y\Delta\emptyset}{2} \right) \left(\emptyset_c - \left(\frac{N_y}{2} - 1 \right)\Delta\emptyset \right) \dots (\emptyset_c) \dots \left(\emptyset_c + \left(\frac{N_y}{2} - 1 \right)\Delta\emptyset \right) \right]_{1\times N_y}.$$

$$\tag{4.60}$$

Then, collect the backscattered electric field for those frequencies and angles as $E^s(f,\emptyset)$.

Step 5: At this final step, we can take the 2D IFT to get the final ISAR image. If the backscattered field data are collected within a small frequency bandwidth and the angles, then IFFT can be readily applied.

Next, we will demonstrate some numeric examples for the construction of ISAR images by applying the steps listed in the above algorithm.

4.5.4.1 ISAR Design Example # 1 The computer-aided design (CAD) view of an airplane model whose ISAR image is going to be constructed is shown in Figure 4.17a. The CAD file is composed of many triangle patches. The dimensions of the plane are 7 m, 11.68 m, and 3.30 m in x, y, and z directions, respectively. We would like to get a 2D ISAR image of the airplane on the x–y plane. Therefore, we will start applying the ISAR design steps:

Step 1: Since the target size is 7 m by 11.68 m, we should choose an image window extend that should cover the whole airplane on the 2D x–y plane. So, we select the size of the image extends as 12 m and 16 m in x and y directions, respectively. The backscattered data are to be collected from the nose-on direction around the center frequency of 6 GHz. Therefore, the range (x) axis will be in the direction from nose

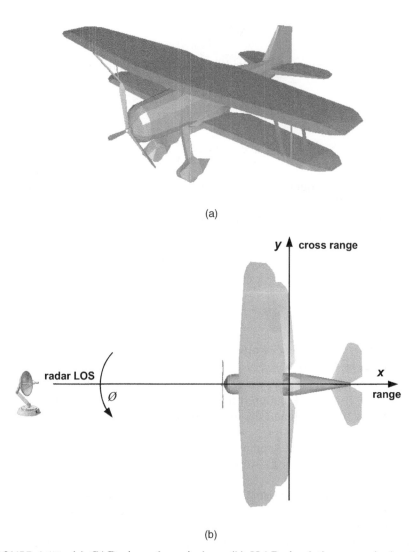

(a)

(b)

FIGURE 4.17 (a) CAD view of an airplane, (b) ISAR simulation scenario for the monostatic situation.

to tail, whereas the cross-range (y) axis will be in the direction from one wing to another. The ISAR scenario for the scene is illustrated in Figure 4.17b.

Step 2: We select the range and cross-range resolutions as $\Delta x = 37.5$ cm and $\Delta y = 25$ cm, respectively. Therefore, the sampling points in range (N_x) and the sampling points in cross range (N_y) will be equal to

$$N_x = \frac{12\,\mathrm{m}}{0,375\,\mathrm{m}}$$
$$= 32$$
$$N_y = \frac{16\,\mathrm{m}}{0,25\,\mathrm{m}}$$
$$= 64.$$

(4.61)

Step 3: Now, we can determine the frequency resolution, Δf, and aspect resolution, $\Delta \varnothing$, as

$$\Delta f = \frac{3 \cdot 10^8 / 2}{12}$$
$$= 12.5\,\mathrm{MHz}$$
$$\Delta \varnothing = \frac{0.05 / 2}{16}$$
$$= 0.0016\,\mathrm{rad}\ (0.09°).$$

(4.62)

The frequency bandwidth and the angular width should then be equal to

$$B = 32 \cdot 125\,\mathrm{MHz}$$
$$= 400\,\mathrm{MHz}$$
$$\Omega = 64 \cdot 0.00016\,\mathrm{rad}$$
$$= 0.1\,\mathrm{rad}\ (5.73°).$$

(4.63)

Notice that the frequency bandwidth is smaller than the one-tenth of the center frequency that satisfies the small-bandwidth approximation. The look angle of radar varies from −0.05 rad to 0.05 rad. Within this angular width, all azimuth angles satisfy $\sin(\varnothing) \cong \varnothing$; therefore, small-angle approximation also holds true.

Step 4: Once the above quantities are defined, the backscattered electric field should be collected for frequencies from 5.80 GHz to 6.1875 GHz for a total of 32 discrete frequencies and angles from −2.86° to 2.78° for 64 distinct angles. Therefore, the simulation of the airplane model is obtained by calculating the backscattered electric field $E^s(f,\varnothing)$ for these frequencies and angles. At the end of simulation, 2D multifrequency multiaspect backscattered field data of size 32 by 64 are obtained.

Step 5: In the last step, the 2D IFT of the collected data is taken to form the ISAR image. The resultant ISAR image is depicted in Figure 4.18.

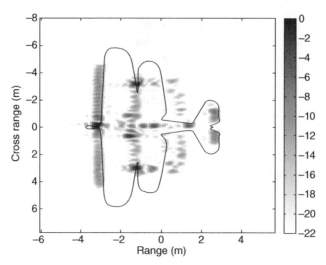

FIGURE 4.18 2D ISAR image of the airplane model.

As stated previously, an ISAR image displays the 2D range–cross-range profile of the target. The ISAR image is plotted in logarithmic scale with the dynamic range of 20 dB. In this image, we observe the main back-scattering centers located around the nose, propellers, wings, and tail of the airplane. The *sinc*-type distortion due to finite frequency bandwidth and the finite angular widths can be easily observed around the scatterer centers in the image. In Chapter 5, we will see how to mitigate these kinds of distortive effects.

4.5.4.2 *ISAR Design Example # 2*   In this example, we will investigate the design parameters for generating the ISAR image of a passenger plane whose CAD file is shown in Figure 4.19a. This airplane has x, y, and z extends of 72.8 m, 61.1748 m, and 22.5324 m, respectively. 2D ISAR image of the plane can be obtained by applying the following ISAR design steps:

Step 1: We would like to get the range and cross-range image of the plane by looking 10° above the nose of the airplane (80° in elevation and 0°in azimuth) at the center frequency of 4 GHz. From this look angle, the airplane has the range extend of 70.5 m and the cross-range extend of 61.1748 m. Therefore, we select the ISAR window size as 80 m by 66 m to be able to cover the whole airplane. The ISAR scenario for the air-plane is shown in Figure 4.19b.

Step 2: We select the range and cross-range resolutions as $\Delta x = 250$ cm and $\Delta y = 103.125$ cm, respectively. Therefore, the sampling points in range (N_x) and the sampling points in cross range (N_y) equal to

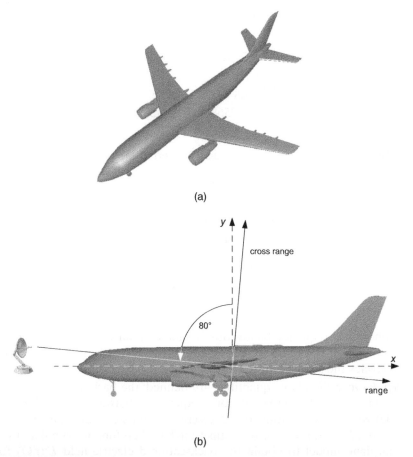

FIGURE 4.19 (a) CAD view of a passenger airplane (b) ISAR simulation scenario for the monostatic situation.

$$N_x = \frac{80}{2.5}$$
$$= 32$$
$$N_y = \frac{66}{1.03125}$$
$$= 64.$$

(4.64)

Step 3: Now, we can determine the resolutions in frequency and azimuth angle domains as

$$\Delta f = \frac{3 \cdot 10^8 / 2}{80}$$
$$= 1875 \text{ KHz}$$
$$\Delta \emptyset = \frac{0.05 / 2}{66}$$
$$= 0.000568 \text{ rad } (0.0326°).$$

(4.65)

Therefore, the frequency bandwidth and the angular width become equal to

$$B = 32 \cdot 1875 \text{ KHz}$$
$$= 60 \text{ MHz}$$
$$\Omega = 64 \cdot 0.000568 \text{ rad}$$
$$= 0.036352 \text{ rad } (2.08°).$$

(4.66)

Since the frequency bandwidth is smaller than the one-tenth of the center frequency and $\sin(\Omega/2) = \sin(0.0184) = 0.0184$, this ISAR setup satisfies the small-bandwidth small-angle ISAR case.

Step 4: Once the above quantities are defined, the backscattered electric field should be gathered for frequencies between 3.97 GHz and 4.0281 Ghz for a total of 32 discrete frequencies and angles from −1.0417° to 1.0092° for 64 distinct angles. Therefore, we simulated the airplane model to obtain the backscattered electric field $E^s(f, \emptyset)$ for those frequencies and angles. At the end of the simulation, the 32 by 64 2D multifrequency multiaspect backscattered field data are obtained.

Step 5: In this last step, the 2D IFT of the collected data is taken to acquire the ISAR image. The resultant image is depicted in Figure 4.20 where the key scattering centers on the nose, tires, engine ducts, and tail can easily be observed. The image also suffers from the sinc-type distortion due to finite bandwidth and finite aspect width, as expected.

4.6 2D ISAR IMAGE FORMATION (WIDE BANDWIDTH, LARGE ANGLES)

ISAR systems generally use narrow angular integration widths that may typically extend only a few degrees while collecting reflectivity data from the target. This is mainly due to the fact that using narrow look-angle apertures provides major simplifications in signal processing and image formation. The

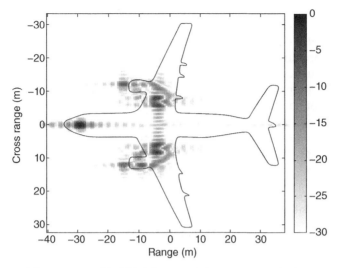

FIGURE 4.20   2D ISAR image of the passenger airplane.

plane-wave illumination assumption can be made under most narrow-angle situations and therefore, direct Fourier transformation procedure can be efficiently employed in forming the final ISAR images as presented in Section 4.5. Besides these benefits, noting that the cross-range resolution is inversely proportional to the look-angle width, high-resolution ISAR images of targets which have large cross-range extents cannot be possible with the narrow-angle data. Therefore, collecting the backscattered data over a wide angular width will improve the cross-range resolution [4]. Furthermore, wide-angle data collections can also provide high range resolutions even with relatively narrow waveform bandwidth [5]. On the other hand, wide-angle systems are faced with the significant problem of unfocused images. Plane-wave illumination assumption is no longer valid in the wide-angle setup and hence, the imaging algorithm must take wavefront curvature effects into account. One possible solution is the sub-aperture approach which assumes planar wavefronts in smaller sub-apertures of the wide-angle data [6]. This procedure exhibits resolution degradations because it does not use the whole angle aperture for the same integration time.

If the frequency bandwidth and the angle width are not small, then the *small-bandwidth and small-angle ISAR* imaging procedure cannot be employed. Then, the double integration in forming the ISAR image should be carried out numerically. There are usually two different ways to acquire the *wide-bandwidth large-angle ISAR* imagery:

1. Direct integration
2. DFT-based integration after polar reformatting

In the former method, the 2D ISAR integral is numerically carried out by applying a numerical integration scheme such as *Simpson's integration* or *Gaussian quadrature integration* procedure. Although the numerical integration method provides better resolved images in range and cross-range domains for wide frequency bandwidth and large angles, the main drawback is the considerable computation time in evaluating the ISAR integral.

In the latter method, the collected data are transformed to a uniformly spaced grid such that the ISAR integral is computed with the help of FFT. This transformation is known as polar formatting in radar literature. Next, we will explore the details of both imaging algorithms with numerical demonstrations.

4.6.1 Direct Integration

This method is based on the fact that the ISAR image is proportional to the following integral:

$$\begin{aligned} ISAR(x,y) &\sim \int_{\varnothing_1}^{\varnothing_2}\int_{k_1}^{k_2}\left\{E^s(k,\varnothing)\right\}\cdot e^{j2(k_x\cdot x+k_y\cdot y)}dk\cdot d\varnothing \\ &= \int_{\varnothing_1}^{\varnothing_2}\int_{k_1}^{k_2}\left\{E^s(k,\varnothing)\right\}\cdot e^{j2(k\cos\varnothing\cdot x+k\sin\varnothing\cdot y)}dk\cdot d\varnothing. \end{aligned} \tag{4.67}$$

Here, the backscattered electric field is assumed to be collected for the spatial frequencies from k_1 to k_2 and for the angles from $\varnothing_1$ to $\varnothing_2$. Let us investigate how this integral is capable of showing the locations of dominant scattering points of a target, that is, the ISAR image: For a single point scatterer at (x_o, y_o), the backscattered electric field can be approximated as

$$E^s(k,\varnothing) \cong A \cdot e^{-j2(k\cos\varnothing\cdot x_o+k\sin\varnothing\cdot y_o)}. \tag{4.68}$$

Substituting Equation 4.68 into Equation 4.67, we get

$$ISAR(x,y) \sim A \cdot \int_{\varnothing_1}^{\varnothing_2}\int_{k_1}^{k_2}\cdot e^{j2k(\cos\varnothing\cdot(x-x_o)+\sin\varnothing\cdot(y-y_o))}dk\cdot d\varnothing. \tag{4.69}$$

While this integration is carried out for different values of x and y, the result of Equation 4.69 is maximized only for $x = x_o$ and $y = y_o$, since the phase of E^s and the phase of the integration argument are fully matched to sum up the energy contained in every pixel on the frequency-aspect domain as the integral is evaluated as shown below:

$$\begin{aligned} ISAR(x_o,y_o) &\sim A \cdot \int_{\varnothing_1}^{\varnothing_2}\int_{k_1}^{k_2}dk\cdot d\varnothing \\ &= A \cdot (k_2-k_1)\cdot(\varnothing_2-\varnothing_1) \\ &= A \cdot BW_k \cdot \Omega, \end{aligned} \tag{4.70}$$

where $BW_k = k_2 - k_1$ is the spatial frequency bandwidth and $\Omega = \varnothing_2 - \varnothing_1$ is the look angle width while collecting the backscattered data. For the other values of x and y different from $x = x_o$ and $y = y_o$, the result of Equation 4.69 is quite small since the phase of E^s and the phase of the integration argument are not matched; therefore, the integration value comes out to be very small when compared to the value at Equation 4.70. This means that the point scatterer at (x_o, y_o) is pinpointed with an appropriate integration routine with the wide-bandwidth large-angle ISAR imaging integral by normalizing Equation 4.67 as

$$ISAR(x,y) = \frac{1}{BW_k \cdot \Omega} \int_{\varnothing_1}^{\varnothing_2} \int_{k_1}^{k_2} \{E_s(k,\varnothing)\} \cdot e^{j2(k\cos\varnothing \cdot x + k\sin\varnothing \cdot y)} dk \cdot d\varnothing. \qquad (4.71)$$

The resolution of this ISAR image is improved by selecting wider bandwidth and larger aspect width. This, of course, in return, necessitates more computation resources as a result of wider integration ranges. The resolution of the ISAR image can also be visually improved by selecting a smaller numerical integration discretization. This is again at the expense of increasing the computational load in terms of the computing memory and the calculation time.

Now, we will demonstrate an application of the wide-bandwidth large-angle ISAR imaging concept via a numerical example. For this purpose, a finite number of perfect point scatterers are assumed to be located as shown in Figure 4.21. A total of 110 point scatterers are placed to emulate the outline of a fictitious airplane. These point scatterers are assumed to scatter the electromagnetic energy in all directions and in all angles with a unitary amplitude.

Before going into the *wide-bandwidth large-angle* ISAR imaging routine, it may be useful to get the *small-bandwidth small-angle ISAR image* of this geometry for comparison reasons. For this purpose, we select an ISAR image

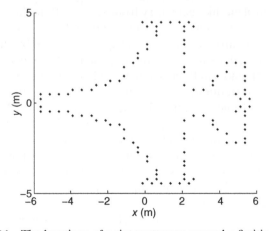

FIGURE 4.21   The locations of point scatterers around a fictitious airplane.

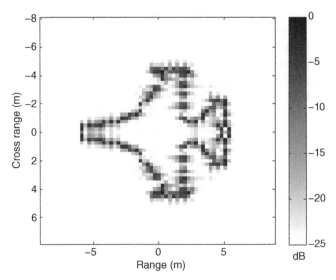

FIGURE 4.22   Small-bandwidth small-angle ISAR image of the hypothetical airplane model.

window of 18 m in the range direction and 16 m in the cross-range direction. Then, applying the basic ISAR design procedure, we end up with getting a frequency bandwidth of 525 MHz around the center frequency of 8 GHz and the angular width of 4.23° around the nose of the airplane-like geometry. After collecting the backscattered electric field for 64 different frequencies and 64 different aspects, the conventional small-bandwidth small-angle ISAR imaging algorithm is applied to obtain the final ISAR image as shown in Figure 4.22.

Now, we will increase both the frequency bandwidth and the aspect width to employ the wide-bandwidth large-angle ISAR imaging procedure. For this purpose, the backscattered electric field from this geometry is collected around the nose of this airplane-like geometry from −30° to 30° in the azimuth angles. While doing this, the frequency is also altered from 6 GHz to 10 GHz, providing a 50% bandwidth around the center frequency of 8 GHz. This data collection setup, of course, does not meet the regular ISAR imagery specifications of small bandwidth and small angles. After collecting the backscattered field for those angles and frequencies, the regular ISAR algorithm based on small bandwidth and small angles is applied using the conventional FFT-based ISAR imaging. The resulting image is depicted in Figure 4.23. Of course, the image is highly distorted since the collected data do not lie on a rectangular grid in the spatial frequency plane and therefore, 2D IFT operations vastly spread the locations and PSRs of the scattering centers on the target. This result clearly shows that wide-bandwidth and/or wide-aspect backscattered data should be treated differently.

The same data are processed through the wide-bandwidth large-angle ISAR imaging integral in Equation 4.71 with the use of Simpson's integration

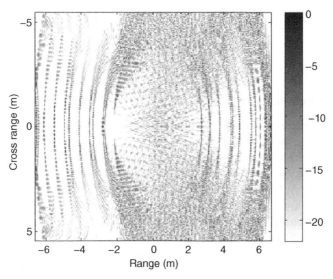

FIGURE 4.23 Aliased ISAR image after applying a 2D IFT to a wide-bandwidth large-angle backscattered field.

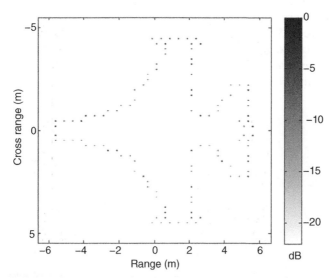

FIGURE 4.24 Wide-bandwidth large-angle ISAR image of the airplane-like geometry (after direct integration).

rule. The resultant ISAR image is plotted in Figure 4.24 where the scattering centers are almost perfectly imaged, thanks to the direct integration method of ISAR. Here, the resolution cell is determined by the integration discretization value which is quite small and which, in turn, provides almost perfect resolution values both in range and cross-range domains. It is worthwhile to mention that there always exists numerical noise due to approximations in the numerical integration scheme. The numerical noise in the ISAR image in Figure 4.24, for instance, can be visible after −25 dB below the maximum pixel in the image.

4.6.2   Polar Reformatting

Another way of treating the wide-bandwidth wide-aspect backscattered data is by utilizing the polar reformatting algorithm. The main idea is to reformat the data in the spatial frequency domain to make use of FFT for fast formation of the ISAR image. In this subsection, wide-bandwidth large-angle ISAR imaging based on the polar reformatting routine will be explored.

Since the data, $E^s(k,\emptyset)$, are collected in the frequency-aspect domain, they are in equally spaced rectangular form in this domain. However, the data are in fact in polar format in the spatial frequency domain on k_x–k_y plane as seen in Figure 4.25. It is obvious from Equation 4.68 that the Fourier relationship exists between k_x and x and between k_y and y. Therefore, fast computation of ISAR integral can be done with the help of FFT only if the data are transformed on a uniform grid on k_x–k_y plane. The use of FFT/IFFT requires the data to be in a discrete, uniformly spaced rectangular form. Therefore, the data should be transformed from the polar format to the Cartesian format as illustrated in Figure 4.25. This process is known as *polar reformatting*. To minimize the error associated with this reformatting procedure, several interpolation schemes such as *four-nearest neighbor approximation* [7, 8] can be employed. After putting the data in their proper format, the ISAR image can be gener-

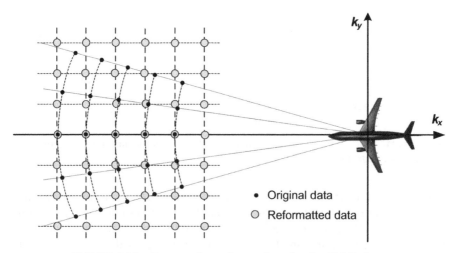

FIGURE 4.25   Rectangular reformatting of polar ISAR data.

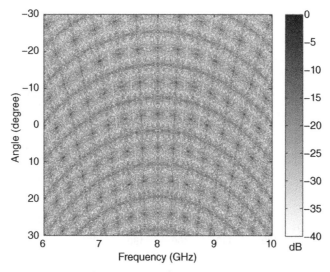

FIGURE 4.26   The backscattered field data in frequency-aspect domain.

ated in a similar manner as explained in Section 4.4 by applying 2D DFT operation.

The polar reformatting algorithm is demonstrated with the same wide-bandwidth large-angle data set collected for the hypothetical target in Figure 4.21. The backscattered field data in the frequency-aspect domain are depicted in Figure 4.26. Then, the data are reformatted in 2D spatial frequency plane of $k_x - k_y$ by using the four-nearest neighbor algorithm. The details of this algorithm will be presented in Chapter 5, Section 5.3. The resultant reformatted data are shown in Figure 4.27. Once the field data are transformed uniformly onto the $k_x - k_y$ plane, FFT processing can then be applied to transform the data onto the $x-y$ plane, that is, the image plane. The result is nothing but the 2D ISAR image as depicted in Figure 4.28. As in the case of direct integration, polar reformatting followed by IFFT processing provides very fine resolutions depending both on range and cross-range dimensions. The numerical noise due to reformatting the data from polar format to the rectangular format is unavoidable. For this particular example, the numerical noise for this ISAR image starts to show up after −30 dB below the maximum point in the image.

4.7   3D ISAR IMAGE FORMATION

In the conventional operation of ISAR, the coherent system tracks the target and collects the frequency-diverse reflected energy for different look angles

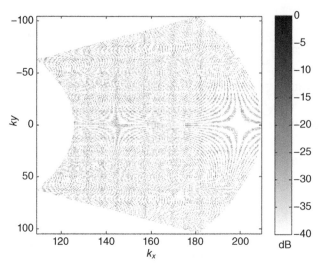

FIGURE 4.27   The backscattered field data on spatial frequency plane of k_x–k_y.

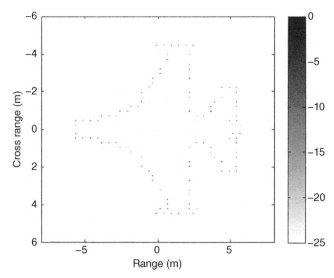

FIGURE 4.28   Wide-bandwidth large-angle ISAR image of the airplane-like geometry (after polar reformatting and FFT).

of target, usually by utilizing the rotational motion of the target. Displaying the target's cross-sectional area over different range points and Doppler frequencies (or cross-range points) yields the 2D ISAR image of the target. In a similar manner, a 3D ISAR image of the target may be obtained if the target's translational and rotational motion components can be measured or at least estimated accurately such that the target's profiles for two different orthogonal look angles are obtained.

Recent studies on 3D ISAR imaging concept have been mainly focused on two different approaches [9–13]: The first approach is based on interferometric ISAR setup, where multiple antennas are used at different heights so that the second cross-range dimension is able to be resolved [14–17]. This approach has many limitations, including its high sensitivity with respect to glint noise of the target [18] and the lack of multiple height (z) information for a selected range cross-range (x,y) point. On the other hand, the second approach that utilizes single ISAR antenna can provide radar cross section (RCS) profiles at multiple heights and multiple range and cross-range points. Some of the 3D ISAR imaging algorithms are listed in References 9–13.

Here, we will present 3D ISAR imaging for small frequency bandwidth and small-angle approximation while collecting the backscattered field data.

A 3D ISAR image, in fact, presents a 3D profile in range (say x) and two cross-range domains (say y and z). Therefore, the scattered field should be collected for various frequencies and different azimuth and frequency angles as depicted in Figure 4.29. As illustrated in this figure, the collected data occupy a space in k_x, k_y, and k_z domains. If the backscattered electric field data are collected within a small bandwidth of frequencies, B (or BW_k), and width of viewing angles, $BW_\varnothing \triangleq \Omega$ and $BW_\theta \triangleq \psi$, the data grid in $k_x - k_y - k_z$ space approaches an equally spaced linear grid. As in the case of the 2D ISAR, this assumption makes it possible to use FFT in forming the 3D ISAR image.

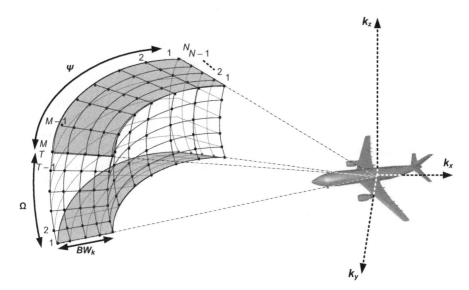

FIGURE 4.29 Collection of raw ISAR data in Fourier space (3D monostatic case).

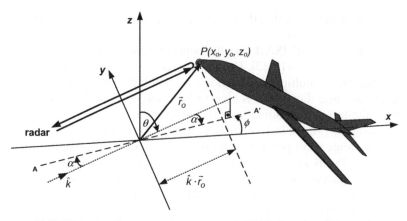

FIGURE 4.30 Geometry for monostatic ISAR imaging (3D case).

The algorithm is based on the model in which a point target is assumed to be present (see Fig. 4.30). The point scatterer to be imaged is located at $P(x_o, y_o, z_o)$. The radar illuminates the target with a wavenumber vector of $\vec{k} = k \cdot \hat{k}$. The direction of illumination makes an angle of θ with the z-axis. It also makes an angle of α with the x–y plane. The line A–A' indicates the projection line of $\vec{k}$ vector on the x–y plane. So, one can easily notice that $\theta + \alpha = 90°$. Taking origin as the phase center of the scenario, the far-field backscattered field from a point scatterer P at an azimuth angle $\emptyset$ and at an elevation angle θ can be written as

$$E^s(k,\emptyset,\theta) = A_o \cdot e^{-j2\vec{k}\cdot\vec{r_o}}. \tag{4.72}$$

Here, A_o is the amplitude of the backscattered electric field intensity, $\vec{k}$ is the vector wave number in k direction, $\vec{r_o}$ is the vector from origin to point P. The multiplier "2" in the exponential accounts for the two-way propagation between the radar and the scatterer. The reason Equation 4.72 has a phase lag of amount $(2\vec{k}\cdot\vec{r_o})$ is that the electromagnetic wave hits point P and reflects back in the same direction will travel an extra the trip distance of $(2\hat{k}\cdot\vec{r_o})$ when compared to reference wave that hits the origin and reflects back. $\vec{k}$ vector can be written in terms of the wave numbers in x, y, and z directions as

$$\begin{aligned}
\vec{k} &= k \cdot \hat{k} \\
&= k \cdot (\hat{x} \cdot \sin\theta \cos\emptyset + \hat{y} \cdot \sin\theta \sin\emptyset + \hat{z} \cdot \cos\theta) \\
&= k \cdot (\hat{x} \cdot \cos\alpha \cos\emptyset + \hat{y} \cdot \cos\alpha \sin\emptyset + \hat{z} \cdot \sin\alpha),
\end{aligned} \tag{4.73}$$

where $\hat{k}, \hat{x}, \hat{y},$ and $\hat{z}$ are the unit vectors in $k, x, y,$ and z directions, respectively. Then, the argument in the phase term of Equation 4.74 can be reorganized to give

$$\vec{k} \cdot \vec{r}_o = k (\hat{x} \cdot cos\alpha cos\emptyset + \hat{y} \cdot cos\alpha sin\emptyset + \hat{z} \cdot sin\alpha) \cdot (\hat{x} \cdot x_o + \hat{y} \cdot y_o + \hat{z} \cdot z_o)$$
$$= kcos\alpha cos\emptyset \cdot x_o + kcos\alpha sin\emptyset \cdot y_o + ksin\alpha \cdot z_o \tag{4.74}$$

Therefore, we can rewrite Equation 4.72 as

$$E^s (k,\emptyset,\theta) = A_o \cdot e^{-j2kcos\theta cos\emptyset \cdot x_o} \cdot e^{-j2kcos\theta sin\emptyset \cdot y_o} \cdot e^{-j2ksin\theta \cdot z_o}. \tag{4.75}$$

This equation offers three separate phase terms as a function of the spatial frequency k and the angles θ and $\emptyset$. At this point we assume that the backscattered signal is collected within a frequency bandwidth, B, that is small compared to the center frequency of operation, f_c. We further assume that the angular bandwidths in both θ (elevation) and $\emptyset$ (azimuth) directions are also small. If the radar is situated around the x-axis, the following assumptions are valid:

$$k \cong k_c$$
$$cos\alpha cos\emptyset \cong 1$$
$$cos\alpha sin\emptyset \cong \emptyset \tag{4.76}$$
$$sin\theta \cong \alpha,$$

where k_c is the wave number corresponding to the center frequency and is given by $k_c = 2\pi f_c/c$. Then, the scattered electric field from point P is approximated as

$$E^s (k,\emptyset,\alpha) = A_o \cdot e^{-j2k \cdot x_o} \cdot e^{-j2k_c\emptyset \cdot y_o} \cdot e^{-j2k_c\alpha \cdot z_o}. \tag{4.77}$$

To be able to use the advantages of FT, we reorganize Equation 4.77 as

$$E^s (k,\emptyset,\alpha) = A_o \cdot e^{-j2\pi\left(\frac{2f}{c}\right)x_o} \cdot e^{-j2\pi\left(\frac{k_c\emptyset}{\pi}\right)y_o} \cdot e^{-j2\pi\left(\frac{k_c\alpha}{\pi}\right)z_o}. \tag{4.78}$$

Then the 3D ISAR image in x–y–z space can be obtained by taking the 3D IFT of Equation 4.78 along $(2f/c)$ and along $(k_c\emptyset/\pi)$ and taking the FT of Equation 4.78 along $(k_c\alpha/\pi)$ as

$$ISAR(x,y,z) = A_o \cdot \mathcal{F}_1^{-1}\left\{e^{-j2\pi\left(\frac{2f}{c}\right)x_o}\right\} \cdot \mathcal{F}_1^{-1}\left\{e^{-j2\pi\left(\frac{k_c\emptyset}{\pi}\right)y_o}\right\} \cdot \mathcal{F}_1\left\{e^{-j2\pi\left(\frac{k_c\alpha}{\pi}\right)z_o}\right\}$$

$$= A_o \cdot \left[\int_{-\infty}^{\infty} e^{-j2\pi\left(\frac{2f}{c}\right)x_o} \cdot e^{j2\pi\left(\frac{2f}{c}\right)x} d\left(\frac{2f}{c}\right)\right] \cdot$$

$$\left[\int_{-\infty}^{\infty} e^{-j2\pi\left(\frac{k_c\emptyset}{\pi}\right)y_o} \cdot e^{j2\pi\left(\frac{k_c\emptyset}{\pi}\right)y} d\left(\frac{k_c\emptyset}{\pi}\right)\right] \cdot \qquad (4.79)$$

$$\left[\int_{-\infty}^{\infty} e^{-j2\pi\left(\frac{k_c\alpha}{\pi}\right)z_o} \cdot e^{j2\pi\left(\frac{k_c\alpha}{\pi}\right)z} d\left(\frac{k_c\alpha}{\pi}\right)\right] \cdot$$

$$= A_o \cdot \delta(x-x_o, y-y_o, z-z_o).$$

Here $\delta(x,y,z)$ represents the 3D impulse function in x–y–z space. As is obvious from Equation 4.79, the point P manifests itself in the ISAR image as the 3D impulse function located at (x_o, y_o, z_o) with the correct reflectivity coefficient of A_o.

The backscattered electric field from a target can be approximated as the sum of scattering from a finite number of single point scatterers, called scattering centers, on the target as shown below:

$$E^s(k,\emptyset,\alpha) \cong \sum_{i=1}^{M} A_i \cdot e^{-j2\vec{k}\cdot\vec{r_i}}. \qquad (4.80)$$

Here, the backscattered electric field from a target is approximated as the sum of backscattered field from M different scattering centers on the target. While A_i represents the complex backscattered field amplitude for the ith scattering center, $\vec{r_i} = x_i \cdot \hat{x} + y_i \cdot \hat{y} + z_i \cdot \hat{z}$ is displacement vector from origin to the location of the ith scattering center. Then, the ISAR image of the target can be found via the following 3D Fourier integral

$$ISAR(x,y,z) = \iiint_{-\infty}^{\infty} \{E^s(k,\emptyset,\alpha)\} \cdot$$

$$\cdot e^{j2\pi\left(\frac{2f}{c}\right)x} e^{j2\pi\left(\frac{k_c\emptyset}{\pi}\right)y} e^{j2\pi\left(\frac{k_c\emptyset}{\pi}\right)y} d\left(\frac{2f}{c}\right) d\left(\frac{k_c\emptyset}{\pi}\right) d\left(\frac{k_c\alpha}{\pi}\right) \qquad (4.81)$$

By entering the scattering center representation for the backscattered field, the small-bandwidth small-angle ISAR image can then be approximated as

$$`ISAR(x,y,z) \cong \sum_{i=1}^{M} A_i \cdot \delta(x-x_i, y-y_i, z-z_i). \qquad (4.82)$$

Therefore, the resulting ISAR image is the sum of M scattering centers with their electromagnetic reflectivity coefficients. Of course, the limits of the integral in Equation 4.81 cannot be infinite in practice due to the fact that the field data can be collected within a finite bandwidth for frequencies and angles. Therefore, the practical ISAR image response distorts from the impulse function to the sinc function.

4.7.1   Range and Cross-Range Resolutions

Range and cross-range resolutions in 3D ISAR are determined in the same way as they are determined in 2D ISAR. Since the same phase terms for the distance variables x and y show up in the 3D ISAR equation of Equation 4.81, the range resolution in x and the cross-range resolution in y is the same as in the case of the 2D case:

$$\Delta x = \frac{c/2}{B}. \tag{4.83}$$

$$\Delta y = \frac{\lambda_c/2}{\Omega}. \tag{4.84}$$

The other cross-range resolution in the z direction can be found by making use of the Fourier relationship between the elevation angle variable, α, and the cross-range distance variable, z. If the backscattered data are collected within finite angular width of $BW_\alpha \triangleq \psi$ in the elevation angles, the cross-range resolution of ISAR in the z direction is equal to

$$\begin{aligned}\Delta z &= \frac{1}{BW\left(\dfrac{k_c \alpha}{\pi}\right)} \\ &= \frac{\pi/k_c}{BW_\alpha} \\ &= \frac{\lambda_c/2}{\psi}. \end{aligned} \tag{4.85}$$

4.7.2   A Design Example

In this example, the 3D ISAR imaging for the airplane whose model is shown in Figure 4.17 will be demonstrated. The model has the extents of 7 m, 11.68 m, and 3.30 m in the x, y, and z planes, respectively.

Step 1: In this example, we will select the center of the radar look angle as $(\theta_c = 60°, \emptyset_c = 0°)$. From this look angle, the airplane's range extent becomes 6.6 m. The cross-range extents become 11.68 m in one direction and 5.65 m in the other direction. Therefore, we select the 3D ISAR window size as 12 m by 16 m by 6 m to be able to cover the whole airplane.

Step 2: We select the range and cross-range resolutions as $\Delta x = 18.75$ cm, $\Delta y = 12.50$ cm, and $\Delta z = 18.75$ cm, respectively. Therefore, the sampling points in range (N_x) and the sampling points in two cross-range dimensions (N_y, N_z) will be equal to

$$N_x = \frac{12}{0.1875} = 64$$

$$N_y = \frac{16}{0.125} = 128 \qquad (4.86)$$

$$N_z = \frac{6}{0.1875} = 32.$$

Step 3: Now, we can determine the resolutions in frequency, Δ_f, elevation angles, Δ_α, and azimuth angles, $\Delta_\emptyset$, as

$$\Delta_f = \frac{3 \cdot 10^8 / 2}{12} = 12.5 \text{ MHz}$$

$$\Delta_\emptyset = \frac{0.05 / 2}{16} = 0.001171875 \text{ rad } (0.0671°) \qquad (4.87)$$

$$\Delta_\alpha = \frac{0.05 / 2}{6} = 0.003125 \text{ rad } (0.179°).$$

Provided that we choose the center frequency of operation as 8 GHz, the figures in Equation 4.83 assure the small-bandwidth, small-angle approximation.

Step 4: Now, we collect the backscattered electric field in 3D frequency-elevation-azimuth domains as follows:

Frequencies: range from 7.60 GHz to 8.3875 GHz for a total of 64 discrete frequencies.

Azimuth angles: vary between −4.30° and 4.23° for a total of 128 distinct angles.

Elevation angles: start from 57.14° to 62.69° for a total of 32 equally spaced angles.

Therefore, the backscattered electric field $E^s(f,\emptyset,\alpha)$ from the airplane model is simulated for those frequencies and angles. At the end of simulation, $64 \times 128 \times 32$ multifrequency multiaspect backscattered field data are obtained.

Step 5: In the final step, the 3D backscattered data $E^s(f,\emptyset,\alpha)$ for the airplane model are inverse Fourier transformed to get the ISAR image. Since the image is 3D, the 2D range (x) cross-range (y) slices of this 3D ISAR image for different values of the other cross-range dimension variable (z) are displayed in Figure 4.31. Although there are 32 slices on the x–y plane overall, only 8 of them are displayed, starting from

$z = -3$ m and rise up to $z = 1.875$ m. By looking at these figures, we can observe different scattering centers for different z values in different 2D ISAR slices. In an alternative representation, the 2D projections of the 3D ISAR image are presented in Figure 4.32, where projections on the principal x–y, x–z, and y–z planes are shown.

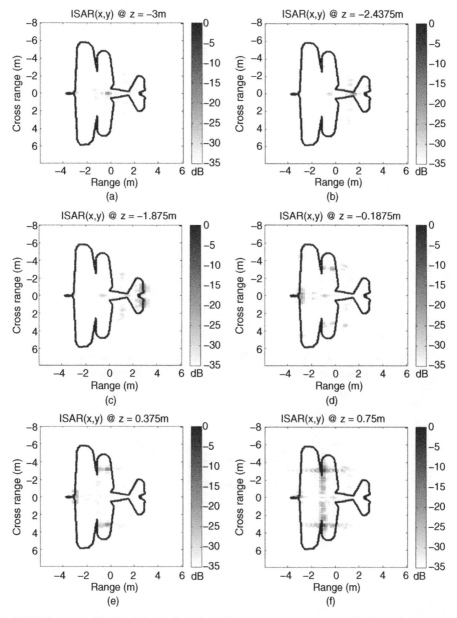

FIGURE 4.31   2D ISAR(x,y) slices for different z values of the 3D ISAR image of the airplane model.

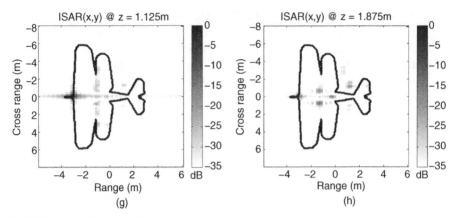

FIGURE 4.31 (*Continued*)

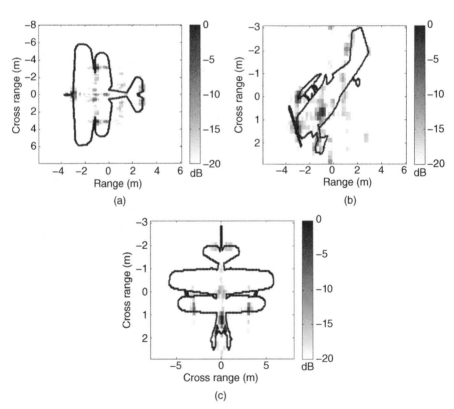

FIGURE 4.32 2D projections of the 3D ISAR image of the airplane model on (a) *x–y* plane, (b) *x–z* plane, and (c) *y–z* plane.

4.8   MATLAB CODES

Below are the Matlab source codes that were used to generate all of the
Matlab-produced Figures in Chapter 4. The codes are also provided in the CD
that accompanies this book.

Matlab code 4.1: Matlab file "Figure4-6.m"

```
%----------------------------------------------------------
% This code can be used to generate Figure 4.6
%----------------------------------------------------------
% This file requires the following files to be present in the
same
% directory:
%
% Es_range.mat
clear all
close all

c = .3; % speed of light
fc = 4; % center frequency
phic = 0*pi/180; % center of azimuth look angles
thc = 80*pi/180; % center of elevation look angles
%_____PRE PROCESSING_____
BWx = 80; % range extend
M = 32; % range sampling
dx = BWx/M; % range resolution
X = -dx*M/2:dx:dx*(M/2-1); % range vector
XX = -dx*M/2:dx/4:-dx*M/2+dx/4*(4*M-1); % range vector (4x
upsampled)
%Form frequency vector
df = c/2/BWx; % frequency resolution
F = fc+[-df*M/2:df:df*(M/2-1)]; % frequency vector
k = 2*pi*F/c; % wavenumber vector

% load backscattered field data for the target
load Es_range
%zero padding (4x);
Enew = E;
Enew(M*4) = 0;

% RANGE PROFILE GENERATION
RP = M*fftshift(ifft(Enew));
h = plot(XX,abs(RP),'k','LineWidth',2);
set(gca,'FontName', 'Arial', 'FontSize',14,'FontWeight','Bold
');
ylabel('Range Profile Intensity'); xlabel('Range [m]');
axis tight
```

Matlab code 4.2: Matlab file "Figure4-8.m"

```
%-----------------------------------------------------------------
% This code can be used to generate Figure 4.8
%-----------------------------------------------------------------
% This file requires the following files to be present in the
same
% directory:
%
% Es_xrange.mat
clear all
close all

c = .3; % speed of light
fc = 4; % center frequency
phic = 0*pi/180; % center of azimuth look angles
%_____PRE PROCESSING_____
BWy = 66; % x-range extend
N = 128; % x-range sampling
dy = BWy/N; % x- range resolution
Y = -dy*N/2:dy:dy*(N/2-1); % x-range vector
YY = -dy*N/2:dy/4:-dy*N/2+dy/4*(4*N-1); % range vector (4x
upsampled)

%Form angle vector
kc = 2*pi*fc/c; % center wavenumber
dphi = pi/(kc*BWy); % azimuth angle resolution
PHI=phic+[-dphi*N/2:dphi:dphi*(N/2-1)]; % azimuth angle
vector

% load backscattered field data for the target
load Es_xrange

%zero padding (4x);
Enew = E;
Enew(N*4) = 0;

% X-RANGE PROFILE GENERATION
XRP = N*fftshift(ifft(Enew));
h = plot(YY,abs(XRP),'k','LineWidth',2);
set(gca,'FontName', 'Arial', 'FontSize',14,'FontWeight','Bold');
ylabel('Range Profile Intensity'); xlabel('Cross-range [m]');
axis tight
```

Matlab code 4.3: Matlab file "Figure4.14.m"

```
%-----------------------------------------------------------------
% This code can be used to generate Figure 4.14
%-----------------------------------------------------------------
% This file requires the following files to be present in the same
% directory:
```

```
%
% Escorner.mat
clear all
close all

c = .3; % speed of light
fc = 10; % center frequency
phic = 180*pi/180; % center of azimuth look angles
%_____PRE PROCESSING OF ISAR_____
BWx = 3; % range extend
M = 16; % range sampling
BWy = 3; % xrange extend
N = 32; % xrange sampling

dx = BWx/M; % range resolution
dy = BWy/N; % xrange resolution

% Form spatial vectors
X = -dx*M/2:dx:dx*(M/2-1);
Y = -dy*N/2:dy:dy*(N/2-1);

%Find resolutions in freq and angle
df = c/(2*BWx); % frequency resolution
dk = 2*pi*df/c; % wavenumber resolution
kc = 2*pi*fc/c;
dphi = pi/(kc*BWy);% azimuth resolution

%Form F and PHI vectors
F=fc+[-df*M/2:df:df*(M/2-1)]; % frequency vector
PHI=phic+[-dphi*N/2:dphi:dphi*(N/2-1)];% azimuth vector
K=2*pi*F/c; % wanenumber vector

%_____GET THE DATA_____
load Escorner

%_____POST PROCESSING OF ISAR_____
ISAR=fftshift(ifft2(Es));
h=figure;
matplot(X,Y,abs(ISAR),50); % form the image
colormap (1-gray);
colorbar
set(gca,'FontName', 'Arial', 'FontSize',12,'FontWeight',
'Bold');
xlabel('Range [m]'); ylabel('Cross-Range [m]');
line([-0.7071 0], [-0.7071 0],'LineWidth',2,'Color','k');
line([-0.7071 0], [0.7071 0],'LineWidth',2,'Color','k');
```

Matlab code 4.4: Matlab file "Figure4-15.m"

```
%-----------------------------------------------------------------
```

```
% This code can be used to generate Figure 4.15
%----------------------------------------------------------------
% This file requires the following files to be present in the same
% directory:
%
% PLANORPHI45_Es.mat
% planorphi45_2_xyout.mat
clear all
close all

c = .3; % speed of light
fc = 6; % center frequency

phic = 45*pi/180; % center of azimuth look angles

%_____PRE PROCESSING OF ISAR_____
BWx = 13; % range extend
M = 32; % range sampling
BWy = 13; % xrange extend
N = 64; % xrange sampling

dx = BWx/M; % range resolution
dy = BWy/N; % xrange resolution

% Form spatial vectors
X = -dx*M/2:dx:dx*(M/2-1);
Y = -dy*N/2:dy:dy*(N/2-1);

%Find resolutions in freq and angle
df = c/(2*BWx); % frequency resolution
dk = 2*pi*df/c; % wavenumber resolution
kc = 2*pi*fc/c;
dphi = pi/(kc*BWy);% azimuth resolution

%Form F and PHI vectors
F = fc+[-df*M/2:df:df*(M/2-1)]; % frequency vector
PHI = phic+[-dphi*N/2:dphi:dphi*(N/2-1)];% azimuth vector
K=2*pi*F/c; % wanenumber vector

%_____GET THE DATA_____
load PLANORPHI45_Es.mat; % load E-scattered
load planorphi45_2_xyout.mat; % load target outline

%_____ POST PROCESSING OF ISAR_____
%windowing;
w=hanning(M)*hanning(N).';
Ess=Es.*w;
```

```
%zero padding;
Enew=Ess;
Enew(M*4,N*4)=0;

% ISAR image formatiom
ISARnew=fftshift(ifft2(Enew));

h=figure;
matplot2(X,Y,abs(ISARnew),22); % form the image
colormap(1-gray);colorbar
line(xyout_yout,xyout_xout,'LineWidth',.25,'LineStyle','.','Co
lor','k');
set(gca,'FontName', 'Arial', 'FontSize',12,'FontWeight',
'Bold');
xlabel('Range [m]'); ylabel('Cross-Range [m]');
```

Matlab code 4.5: Matlab file "Figure4-18.m"
```
%----------------------------------------------------------------
% This code can be used to generate Figure 4.18
%----------------------------------------------------------------
% This file requires the following files to be present in the
same
% directory:
%
% Esplanorteta60.mat
% planorteta60_2_xyout.mat
clear all
close all

c = .3; % speed of light
fc = 6; % center frequency
phic = 0*pi/180; % center of azimuth look angles
thc = 80*pi/180; % center of elevation look angles

%_____PRE PROCESSING OF ISAR_____
BWx = 12; % range extend
M = 32; % range sampling
BWy = 16; % x-range extend
N = 64; % x-range sampling

dx = BWx/M; % range resolution
dy = BWy/N; % xrange resolution

% Form spatial vectors
X = -dx*M/2:dx:dx*(M/2-1);% range vector
XX = -dx*M/2:dx/4:-dx*M/2+dx/4*(4*M-1); % range vector (4x
upsampled)
Y = -dy*N/2:dy:dy*(N/2-1); % x-range vector
```

```
YY = -dy*N/2:dy/4:-dy*N/2+dy/4*(4*N-1); % range vector (4x
upsampled)


%_____GET THE DATA_____
load Esplanorteta60 % load E-scattered
load planorteta60_2_xyout % load target outline


%_____ POST PROCESSING OF ISAR_____
%zero padding;
Enew = Es;
Enew(M*4,N*4)=0;

% ISAR image formatiom
h = figure;
ISARnew = fftshift(ifft2(Enew));
matplot2(X(M:-1:1),Y,abs(ISARnew.'),22); % form the image
colormap(1-gray);colorbar
line(-xyout_xout,xyout_yout,'LineWidth',.25,'LineStyle','.','C
olor','k');
set(gca,'FontName', 'Arial', 'FontSize',14,'FontWeight',
'Bold');
xlabel('Range [m]'); ylabel('Cross-Range [m]');
```

Matlab code 4.6: Matlab file "Figure4-20.m"
```
%----------------------------------------------------------------
% This code can be used to generate Figure 4.20
%----------------------------------------------------------------
% This file requires the following files to be present in the
same
% directory:
%
% Esairbus.mat
% airbusteta80_2_xyout.mat
clear all
close all

c = .3; % speed of light
fc = 4; % center frequency
phic = 0*pi/180; % center of azimuth look angles
thc = 80*pi/180; % center of elevation look angles

%_____PRE PROCESSING OF ISAR_____
BWx = 80; % range extend
M = 32; % range sampling
BWy = 66; % x-range extend
N = 64; % x-range sampling


dx = BWx/M; % range resolution
```

```
dy = BWy/N; % xrange resolution

% Form spatial vectors
X = -dx*M/2:dx:dx*(M/2-1);% range vector
XX = -dx*M/2:dx/4:-dx*M/2+dx/4*(4*M-1); % range vector (4x
upsampled)
Y = -dy*N/2:dy:dy*(N/2-1); % x-range vector
YY = -dy*N/2:dy/4:-dy*N/2+dy/4*(4*N-1); % range vector (4x
upsampled)

%_____GET THE DATA_____
load Esairbus % load E-scattered
load airbusteta80_2_xyout.mat % load target outline

% ISAR 4x UPSAMPLED-------------------
%zero padding;
Enew = Es;
Enew(M*4,N*4)=0;

% ISAR image formatiom
h = figure;
ISARnew = fftshift(ifft2(Enew));
matplot2(X,Y,abs(ISARnew.'),30); % form the image
colormap(1-gray);colorbar
line(-xyout_xout,xyout_yout,'LineWidth',.25,'LineStyle','.','Color
','k');
set(gca,'FontName', 'Arial', 'FontSize',14,'FontWeight',
'Bold');
xlabel('Range [m]'); ylabel('Cross-Range [m]');
```

Matlab code 4.7: Matlab file "Figure4-21and4-22.m"
```
%---------------------------------------------------------------
% This code can be used to generate Figure 4-21 and 4-22
%---------------------------------------------------------------
clear all
close all

c = .3; % speed of light
fc = 8; % center frequency
phic = 0*pi/180; % center of azimuth look angles

%_____PRE PROCESSING OF ISAR_____
BWx = 18; % range extend
M = 64; % range sampling
BWy = 16; % xrange extend
N = 64; % xrange sampling

dx = BWx/M; % range resolution
dy = BWy/N; % xrange resolution
```

```
% Form spatial vectors
X = -dx*M/2:dx:dx*(M/2-1);
Y = -dy*N/2:dy:dy*(N/2-1);

%Find resoltions in freq and angle
df = c/(2*BWx); % frequency resolution
dk = 2*pi*df/c; % wavenumber resolution
kc = 2*pi*fc/c;
dphi = pi/(kc*BWy);% azimuth resolution

%Form F and PHI vectors
F = fc+[-df*M/2:df:df*(M/2-1)]; % frequency vector
PHI = phic+[-dphi*N/2:dphi:dphi*(N/2-1)];% azimuth vector
K = 2*pi*F/c; % wavenumber vector

%_____ FORM RAW BACKSCATTERED DATA_____
%load scattering centers
load fighterSC
l = length(xx);
%---Figure 4.21--------------------------------------------------
h = figure;
plot(xx,yy,'.')
set(gca,'FontName', 'Arial', 'FontSize',12,'FontWeight',
'Bold');
xlabel('Range [m]'); ylabel('Cross - range [m]');
colormap(1-gray);
xlabel('X [m]');
ylabel('Y [m]');
saveas(h,'Figure4-21.png','png');

%form backscattered E-field from scattering centers
Es = zeros(M,N);
for m=1:l;
 Es = Es+1.0*exp(-j*2*K'*(cos(PHI)*xx(m)+sin(PHI)*yy(m)));
end

%_____ POST PROCESSING OF ISAR (small-BW small
angles)_____
ISAR = fftshift(ifft2(Es.'));

%---Figure 4.22--------------------------------------------------
h = figure;
matplot2(X,Y,ISAR,25); colormap(1-gray); colorbar
set(gca,'FontName', 'Arial', 'FontSize',12,'FontWeight',
'Bold');
xlabel('Range [m]'); ylabel('Cross - range [m]');
colormap(1-gray);
saveas(h,'Figure4-22.png','png');
```

Matlab code 4.8: Matlab file "Figure4-23and24.m"

```
%-----------------------------------------------------------------
% This code can be used to generate Figure 4.23 and 4.24
%-----------------------------------------------------------------
% This file requires the following files to be present in the same
% directory:
%
% fighterSC.mat
clear all
close all

c = .3; % speed of light
fc = 8; % center frequency
fMin = 6; % lowest frequency
fMax = 10; % highest frequency

phic = 0*pi/180; % center of azimuth look angles
phiMin = -30*pi/180; % lowest angle
phiMax = 30*pi/180; % highest angle
%-------------------------------------------------
% WIDE-BW AND LARGE ANGLES ISAR
%-------------------------------------------------
% A- INTEGRATION
%-------------------------------------------------
nSampling = 300; % sampling number for integration

% Define Arrays
f = fMin:(fMax-fMin)/(nSampling-1):fMax;
k = 2*pi*f/.3;
kMax = max(k);
kMin = min(k);
kc = (max(k)+min(k))/2;
phi = phiMin:(phiMax-phiMin)/(nSampling-1):phiMax;

% resolutions
dx = pi/(max(k)-min(k)); % range resolution
dy = pi/kc/(max(phi*pi/180)-min(phi*pi/180)); % xrange
resolution

% Form spatial vectors
X = -nSampling*dx/2:dx:nSampling*dx/2;
Y = -nSampling*dy/2:dy:nSampling*dy/2;

%_____ FORM RAW BACKSCATTERED DATA_____
%load scattering centers
load fighterSC
l = length(xx);
```

```
%form backscattered E-field from scattering centers
clear Es;
Es = zeros((nSampling),(nSampling));
for m=1:1;
 Es = Es+1.0*exp(-j*2*k.'*cos(phi)*xx(m)).*exp(-j*2*k.'*sin(p
hi)*yy(m));
end


axisX = min(xx)-1:0.05:max(xx)+1;
axisY = min(yy)-1:0.05:max(yy)+1;

% take a look at what happens when DFT is used
%---Figure 4.23-----------------------------------------------
ISAR1 = fftshift(ifft2(Es.'));
matplot2(axisX,axisY,ISAR1,22);
colormap(1-gray); colorbar
set(gca,'FontName', 'Arial', 'FontSize',12,'FontWeight',
'Bold');
xlabel('Range [m]'); ylabel('Cross - range [m]');

% INTEGRATION STARTS HERE

% Building Simpson Nodes; Sampling Rate is nSampling
% Weights over k
h = (kMax-kMin)/(nSampling-1);
k1 = (kMin:h:kMax).';
wk1 = ones(1,nSampling);
wk1(2:2:nSampling-1) = 4;
wk1(3:2:nSampling-2) = 2;
wk1 = wk1*h/3;


% Weights over phi
h = (phiMax-phiMin)/(nSampling-1);
phi1 = (phiMin:h:phiMax).';
wphi1 = ones(1,nSampling);
wphi1(2:2:nSampling-1) = 4;
wphi1(3:2:nSampling-2) = 2;

wphi1 = wphi1*h/3;
% Combine for two dimensional integration
[phi1,k1] = meshgrid(phi1,k1);
phi1 = phi1(:);
k1 = k1(:);
w = wk1.'*wphi1;
w = w(:).';


newEs = Es(:).';
newW = w.*newEs;
```

```
% Integrate
b = 2j;
ISAR2 =
zeros((max(xx)-min(xx)+2)/0.05+1,(max(yy)-min(yy)+2)/0.05+1);

k1 = k1.*b;
cosPhi = cos(phi1);
sinPhi = sin(phi1);

tic;
x1 = 0;
for X1 = axisX
 x1 = x1+1;
 y1 = 0;
 for Y1 = axisY
 y1 = y1+1;
 ISAR2(x1,y1) = newW*(exp(k1.*(cosPhi.*X1+sinPhi.*Y1)));
 end
end
time1 = toc;

%---Figure 4.24-----------------------------------------------
matplot2(axisX,axisY,ISAR2.',22);
colormap(1-gray); colorbar
set(gca,'FontName', 'Arial', 'FontSize',12,'FontWeight',
'Bold');
xlabel('Range [m]');
ylabel('Cross - range [m]');
```

Matlab code 4.9: Matlab file "Figure4-26thru4-28.m"
```
%-------------------------------------------------------------
% This code can be used to generate Figure 4.26 thru 4.28
%-------------------------------------------------------------
% This file requires the following files to be present in the
same
% directory:
%
% fighterSC.mat
clear all
close all

c = .3; % speed of light
fc = 8; % center frequency
fMin = 6; % lowest frequency
fMax = 10; % highest frequency

phic = 0*pi/180; % center of azimuth look angles
phiMin = -30*pi/180; % lowest angle
phiMax = 30*pi/180; % highest angle
```

```
%-------------------------------------------------
% WIDE BW AND WIDE ANGLE ISAR
%-------------------------------------------------
% B- POLAR REFORMATTING
%-------------------------------------------------
nSampling = 1500; % sampling number for integration

% Define Bandwidth
f = fMin:(fMax-fMin)/(nSampling):fMax;
k = 2*pi*f/.3;
kMax = max(k);
kMin = min(k);

% Define Angle
phi = phiMin:(phiMax-phiMin)/(nSampling):phiMax;

kc = (max(k)+min(k))/2;

kx=k.'*cos(phi);
ky=k.'*sin(phi);

kxMax = max(max(kx));
kxMin = min(min(kx));
kyMax = max(max(ky));
kyMin = min(min(ky));

MM=4; % up sampling ratio
clear kx ky;
kxSteps = (kxMax-kxMin)/(MM*(nSampling+1)-1);
kySteps = (kyMax-kyMin)/(MM*(nSampling+1)-1);
kx = kxMin:kxSteps:kxMax; Nx=length(kx);
ky = kyMin:kySteps:kyMax; Ny=length(ky);
kx(MM*(nSampling+1)+1) = 0;
ky(MM*(nSampling+1)+1) = 0;

%_____ FORM RAW BACKSCATTERED DATA_____
%load scattering centers
load fighterSC
l = length(xx);

%form backscattered E-field from scattering centers
Es = zeros((nSampling+1),(nSampling+1));
for n=1:length(xx);
 Es = Es+exp(-j*2*k.'*cos(phi)*xx(n)).*exp(-j*2*k.'*sin(phi)*
yy(n));
end

%---Figure 4.24-------------------------------------------------
matplot2(f,phi*180/pi,Es,40);
```

```
colormap(1-gray); colorbar
set(gca,'FontName', 'Arial', 'FontSize',12,'FontWeight',
'Bold');
xlabel('Frequency [GHz]');
ylabel('Angle [Degree]');

newEs = zeros(MM*(nSampling+1)+1,MM*(nSampling+1)+1);
t = 0;
v = 0;
for tmpk = k
 t = t+1;
 v = 0;
 for tmpPhi = phi
 v = v+1;
 tmpkx = tmpk*cos(tmpPhi);
 tmpky = tmpk*sin(tmpPhi);
 indexX = floor((tmpkx-kxMin)/kxSteps)+1;
 indexY = floor((tmpky-kyMin)/kySteps)+1;

 r1 = sqrt(abs(kx(indexX)-tmpkx)^2+abs(ky(indexY)-tmpky)^2);
 r2 =
sqrt(abs(kx(indexX+1)-tmpkx)^2+abs(ky(indexY)-tmpky)^2);
 r3 = sqrt(abs(kx(indexX)-tmpkx)^2+abs(ky(ind
exY+1)-tmpky)^2);
 r4 = sqrt(abs(kx(indexX+1)-tmpkx)^2+abs(ky(ind
exY+1)-tmpky)^2);

 R = 1/r1+1/r2+1/r3+1/r4;

 A1 = Es(t,v)/(r1*R);
 A2 = Es(t,v)/(r2*R);
 A3 = Es(t,v)/(r3*R);
 A4 = Es(t,v)/(r4*R);
 newEs(indexY,indexX) = newEs(indexY,indexX)+A1;
 newEs(indexY,indexX+1) = newEs(indexY,indexX+1)+A2;
 newEs(indexY+1,indexX) = newEs(indexY+1,indexX)+A3;
 newEs(indexY+1,indexX+1) = newEs(indexY+1,indexX+1)+A4;
 end
end

% down sample newEs by MM times
newEs=newEs(1:MM: size(newEs),1:MM: size(newEs));

%---Figure 4.25-----------------------------------------------
% reformatted data
h = figure;
Kx = kx(1:Nx-1);
Ky = ky(1:Ny-1);
matplot2(Kx,Ky,newEs,40);
```

```
colormap(1-gray);
colorbar
set(gca,'FontName', 'Arial', 'FontSize',12,'FontWeight',
'Bold');
xlabel('kx [rad/m]'); ylabel('ky [rad/m]');

% Find Corresponding ISAR window in Range and X-Range
kxMax = max(max(kx));
kxMin = min(min(kx));
kyMax = max(max(ky));
kyMin = min(min(ky));

BWKx = kxMax-kxMin;
BWKy = kyMax-kyMin;

dx = pi/BWKx;
dy = pi/BWKy;
X = dx*(-nSampling/2:nSampling/2);
Y = dy*(-nSampling/2:nSampling/2);

%---Figure 4.26------------------------------------------------
% Plot the resultant ISAR image
h = figure;
tt = nSampling/4:3*nSampling/4;
ISAR3 = fftshift(ifft2(newEs));
matplot2(X,Y,ISAR3(:,tt),25);
axis([-8 8 -6 6])
colormap(1-gray);
colorbar
set(gca,'FontName', 'Arial', 'FontSize',12,'FontWeight',
'Bold');
xlabel('Range [m]');
ylabel('Cross - range [m]');
```

Matlab code 4.10: Matlab file "Figure4-31and4-32.m"
```
%-------------------------------------------------------------
% This code can be used to generate Figure 4.31 and 4.32
%-------------------------------------------------------------
% This file requires the following files to be present in the
same
% directory:
%
% E_field.mat
% planorteta60_2_xyout.mat
% planorteta60xzout.mat

clear all
close all
```

```
c = .3; % speed of light
fc = 8; % center frequency
phic = 0*pi/180; % center of azimuth look angles
thc = 60*pi/180; % center of elevation look angles


%_____PRE PROCESSING OF ISAR_____
BWx = 12; % range extend
M = 64; % range sampling
BWy = 16; % x-range1 extend
N = 128; % x-range1 sampling
BWz = 6; % x-range2 extend
P = 32; % x-range2 sampling

%Find spatial resolutions
dx = BWx/M; % range resolution
dy = BWy/N; % xrange1 resolution
dz = BWz/P; % xrange1 resolution

% Form spatial vectors
X = -dx*M/2:dx:dx*(M/2-1);% range vector
XX = -dx*M/2:dx/4:-dx*M/2+dx/4*(4*M-1); % range vector (4x
upsampled)
Y = -dy*N/2:dy:dy*(N/2-1); % x-range1 vector
YY = -dy*N/2:dy/4:-dy*N/2+dy/4*(4*N-1); % x-range1 vector (4x
upsampled)
Z = -dz*P/2:dz:dz*(P/2-1); % x-range2 vector
ZZ = -dz*P/2:dz/4:-dz*P/2+dz/4*(4*P-1);% x-range2 vector (4x
upsampled)

%Find resoltions in freq and angle
df = c/(2*BWx);
dk = 2*pi*df/c;
kc = 2*pi*fc/c;

dphi = pi/(kc*BWy);
dth = pi/(kc*BWz);

%Form F and PHI vectors
F = fc+[-df*M/2:df:df*(M/2-1)];
PHI = phic+[-dphi*N/2:dphi:dphi*(N/2-1)];
TET = thc+[-dth*P/2:dth:dth*(P/2-1)];

%_____GET THE DATA_____
load E_field % load E-scattered
load planorteta60_2_xyout % load target outline

% ISAR
ISAR=fftshift(ifftn(E3d));
```

```
%-------ISAR(x,y) Slices-------------
A = max(max(max(ISAR)));
for m=1:P;
 EE=ISAR(:,:,m);
 EE(1,1,1)=A;
 zp = num2str(Z(m));
 zpp = ['ISAR(x,y) @ z = ' zp 'm'];
%---Figure 4.31----------------------------------------------
 matplot2(-X,Y, EE.',35);colorbar; colormap(1-gray);
 set(gca,'FontName', 'Arial', 'FontSize',14,'FontWeight',
 'Bold');
 xlabel('Range [m]'); ylabel('X-Range [m]');
 drawnow;
 title(zpp);
 h = line(-xyout_xout,xyout_yout,'Color','k','LineStyle','.',
 'MarkerSize',5);
 pause
end

%---Figure 4.32----------------------------------------------
%-------XY Projection-------------
figure;
EExy = zeros(M,N);
load planorteta60_2_xyout

for m = 1:P;
 EExy = EExy+ISAR(:,:,m);
end
 matplot2(-X,Y,EExy.',20);
 colorbar;
 colormap(1-gray);
 set(gca,'FontName', 'Arial', 'FontSize',14,'FontWeight',
 'Bold');
 xlabel('Range [m]');
 ylabel('X-Range [m]');
h=line(-xyout_xout,xyout_yout,'Color','k','LineStyle','.','Mar
kerSize',5);

 %-------XZ Projection-------------
figure;
load planorteta60xzout.mat
for m = 1:M;
 for n = 1:P
 EExz(m,n) = sum(ISAR(m,:,n));
 end
end
 matplot2(-X,Z,EExz.',20);colorbar;
 colormap(1-gray);
```

```
set(gca,'FontName', 'Arial', 'FontSize',14,'FontWeight',
'Bold');
xlabel('Range [m]');
ylabel('X-Range [m]');
h=line(-xzout_xout,-xzout_zout,'Color','k','LineStyle','.','Ma
rkerSize',5);

%-------YZ Projection------------
figure;
load planorteta60yzout.mat
for m = 1:N;
 for n = 1:P
 EEyz(m,n) = sum(ISAR(:,m,n));
 end
end
 matplot2(-Y,Z,EEyz.',20);colorbar;
 colormap(1-gray);
 set(gca,'FontName', 'Arial', 'FontSize',14,'FontWeight',
'Bold');
 xlabel('X-Range [m]');
 ylabel('X-Range [m]');
h=line(-yzout_yout,-yzout_zout,'Color','k','LineStyle','.','Ma
rkerSize',5);
```

REFERENCES

1 T. H. Chu and D.-B. Lin, Y.-W. Kiang. Microwave diversity imaging of perfectly conducting objects in close near field region. *Antennas and Propagation Society International Symposium*, 1989. AP-S. Digest, vol. 1, (1995) 82–85.

2 R. Bhalla and H. Ling. ISAR image formation using bistatic data computed from the shooting and bouncing ray technique. *Journal of Electromagnetic Waves and Applications* 7(9) (1993) 1271–1287.

3 H. Ling, R. Chou, and S. W. Lee. Shooting and bouncing rays: Calculating the RCS of an arbitrary shaped cavity. *IEEE Transactions on Antennas and Propagation* 37 (1989) 194–205.

4 D. R. Wehner. *High resolution radar*. Artech House, Norwood, MA, 1997.

5 J. E. Luminati. *Wide-angle multistatic synthetic aperture radar: Focused image formation and aliasing artifact mitigation*, PhD thesis, Air Force Institute of Technology, Wright-Patterson Air Force Base, Ohio, USA, 2005.

6 C. Ozdemir, O. Kirik, and B. Yilmaz. Sub-aperture method for the wide-bandwidth wide-angle Inverse Synthetic Aperture Radar imaging. International Conference on Electrical and Electronics Engineering-ELECO'2009, Bursa, Turkey, December 2009, 288–292.

7 C. Özdemir, R. Bhalla, L. C. Trintinalia, and H. Ling. ASAR—Antenna synthetic aperture radar imaging. *IEEE Transactions on Antennas and Propagation* 46(12) (1998) 1845–1852.

8 J. Li, et al. Comparison of high-resolution ISAR imageries from measurement data and synthetic signatures. SPIE Proceedings on Radar Processing, Technology, and Applications IV, vol. 3810: 170–179, 1999.

9 F. E. McFadden. Three dimensional reconstruction from ISAR sequences. Proceedings of SPIE, vol. 4744, 2002, pp. 58–67.

10 J. T. Mayhan, M. L. Burrows, and K. M. Cuomo. "High resolution 3D snapshot" ISAR imaging and feature extraction. *IEEE Transactions on Aerospace and Electronic Systems* 37(2) (2001) 630–642.

11 F. Fortuny. An efficient 3-D near field ISAR algorithm. *IEEE Transaction on Aerospace and Electronic Systems* 34 (1998) 1261–1270.

12 K. K. Knoell and G. P. Cardillo. Radar tomography for the generation of three-dimensional images. *IEE Proceedings on Radar, Sonar and Navigation* 142(2) (1995) 54–60.

13 R. T. Lord, W. A. J. Nel, and M. Y. Abdul Gaffar. Investigation of 3-D RCS image formation of ships using ISAR. 6th European Conference on Synthetic Aperture Radar—EUSAR 2006, Dresden, Germany, 2006, pp 4–7.

14 X. Xu and R. M. Narayanan. 3-D interferometric ISAR images for scattering diagnosis of complex radar targets. IEEE Radar Conference, April 1999, pp. 237–241.

15 G. Wang, X. G. Xia, and V. C. Chen. Three-dimensional ISAR imaging of maneuvering targets using three receivers. *IEEE Transactions on Image Processing* 10(3) (2001) 436–447.

16 X. Xu and R. M. Narayanan. Three-dimensional interferometric ISAR imaging for target scattering diagnosis and modeling. *IEEE Transactions on Image Processing* 10(7) (2001) 1094–1102.

17 Q. Zhang, T. S. Yeo, G. Du, and S. Zhang. Estimation of three dimensional motion parameters in interferometric ISAR imaging. *IEEE Transactions on Geoscience and Remote Sensing* 42(2) (2004) 292–300.

18 W.-R. Wu. Target tracking with glint noise. *IEEE Transactions on Aerospace and Electronic Systems* 29(1) (1993) 174–185.

CHAPTER FIVE

Imaging Issues in Inverse Synthetic Aperture Radar

While applying the inverse synthetic aperture radar (ISAR) imaging algorithms, there are some design parameters, such as reformatting the raw ISAR data and applying window functions, that should be carefully handled such that the resultant ISAR image quality can be improved. In this chapter, we will describe and discuss how to tweak these parameters to obtain a good quality ISAR image.

5.1 FOURIER-RELATED ISSUES

The concept of Fourier transform (FT) is very important in synthetic aperture radar (SAR)/ISAR imaging as has been demonstrated in many different places in this book. The classic way of forming the ISAR image is accomplished by transforming the scattered field data from the frequency-angle domain to the image domain by applying the FT operation.

In SAR/ISAR processing, the collected data sequence is usually digitized such that the digital signal processing algorithms are used to form the final image [1, 2]. The process of digitizing the signal has already been shown in Chapter 1, Section 1.6. After the data are converted to the digital form, the Fourier processing can be readily applied with the help of discrete Fourier transform (DFT) operations. Next, we will briefly explain the very basic but also very critical issues regarding getting a well-implemented and good quality ISAR image.

Inverse Synthetic Aperture Radar Imaging with MATLAB Algorithms, First Edition.
Caner Özdemir.
© 2012 John Wiley & Sons, Inc. Published 2012 by John Wiley & Sons, Inc.

5.1.1   DFT Revisited

The sampling and digitizing processes of an analog signal are demonstrated in Figure 5.1. Let $x(t)$ be the continuous time-domain signal as shown in Figure 5.1a, and let $s(t)$ be the observed (or recorded) time-domain signal with a duration of T as depicted in Figure 5.1b. Therefore, $s(t)$ may represent a portion of the original signal $x(t)$. After digitizing $s(t)$ with a sampling interval of $T_s = 1/f_s$, a discrete set of $s[n]$ is obtained, as in Figure 5.1c. Here, f_s is the sampling frequency and also represents the fundamental frequency of a periodic version of the observed signal $s(t)$. In fact, as demonstrated in Figure 5.1d, the resultant discrete-time signal $s[n]$ is periodic in time. As its DFT will be shown later, $S[k]$ is periodic in frequency as well. It can be easily observed from Figure 5.1d that the period is equal to N times the sampling interval as

$$T = N \cdot T_s. \tag{5.1}$$

It is important to note that only the $s[n]$ values for $n = 0,1,2,\ldots,N-1$ is included in the discrete-time base signal. The Nth datum, $s[N]$, belongs to the next period and is the same as $s[0]$. This observation is important when selecting the correct duration (or bandwidth) of the signal in time (or frequency). This phenomenon is clarified with the following examples.

Example 5.1:   Suppose that a discrete time-domain signal is collected for a time interval of 1 ms with 128 samples. Here, the first sample is taken at $t = 0$ *seconds* and the last sample is taken at $t = 1$ ms for a total of 128 points. Therefore, the sampling interval is equal to

$$\begin{aligned} T_s &= \frac{1\,\mathrm{ms}}{127} \\ &= 7.874\ \mu\mathrm{s}. \end{aligned} \tag{5.2}$$

Therefore, the total time interval is equal to

$$\begin{aligned} T &= N \cdot T_s \\ &= 128 \cdot 7.874\ \mu\mathrm{s} \\ &= 1.0078\ \mathrm{ms}, \end{aligned} \tag{5.3}$$

which is greater than the observation time of 1 ms. This is because of the fact that DFT is periodic, and the next period starts one sampling interval later than the last sampling point. This kind of calculation becomes important when dealing with the scaling in the other domain, especially when finding the resolutions and the bandwidths. The frequency resolution for this example is therefore

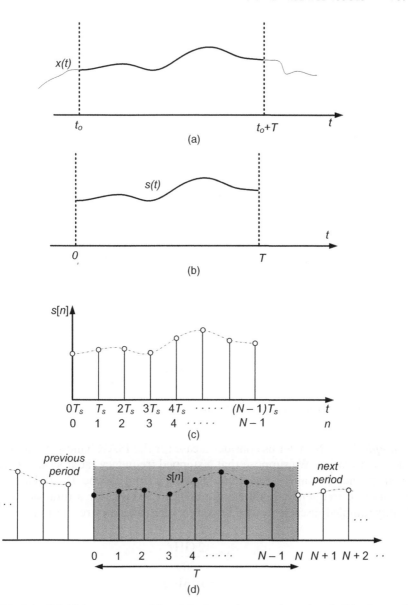

FIGURE 5.1 Digitizing process: (a) original continuous time signal, (b) observed signal with a duration of T, (c) sampled signal with N discrete points, (d) corresponding discrete-time signal $s[n]$ is periodic.

$$\Delta f = \frac{1}{T}$$

$$= \frac{1}{1.0078 \text{ ms}} \tag{5.4}$$

$$= 992.1875 \text{ Hz}.$$

Then, the frequency bandwidth becomes equal to

$$B = N \cdot \Delta f$$

$$= 128 \cdot 992.1875 \tag{5.5}$$

$$= 127 \text{ KHz}.$$

To check the accuracy of the calculations, let us find the time resolution as

$$\Delta t = \frac{1}{B}$$

$$= \frac{1}{127 \text{ KHz}} \tag{5.6}$$

$$= 7.874 \, \mu\text{s},$$

which is exactly the same as a sampling interval of T_s.

Example 5.2: Now, let us consider a case for the ISAR bandwidth and resolution calculation. We suppose that a stepped frequency radar system collects the frequency-domain backscattered field data from a scene. The radar records the discrete frequency signal from 8 GHz to 10 GHz for a total of 201 uniformly sampled discrete points. Therefore, the frequency resolution is

$$\Delta f = \frac{2 \text{ GHz}}{200} \tag{5.7}$$

$$= 10 \text{ MHz}.$$

Therefore, the total frequency bandwidth becomes

$$B = N \cdot \Delta f$$

$$= 201 \cdot 10 \text{ MHz} \tag{5.8}$$

$$= 2.01 \text{ GHz}.$$

This represents a range resolution of

$$\Delta r = \frac{c}{2 \cdot B}$$
$$= \frac{0.3}{2 \cdot 2.01} \tag{5.9}$$
$$= 7.4627 \text{ cm.}$$

The total range can be calculated by multiplying the range resolution by the number of samples as

$$R = N \cdot \Delta r$$
$$= 201 \cdot 7.4627 \text{ cm} \tag{5.10}$$
$$= 15 \text{ m.}$$

To check the accuracy of the calculations, let us determine the frequency resolution via the following Fourier equality:

$$\Delta f = \frac{c}{2 \cdot R}$$
$$= \frac{0.3 \cdot 10^9 \text{ m/s}}{2 \cdot 15} \tag{5.11}$$
$$= 10 \text{ MHz,}$$

which is exactly the same as in Equation 5.7.

5.1.2 Positive and Negative Frequencies in DFT

The definition and the formulation of DFT have been already given in Chapter 1. The forward and inverse DFT pair is defined as given below:

$$S[k] = \sum_{n=0}^{N-1} s[n] \cdot e^{-j2\pi \frac{k}{N} n} \quad k = 0,1,2,\dots,N-1, \tag{5.12a}$$

$$s[n] = \sum_{k=0}^{N-1} S[k] \cdot e^{j2\pi \frac{n}{N} k} \quad n = 0,1,2,\dots,N-1, \tag{5.12b}$$

where $s[n]$ and $S[k]$ present the time-domain and the frequency-domain signals, respectively. While each n step in Equation 5.12b represents a T_s increment in the time axis, each k step in Equation 5.12a represents an $f_s = 1/T$ increment in the frequency axis. It is important to note that DFT representation includes both the positive and the negative frequencies or the harmonics. Table 5.1 explains the indexing of positive and negative frequencies. While the first $N/2$ index entries are for the positive Fourier harmonics, the next $N/2$ index entries stand for the negative Fourier harmonics. This phenomenon can

TABLE 5.1 Index Allocation of Positive and Negative Frequencies in DFT Representation

Indexing	Corresponding Harmonic Frequencies
$0 \leq k \leq \dfrac{N}{2} - 1$	Positive frequencies: $k \cdot f_s$
$\dfrac{N}{2} \leq k \leq N - 1$	Negative frequencies: $(k - N) \cdot f_s$

be obtained from the definition of DFT as follows: Let us investigate the index terms in the second $N/2$ index entries by putting $(N - k)$ instead of k in Equation 5.12a as

$$
\begin{aligned}
S[N-k] &= \sum_{n=0}^{N-1} s[n] \cdot e^{-j2\pi \frac{(N-k)}{N} n} \\
&= \sum_{n=0}^{N-1} s[n] \cdot e^{-j2\pi n} \cdot e^{-j2\pi \frac{(-k)}{N} n} \\
&= \sum_{n=0}^{N-1} s[n] \cdot e^{-j2\pi \frac{(-k)}{N} n} \\
&\triangleq S[-k] .
\end{aligned}
\tag{5.13}
$$

Therefore, $S[N - k]$ is identical to $S[-k]$. This phenomenon is demonstrated through Figure 5.2. Since negative frequencies come after the positives ones, they have to be interchanged to correctly arrange the frequency axis.

If the data are collected in the frequency domain, the signal in time domain, after inverse DFT, includes both the positive and the negative time indices, analogously. This can be easily observed from the definition of inverse DFT by testing the following. Let us investigate the index terms in the second $N/2$ index entries by putting $(N - n)$ instead of n in Equation 5.12b as

$$
\begin{aligned}
S[N-n] &= \sum_{k=0}^{N-1} S[k] \cdot e^{j2\pi \frac{(N-n)}{N} k} \\
&= \sum_{k=0}^{N-1} S[k] \cdot e^{j2\pi k} \cdot e^{j2\pi \frac{(-k)}{N} n} \\
&= \sum_{k=0}^{N-1} S[k] \cdot e^{j2\pi \frac{(-n)}{N} k} \\
&\triangleq S[-n].
\end{aligned}
\tag{5.14}
$$

Therefore, Equation 5.14 clearly shows that indices in the second $N/2$ entries in fact correspond to negative time indices.

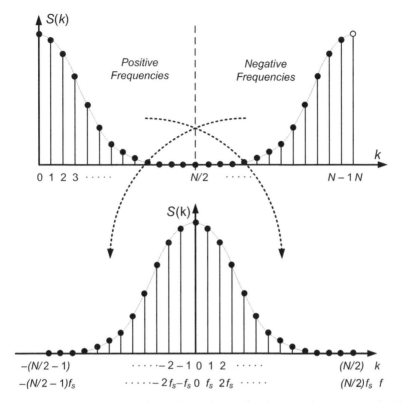

FIGURE 5.2 Demonstration of positive and negative frequencies in DFT: First half and the second half of the DFT sequence should be swapped to organize the frequency axis.

In ISAR imaging, the raw data are usually collected in the frequency-angle axis and are transformed to range cross-range (or range-Doppler) domain after Fourier transforming. To correctly position the ISAR image, the data indices corresponding to positive and negative range cross-range quantities should be swapped.

To demonstrate the use of this property in ISAR imagery, let us reconsider the plane example in Chapter 4, Figure 4.22. As listed in the Matlab script of that example, the ISAR image is swapped by the use of "fftshift" command to correctly display the ISAR image. This is because of the fact that DFT and inverse discrete Fourier transform (IDFT) operations are periodic, which means that the image repeats itself in every image window size of $(\Delta x \cdot N_x)$-by-$(\Delta y \cdot N_y)$, as shown in Figure 5.3a. After DFT operation, the discrete data entry for the first indices in range and cross-range domains corresponds to "0 m" as shown in the Figure 5.3b. After swapping the negative and positive range and cross-range distances as demonstrated in Figure 5.3, the final image is correctly positioned in the image domain as shown in Figure 5.3c.

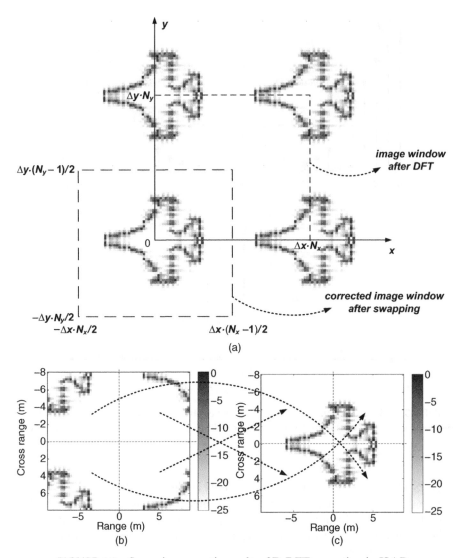

FIGURE 5.3   Swapping operations after 2D DFT operation in ISAR.

5.2   IMAGE ALIASING

ISAR image aliasing occurs when the backscattered field data are not collected with the minimum required frequency and/or angle sampling rates. This
is, of course, the nature of the DFT and shows up when applying the ISAR
signal processing. This phenomenon will be explained with an example.

Let us consider the point scatterers whose locations are already plotted in
Chapter 4, Figure 4.21. To be able to get the ISAR image of these scatterers,
the ISAR image size is selected as [18 m · 16 m]. This selection of ISAR image

size imposes the following sampling ratios (or resolutions) in frequency and aspect domains:

$$\Delta_f = \frac{c/2}{X_{max}}$$
$$= \frac{3 \cdot 10^8 / 2}{18} \tag{5.15a}$$
$$= 8.333 \, \text{MHz.}$$

$$\Delta_\emptyset = \frac{\lambda_c / 2}{Y_{max}}$$
$$= \frac{0.0375}{16} \tag{5.15b}$$
$$= 0.001171875 \, \text{rad} \, (0.0671°).$$

If the data were collected with the above sampling ratios, the correct ISAR image as seen in Chapter 4, Figure 4.22 would be obtained. However, if the data are undersampled, then we will have an aliased ISAR image. To demonstrate this property, the data are sampled in frequency and aspect with twice the ratio that is calculated in Equation 5.15, selecting $\Delta_f = 16.666$ MHz and $\Delta_\emptyset = 0.1342°$. This selection of resolutions in frequency and angle will correspond to a smaller ISAR window. In Figure 5.4, the corresponding ISAR image of 9 m · 8 m in size is plotted. As is obvious from the figure, the image suffers from the aliasing phenomenon due to undersampling of the collected data in the Fourier domain. This is due to the fact that the new Δf and Δø values correspond to an ISAR image size of 8 m by 8 m. Therefore, any

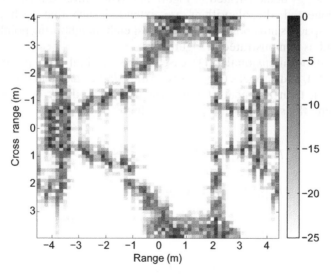

FIGURE 5.4   An example of aliased ISAR image.

scattering center beyond this image frame leaks back to the image from the opposite side of the frame according to the Fourier theory.

5.3   POLAR REFORMATTING REVISITED

Polar reformatting is the common mapping technique that is widely used in SAR/ISAR processing [3, 4]. The problem of polar reformatting was already studied in Chapter 4, Section 4.6.2. What is mainly done in polar reformatting is to reformat the collected backscattered field data that correspond to polar data in spatial frequency onto a rectangular grid so that DFT can be applied for fast formation of the ISAR image, as depicted in Figure 5.5.

In this section, we will look into the polar reformatting process more closely by investigating some common interpolation schemes.

5.3.1   Nearest Neighbor Interpolation

Since the collected frequency-aspect data do not lie on a rectangular grid on the $k_x - k_y$ plane, most of these data points will not coincide with the grid points in the Fourier domain as seen in Figure 5.5.

One of the most popular interpolation techniques used in polar reformatting is the *nearest neighbor scheme* [3, 5]. It is a very general technique and can be applied to any data that is not required to be in the form of a uniform grid. For SAR/ISAR applications, however, it is commonly used to interpolate the polar formatted data to a uniformly sampled rectangular grid. In the first-order interpolation scheme, the closest four neighboring points on the uniform grid are linearly updated depending on their distances from the original data point at k_{x_i}, k_{y_i} as demonstrated in Figure 5.5. When three-dimensional (3D) data are collected, the algorithms are extended such that each point in $k_x - k_y - k_z$ space is interpolated to the closest eight neighboring points on the uniform grid as demonstrated in Figure 5.6.

For the two-dimensional (2D) case, if the original data point has an amplitude of A_i and happens to be between $[n \cdot \Delta k_x, m \cdot \Delta k_y]$ and $[(n + 1) \cdot \Delta k_x, (m + 1) \cdot \Delta k_y]$, then the nearest four grid points on the $k_x - k_y$ plane are updated using the standard interpolation scheme as follows:

$$
\begin{aligned}
\tilde{E}^s \left[n \cdot \Delta k_x, m \cdot \Delta k_y \right] &= A_i \cdot \frac{R}{r_1} \\[2mm]
\tilde{E}^s \left[n \cdot \Delta k_x, (m+1) \cdot \Delta k_y \right] &= A_i \cdot \frac{R}{r_2} \\[2mm]
\tilde{E}^s \left[(n+1) \cdot \Delta k_x, (m+1) \cdot \Delta k_y \right] &= A_i \cdot \frac{R}{r_3} \\[2mm]
\tilde{E}^s \left[(n+1) \cdot \Delta k_x, m \cdot \Delta k_y \right] &= A_i \cdot \frac{R}{r_4},
\end{aligned}
\tag{5.16}
$$

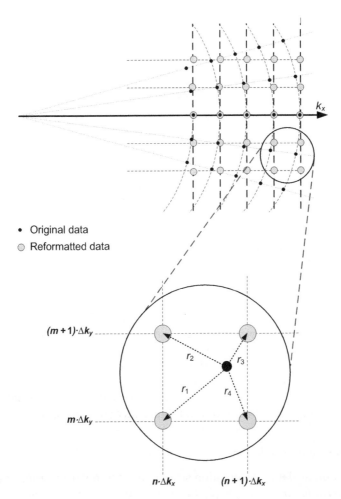

FIGURE 5.5 Interpolation can be employed in different ways to reformat the polar ISAR data to rectangular data (2D case).

where

$$R = \frac{1}{\dfrac{1}{r_1} + \dfrac{1}{r_2} + \dfrac{1}{r_3} + \dfrac{1}{r_4}} \tag{5.17}$$

and r_k's ($k = 1, 2, 3, 4$) are the distances from the four grid points to the original data position as shown in Figure 5.5. $\tilde{E}^s$ is the uniformly sampled reformat-

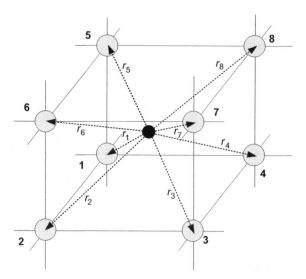

FIGURE 5.6   First-order nearest neighbor interpolation (3D case): Eight nearest data points are updated.

ted data on the $k_x - k_y$ plane. It can be deduced from Equations 5.16 and 5.17 that the amplitude share of a grid point is inversely proportional to its distance to the original point. If it is close to the original data point, it will have a large portion of the amplitude of the original data. If not, it will share only a small fraction of the amplitude of the original data.

In the second-order interpolation scheme, 16 nearest neighbor data points on the uniform grid are updated in the same way (see Fig. 5.7).

5.3.2   Bilinear Interpolation

Another popular data interpolation scheme used in polar reformatting is the *bilinear interpolation* technique [6]. In contrast to the nearest neighbor scheme, the original data grid should be uniformly sampled in this type of interpolation technique. Since the collected raw backscattered data are in rectangular format on the frequency-aspect plane but in polar format in Fourier space, the interpolation should be done in frequency-aspect space. The implementation of bilinear interpolation is illustrated in Figure 5.8 where the original data (shown as black dots) are collected uniformly on the $f - \emptyset$ plane. To be able to apply the FFT, the data should be uniformly sampled on the $k_x - k_y$ plane. If this uniformly sampled grid on the $k_x - k_y$ plane is transformed to the $f - \emptyset$ plane, it corresponds to a nonuniform, polar grid, demonstrated in Figure 5.8 as empty dots.

According to the bilinear interpolation, any point in between the uniformly sampled data points can be interpolated in the following way:

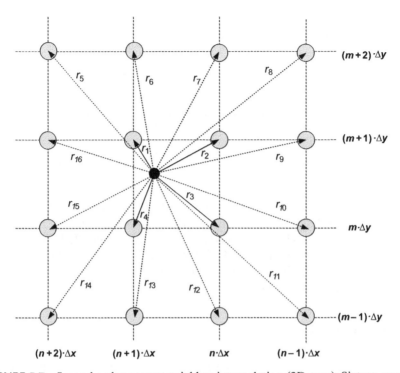

FIGURE 5.7   Second-order nearest neighbor interpolation (2D case): Sixteen nearest data points are updated.

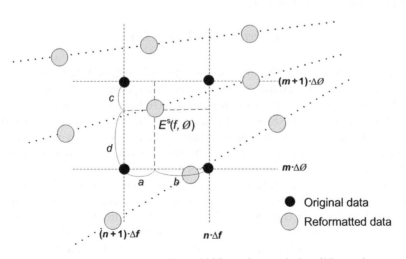

FIGURE 5.8   Implementation of bilinear interpolation (2D case).

$$\tilde{E}^s(f,\emptyset) = E^s(n \cdot \Delta f, m \cdot \Delta\emptyset) \cdot \frac{b \cdot d}{(a+b) \cdot (c+d)}$$

$$+ E^s((n+1) \cdot \Delta f, m \cdot \Delta\emptyset) \cdot \frac{a \cdot d}{(a+b) \cdot (c+d)}$$

$$+ E^s(n \cdot \Delta f,(m+1) \cdot \Delta\emptyset) \cdot \frac{b \cdot c}{(a+b) \cdot (c+d)} \qquad (5.18)$$

$$+ E^s((n+1) \cdot \Delta f,(m+1) \cdot \Delta\emptyset) \cdot \frac{a \cdot c}{(a+b) \cdot (c+d)},$$

where $a = (n+1) \cdot \Delta f - f$, $b = f - n\Delta f$, $c = (m+1) \cdot \Delta\emptyset - \emptyset$ and $d = \emptyset - m \cdot \Delta\emptyset$.

For the interpolation of 3D uniform data onto a nonuniform 3D grid, a suitable interpolation scheme is called *trilinear interpolation* [7]. It is also possible to use cubic approximation to interpolate the points in between the regular grid points. If this is the case, bicubic [8] and tricubic [9] interpolation schemes are applied for interpolating 2D and 3D data sets, respectively.

5.4 ZERO PADDING

Zero padding is also regarded as an interpolation technique that is used to enhance the visual image quality in SAR/ISAR imaging. To express how zero padding in one domain means interpolation in the other domain, we start with the FT pair shown in Figure 5.9. It is well known that a rectangular pulse of duration T has the FT of a *sinc* waveform that has a zero crossing at integer multiples of $1/T$, as depicted in Figure 5.9. To be more precise, the FT pair can be represented as the following:

$$FT\{rect(t/T)\} = T \cdot sinc(fT). \qquad (5.19)$$

Now, let us consider that the time-domain pulse in Figure 5.9a is sampled to have a discrete signal such that it forms a *comb waveform* as depicted in Figure 5.9c. The DFT of this discrete sequence happens to have a single nonzero entry in the Fourier domain as shown in Figure 5.9d. As mentioned in previous sections, if the bandwidth of the time-domain signal is T, then the resolution in the frequency domain is $1/T$, which is the same as the zero crossing interval of the continuous wave *sinc* function seen in Figure 5.9b. This situation is illustrated in Figure 5.9d where the critically sampled *sinc* signal is plotted. Next, this discrete signal is zero-padded to have time duration of $3T$ as shown in Figure 5.9e. Its DFT is drawn in Figure 5.9f where the frequency sampling interval is now three times smaller than the original one. With this construct, *sinc* waveform can now be visible after DFT processing, provided that the new frequency-domain signal has a DFT length three times larger and therefore

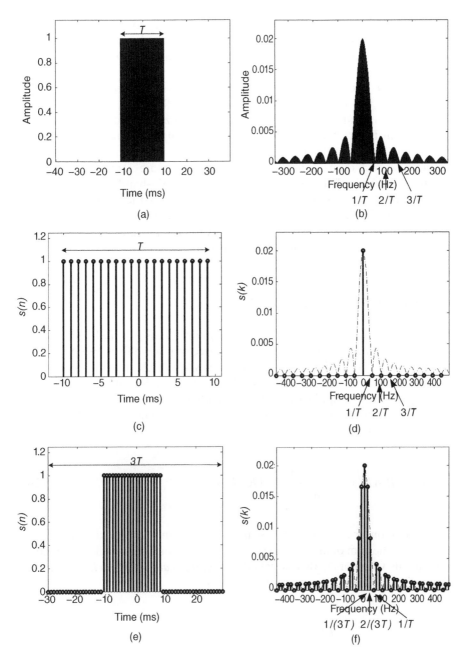

FIGURE 5.9 Illustration of interpolation with zero padding: (a) a rectangular pulse in time, (b) its Fourier transform, a *sinc*, (c) discrete time-domain pulse, (d) its DFT which is a critically sampled *sinc*, (e) zero-padded version of the discrete time-domain pulse, (b) its Fourier transform, an interpolated *sinc*.

has three times more samples of the original frequency signal. This whole process of zero padding in one domain and therefore having more sampled version of the original signal in the other domain is often called *FFT interpolation*.

The use of interpolation via zero padding is often used in SAR/ISAR imagery. Since the data in a SAR/ISAR image are 2D or 3D, the zero padding should be done in all directions. Some examples of zero-padded ISAR images are shown in Figure 5.10. On the left side of the figure, some original ISAR images are shown. A four-times zero-padding procedure is applied in the frequency-aspect domain to obtain images with more sampling points, as shown on the right. It is easily noticed from the right-hand side images that the distribution of the scattering centers are much smoother, to produce more visually satisfying images. It is important to note that zero padding does not really improve the resolution of the image. Rather, it just interpolates the data in between to show a smoother data transition.

5.5 POINT SPREAD FUNCTION (PSF)

In its general usage, the *point spread response* (PSR) or PSF defines the response of an imaging system to a point source or point object. In SAR/ISAR nomenclature, it is the impulse response of a SAR/ISAR imaging system to a point scatterer. As will be demonstrated in this section, its effect is more meaningful with the application of zero padding.

As defined in the previous chapter, a 2D ISAR image is described as the following double integral:

$$ISAR(x,y) = \frac{1}{\pi^2} \cdot \int_{-\infty}^{\infty} \int_{-\infty}^{\infty} \left\{ E^s(k_x, k_y) \right\} \cdot e^{j2(k_x \cdot x + k_y \cdot y)} dk_x \cdot dk_y. \quad (5.20)$$

It is a common practice that any scattered field can be approximated as the response of the sum of finite point scatterers on the target. As will be explained in detail in Chapter 7, this is called the *point-scatterer model*. Under this assumption, $E^s(k_x, k_x)$ can be represented as

$$E^s(k_x, k_y) \cong \sum_{i=1}^{N} A_i \cdot e^{-j2(k_x \cdot x_i + k_y \cdot y_i)}, \quad (5.21)$$

where N represents the total number of point scatterers, A_i is the scattered field amplitude, and (x_i, y_i) is the spatial location for the ith point scatterer.

Ideally, perfect ISAR imaging of these scatterers is possible, assuming that the scattered field is collected for the infinite values of k_x and k_y (or frequency

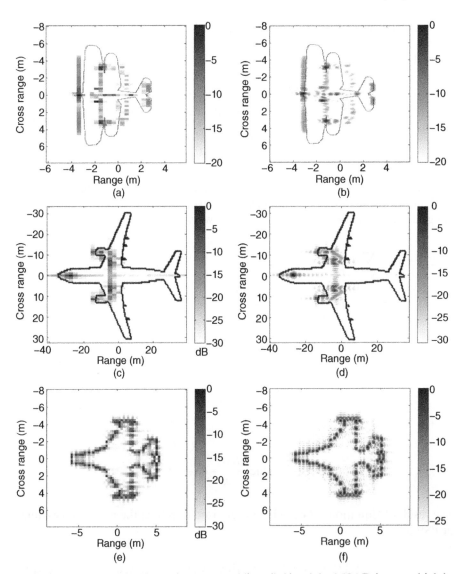

FIGURE 5.10 Interpolation using zero padding: (left) original ISAR images, (right) interpolated ISAR images after applying four-times zero-padding procedure in the Fourier domain.

and aspect). Then, the ISAR image converts to 2D ideal impulse functions, as demonstrated below:

$$ISAR(x,y) = \sum_{i=1}^{N} \frac{1}{\pi^2} \cdot \int_{-\infty}^{\infty} \int_{-\infty}^{\infty} A_i \cdot e^{j2(k_x \cdot (x-x_i) + k_y \cdot (y-y_i))} dk_x \cdot dk_y$$

$$= \sum_{i=1}^{N} A_i \cdot \delta(x - x_i, y - y_i).$$

(5.22)

In practice, however, the frequencies and angles (or k_x and k_y) are finite. Let k_x^L, k_x^H, k_y^L, and k_y^H be the lower and upper limits of the spatial frequencies defined as

$$
\begin{aligned}
k_x^H &= k_{xo} + BW_{k_x}/2 \\
k_x^L &= k_{xo} - BW_{k_x}/2 \\
k_y^H &= k_{yo} + BW_{k_y}/2 \\
k_y^L &= k_{yo} - BW_{k_y}/2.
\end{aligned}
\tag{5.23}
$$

Here, k_{xo} and k_{yo} are the center spatial frequencies, and BW_{k_x} and BW_{k_y} are the bandwidths in k_x and k_y domains, respectively. Then the result of the ISAR integral can be calculated in the following way:

$$
\begin{aligned}
ISAR(x,y) &= \sum_{i=1}^{N} \frac{1}{\pi^2} \cdot \int_{k_x^L}^{k_x^H} \int_{k_y^L}^{k_y^H} A_i \cdot e^{j2(k_x \cdot (x-x_i)+k_y \cdot (y-y_i))} dk_x \cdot dk_y \\
&= \sum_{i=1}^{N} A_i \cdot \left(\frac{e^{j2k_x^H (x-x_i)} - e^{j2k_x^L (x-x_i)}}{j2\pi(x-x_i)} \right) \left(\frac{e^{j2k_y^H (y-y_i)} - e^{j2k_y^L (y-y_i)}}{j2\pi(x-x_i)} \right) \\
&= \sum_{i=1}^{N} A_i \cdot \left(e^{j2k_{xo}(x-x_i)} \cdot \frac{e^{jBW_{k_x}(x-x_i)} - e^{-jBW_{k_y}(x-x_i)}}{j2\pi(x-x_i)} \right) \\
&\qquad \cdot \left(e^{j2k_{yo}(y-y_i)} \cdot \frac{e^{jBW_{k_x}(y-y_i)} - e^{jBW_{k_y}(y-y_i)}}{j2\pi(y-y_i)} \right) \\
&= \sum_{i=1}^{N} A_i \cdot \left(e^{j2k_{xo}(x-x_i)} \cdot \frac{BW_{k_x}}{\pi} \cdot \mathrm{sinc}\left(\frac{BW_{k_x}}{\pi}(x-x_i) \right) \right) \\
&\qquad \cdot \left(e^{j2k_{yo}(y-y_i)} \cdot \frac{BW_{k_y}}{\pi} \cdot \mathrm{sinc}\left(\frac{BW_{k_y}}{\pi}(y-y_i) \right) \right).
\end{aligned}
\tag{5.24}
$$

The result of Equation 5.24 can be reorganized to reflect the physical meaning of PSF as

$$
ISAR(x,y) = \sum_{i=1}^{N} A_i \cdot \delta(x-x_i, y-y_i) * h(x,y),
\tag{5.25}
$$

where $h(x,y)$ is the so-called PSF and is given by

$$
h(x,y) = \left(e^{j2k_{xo} \cdot x} \cdot \frac{BW_{k_x}}{\pi} \cdot \mathrm{sinc}\left(\frac{BW_{k_x}}{\pi} x \right) \right) \left(e^{j2k_{yo} \cdot y} \cdot \frac{BW_{k_y}}{\pi} \cdot \mathrm{sinc}\left(\frac{BW_{k_y}}{\pi} y \right) \right).
\tag{5.26}
$$

According to Equation 5.25, PSF can be regarded as the impulse response of the ISAR imaging arrangement to any point scatterer on the target. This physi-

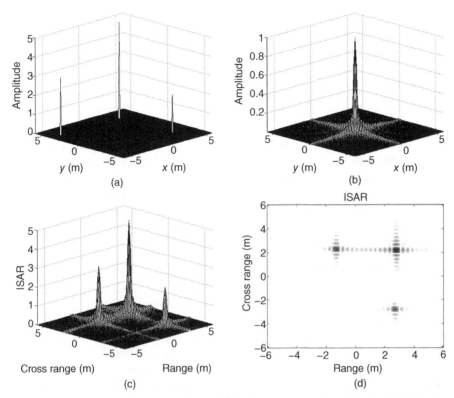

FIGURE 5.11   The physical meaning of PSF: (a) point scatterers, (b) the PSF, (c) the ISAR image is constructed by the convolution of point scatter with the PSF, (d) the effect of PSF: *sinc* sidelobes are noticeable in the 2D image plane.

cal meaning of PSF is illustrated in Figure 5.11. A finite number of point scatterers that have different scattering amplitudes exist in 2D $x - y$ space as shown in Figure 5.11a. The resultant ISAR image (Fig. 5.11c) is nothing but the convolution of the point scatterers with the 2D PSF function in Figure 5.11b. In the common display format of ISAR, the resultant image is displayed in the 2D range and cross-range planes, as shown in Figure 5.11d. Because of the finite bandwidths in k_x and k_y domains, the PSF has tails that correspond to sidelobes of *sinc* functions in both range and cross-range directions. The common way of suppressing the sidelobes of PSF is to use windowing, as we shall explore next.

5.6   WINDOWING

5.6.1   Common Windowing Functions

In radar imaging applications, windowing is a common practice to tone down the magnitudes of the sidelobes of the PSF such that the resultant SAR/ISAR image looks much smoother and more localized in terms of the point scatterers. A *windowing* (or *tapering*) function has nonzero entries within a chosen interval and zero outside this interval as defined below:

$$w[n] = \begin{cases} h[n]; & n = 0,1,2, \ N-1 \\ 0 & else\,where, \end{cases} \tag{5.27}$$

where the entries of $h[n]$ are equal to or less than 1. Here, the N-point $w[n]$ is to be applied to a signal length of T_o. In many practical applications, the window functions that have smooth "bell-shaped" characteristics are used to successfully suppress the undesired sidelobes. *Hanning, Hamming, Kaiser, Blackman*, and *Chebyshev* windows fall into this category. While these windows provide better focused images due to decreased sidelobe levels (SLLs), they unfortunately present lower resolution compared to windows types such as *rectangular* or *triangular* due to increased width of main lobe.

5.6.1.1 Rectangular Window If $h[n] = 1$ for the chosen interval, this window is commonly known as a rectangular window, as depicted in Figure 5.12a. Applying a rectangular window produces the same data with the case when truncating the same portion of the data of the chosen interval. The spectrum of rectangular window has the maximum SLL of about −13 dB, which is the highest when compared with any other windowing function (see Fig. 5.12b). It has, however, the narrowest main lobe width compared with the others. While the −3 dB width of the main lobe is about 88% of the one FFT bin, single FFT bin corresponds to the −4 dB width. This single frequency bin is equal to one fundamental frequency resolution of $1/(T_oN)$.

5.6.1.2 Triangular Window The triangular window function (Fig. 5.13a) has the following formula within the chosen integral:

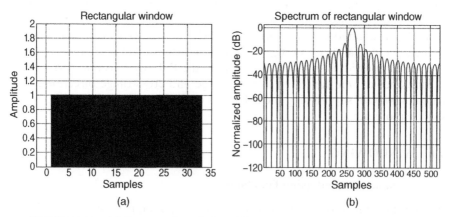

FIGURE 5.12 (a) Rectangular window, (b) spectrum of rectangular window.

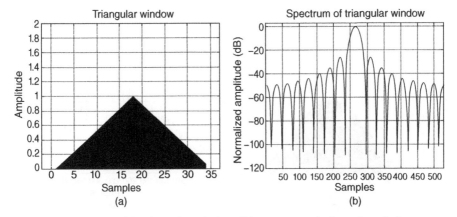

FIGURE 5.13 (a) Triangular window, (b) spectrum of triangular window.

$$h[n] = 1 - \frac{2}{N} \cdot \left| n - \frac{N-1}{2} \right|. \tag{5.28}$$

Since a triangular window of length N can be obtained by convolving two identical rectangular windows of length $N/2$, the spectrum of a triangular window of length N is equal to the square of the spectrum of a rectangular window of length $N/2$. Therefore, the main lobe width of a triangular window is twice as big as the main lobe width of a rectangular window of the same length (see Fig. 5.13b). The spectrum of triangular window has the maximum SLL of about −26 dB that is twice to that of decibel value of the rectangular window as expected.

5.6.1.3 *Hanning Window*   The shape of the Hanning or *Hann* window looks like half of a cycle of a cosine waveform [10]. The equation for the definition of this windowing function is equal to

$$h[n] = 0.5 \cdot \left(1 - cos \left[\frac{2\pi n}{N-1} \right] \right). \tag{5.29}$$

As is obvious from Equation 5.29, the Hanning window function is in fact the normalized version of an up-shifted cosine waveform as plotted in Figure 5.14a. That is why the Hanning window is also known as *raised cosine* window. The spectrum of this window has the maximum SLL of −32 dB.

5.6.1.4 *Hamming Window*   A Hamming window (Fig. 5.15a) is a modified version of the Hanning window as its wave equation is as below:

$$h[n] = 0.53836 - 0.46164 \cdot cos \left[\frac{2\pi n}{N-1} \right]. \tag{5.30}$$

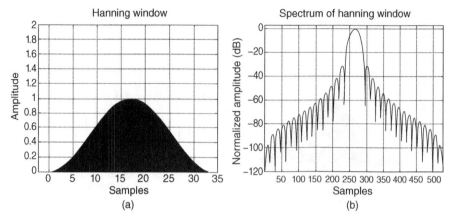

FIGURE 5.14   (a) Hanning window, (b) spectrum of Hanning window.

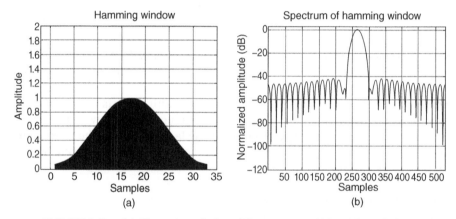

FIGURE 5.15   (a) Hamming window, (b) spectrum of Hamming window.

Richard W. Hamming [11] proposed the above particular coefficients for the two terms of the original Hanning equation that results in the maximum SLL of −43 dB in the spectrum of the window. This proper selection of coefficients helps to decrease the sidelobes closer to the main lobe and increase the sidelobes that are far to the main lobe as plotted in Figure 5.15b.

5.6.1.5   *Kaiser Window*   The equation for the Kaiser window [12] is given below:

$$h[n] = \frac{I_o\left(\alpha\left(1-\left(\frac{2n}{N-1}-1\right)^2\right)^{1/2}\right)}{I_o(\alpha)}, \tag{5.31}$$

where $I_o(\alpha)$ is the zeroth order modified Bessel function of first kind. The spectrum of the Kaiser window has the maximum SLL of −36 dB for $\alpha = 1.5\ \pi$. The window waveform and its spectrum for this windowing function is drawn in Figure 5.16.

5.6.1.6 Blackman Window   The equation for the Blackman window [13] is given below:

$$h[n] = 0.42 - 0.5 \cdot \cos\left[\frac{2\pi n}{N-1}\right] + 0.08 \cdot cos\left[\frac{4\pi n}{N-1}\right]. \tag{5.32}$$

The spectrum of the Blackman window has the maximum SLL of −58 dB. The window function and its spectrum for the Blackman is plotted in Figure 5.17.

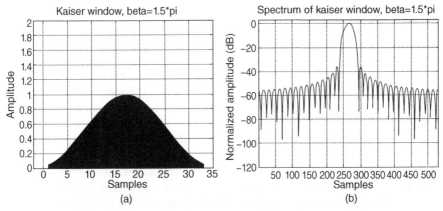

FIGURE 5.16   (a) Kaiser window, (b) spectrum of Kaiser window.

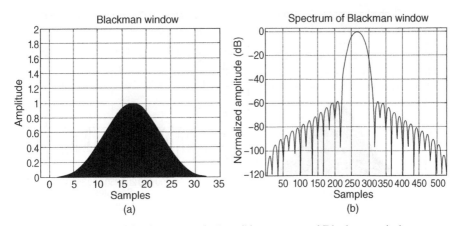

FIGURE 5.17   (a) Blackman window, (b) spectrum of Blackman window.

5.6.1.7 *Chebyshev Window*   Chebyshev window function [14] is characterized by the following equation:

$$h[n] = \frac{(r+2)}{N} \sum_{k=1}^{(N-1)/2} C_{N-1}\left(t_o \cos\left(\frac{k\pi}{N}\right)\right) \cos\left(\frac{2k\pi(n-(N-1)/2)}{N}\right)$$ (5.33)

where r is the ratio of the main lobe to sidelobes in decibel, $C_k(\alpha)$ is the kth order Chebyshev polynomial given below,

$$C_k(\alpha) = \begin{cases} \cos(k\cos^{-1}(\alpha)) & (\alpha) \leq 1 \\ \cosh(k\cosh^{-1}(\alpha)) & (\alpha) > 1 \end{cases}$$ (5.34)

and the parameter t_o is given as

$$t_o = \cosh\left[\frac{\cosh^{-1}(r)}{N-1}\right].$$ (5.35)

The Chebyshev window waveform and its spectrum are shown in Figure 5.18. For this example, r is selected as 80 dB. As is shown in Figure 5.18b, the spectrum makes many ripples providing that the sidelobes have a maximum value of −80 dB.

Table 5.2 summarizes the characteristics of several windowing functions used for smoothing the data of interest. Since higher resolution means greater sidelobes, a trade-off between the main lobe width and the maximum SLL should be taken into account when applying any smoothing window to the data. Table 5.2 presents the −3 dB width of the main lobe and the maximum SLL for the windowing functions that are considered in this book.

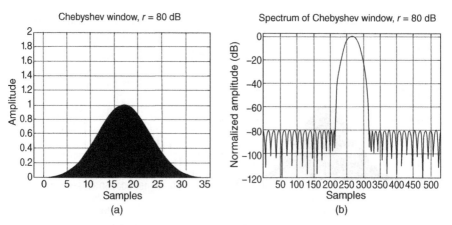

FIGURE 5.18   (a) Chebyshev window, (b) spectrum of Chebyshev window.

TABLE 5.2 Comparative Characteristics of Different Windowing Functions

Window Function	Expression	−3 dB Main Lobe Width (FFT Bins)	Maximum SLL (dB)
Rectangular	$h[n] = 1$	0.88	−13 dB
Triangular	$h[n] = 1 - \dfrac{2}{N} \cdot \left\lvert n - \dfrac{N-1}{2} \right\rvert$	1.24	−26 dB
Hanning	$h[n] = 0.5 \cdot \left(1 - \cos\left[\dfrac{2\pi n}{N-1} \right] \right)$	1.40	−32 dB
Hamming	$h[n] = 0.53836 - 0.46164 \cdot \cos\left[\dfrac{2\pi n}{N-1} \right]$	1.33	−43 dB
Kaiser	$h[n] = \dfrac{I_o\left(\alpha\left(1 - \left(\dfrac{2n}{N-1} - 1 \right)^2 \right)^{1/2} \right)}{I_o(\alpha)}$	1.30	−36 dB (for $\alpha = 1.5\,\pi$)
Blackman	$h[n] = 0.42 - 0.5 \cdot \cos\left[\dfrac{2\pi n}{N-1} \right] + 0.08 \cdot \cos\left[\dfrac{4\pi n}{N-1} \right]$	1.69	−58 dB
Chebyshev	$h[n] = \dfrac{(r+2)}{N} \sum_{k=1}^{(N-1)/2} C_{N-1}\left(t_o \cos\left(\dfrac{k\pi}{N} \right) \right)$ $\cdot \cos\left(\dfrac{2k\pi(n - (N-1)/2)}{N} \right)$	1.68	−80 dB ($r = 80$)

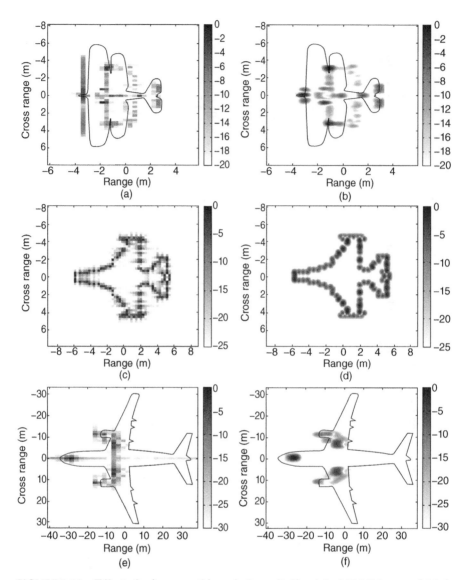

FIGURE 5.19 Effect of using smoothing windows: (left) original ISAR images, (right) interpolated ISAR images after applying four-times zero-padding procedure and Hamming window in the Fourier domain.

5.6.2 ISAR Image Smoothing via Windowing

Applying a window before displaying the final ISAR image is a common procedure for smoothing the PSF around the scattering centers on the image. This process enhances the image visual appearance by suppressing the sidelobes around the scattering centers at the price of losing some resolution.

To demonstrate the use of windowing functions in ISAR imaging, the ISAR images that are given in Figure 5.10 are used. The original ISAR images are shown along the left column in Figure 5.19. Then, a 2D Hamming window is applied to these images in the Fourier domain. A four-times zero-padding scheme is also applied to interpolate the data points. The resultant smoothed 2D ISAR images are plotted along the right column of Figure 5.19. New images do not suffer from the tails (sidelobes) of the 2D PSF around the scattering centers. Since the tails of the scattering centers are suppressed, scattering mechanisms that fall beneath the sidelobes of the nearby scattering centers' PSF are now visible in the new windowed ISAR images. This feature can also be seen from the first and third ISAR images. On the other hand, the decrease in the image resolution is also apparent in the new ISAR images, as we discussed before.

5.7   MATLAB CODES

Below are the Matlab source codes that were used to generate all of the Matlab-produced Figures in Chapter 5. The codes are also provided in the CD that accompanies this book.

Matlab code 5.1: Matlab file "Figure5-9.m"

```
%------------------------------------------------------------
% This code can be used to generate Figure 5.9
%------------------------------------------------------------
clear all
close all
%_____Implementation OF FT Window/
Sinc_____
M = 500;
t = (-M:M)*1e-3/5;
E(450:550) = 1;E(1001)=0;
T = t(550)-t(450);
index=300:700;
%---Figure 5.9(a)--------------------------------------------
area(t(index)*1e3,E(index)); axis([min(t(index))*1e3
max(t(index))*1e3 0 1.15])
set(gca,'FontName', 'Arial', 'FontSize',14,'FontWeight',
'Bold');
xlabel('Time (ms)'); ylabel('Amplitude');
colormap(gray);
%---Figure 5.9(b)--------------------------------------------
index = 430:570;
d = 1/(max(t)-min(t));
f = (-M:M)*df;
Ef = T*fftshift(fft(E))/length(450:550);
figure;
area(f(index),abs(Ef(index)));
axis([min(f(index)) max(f(index)) 0 .023])
set(gca,'FontName', 'Arial', 'FontSize',14,'FontWeight',
'Bold');
xlabel('Frequency (Hz)'); ylabel('Amplitude');%grid on;
```

```
colormap(gray);
%_____Implementation OF DFT_____
clear all;
% TIME DOMAIN SIGNAL
t = (-10:9)*1e-3; N=length(t);
En(1:N) = 1;
%---Figure 5.9(c)----------------------------------------------
figure;
stem(t*1e3,En,'k','LineWidth',3); axis([min(t)*1.2e3
max(t)*1.2e3 0 1.25])
set(gca,'FontName', 'Arial', 'FontSize',12,'FontWeight',
'Bold');
xlabel('Time [ms]'); ylabel('s[n]');%grid on;
%----------FREQ DOMAIN SIGNAL---
dt = t(2)-t(1);
BWt = max(t)-min(t)+dt;
df = 1/BWt;
f = (-10:9)*df;
Efn = BWt*fftshift(fft(En))/length(En);
%---Figure 5.9(d)----------------------------------------------
figure;
stem(f,abs(Efn),'k','LineWidth',3);
axis([min(f) max(f) 0 1.15])
set(gca,'FontName', 'Arial', 'FontSize',14,'FontWeight',
'Bold');
xlabel('Frequency [Hz]'); ylabel('S[k]');%grid on;
colormap(gray);hold on
%----this part for the sinc template
clear En2;
En2(91:110) = En;
En2(200) = 0;
Efn2 = BWt*fftshift(fft(En2))/length(En);
f2 = min(f):df/10:(min(f)+df/10*199);
plot(f2,abs(Efn2),'k-.','LineWidth',1);
axis([min(f2) max(f2) 0 .023]); hold off
%---------ZERO PADDING ----------
%TIME DOMAIN
clear En_zero;
En_zero(20:39) = En;
En_zero(60) = 0;
dt = 1e-3;
t2 = dt*(-30:29);
%---Figure 5.9(e)----------------------------------------------
figure;
stem(t2*1e3,En_zero,'k','LineWidth',3); axis([-dt*30e3
dt*29e3 0 1.25])
set(gca,'FontName', 'Arial', 'FontSize',14,'FontWeight',
'Bold');
xlabel('Time [ms]'); ylabel('s[n]');%grid on;
```

```
%FREQUENCY DOMAIN
Efn2_zero = BWt*fftshift(fft(En_zero))/length(En);
f2 = min(f):df/3:(min(f)+df/3*59);
%---Figure5.9(f)-------------------------------------------------
figure;
plot(f2,abs(Efn2_zero),'k-.','LineWidth',1);hold on
stem(f2,abs(Efn2_zero),'k','LineWidth',3);
set(gca,'FontName', 'Arial', 'FontSize',12,'FontWeight',
'Bold');
xlabel('Frequency [Hz]'); ylabel('S[k]');%grid on;
axis([min(f2) max(f2) 0 0.023]); hold off
```

Matlab code 5.2: Matlab file "Figure5-10ab.m"

```
%----------------------------------------------------------------
% This code can be used to generate Figure 5.10 (a-b)
%----------------------------------------------------------------
% This file requires the following files to be present in the
same
% directory:
%
% Esplanorteta60.mat
% planorteta60_2_xyout.mat
clear all
close all
c=.3; % speed of light
%_____PRE PROCESSING OF ISAR_____
%Find spatial resolutions
BWx = 12;
BWy = 16;
M = 32;
N = 64;
fc = 6;
phic=0;
dx = BWx/M;
dy = BWy/N;
% Form spatial vectors
X = -dx*M/2:dx:dx*(M/2-1);
Y = -dy*N/2:dy:dy*(N/2-1);
%Find resoltions in freq and angle
df = c/(2*BWx);
dk = 2*pi*df/c;
kc = 2*pi*fc/c;
dphi = pi/(kc*BWy);
%Form F and PHI vectors
F = fc+[-df*M/2:df:df*(M/2-1)];
PHI = phic+[-dphi*N/2:dphi:dphi*(N/2-1)];
k = 2*pi*F/c;
% Load the backscattered data
load Esplanorteta60
```

```
load planorteta60_2_xyout
%_____ POST PROCESSING OF ISAR_____
ISAR = fftshift(fft2(Es.'));
ISAR = ISAR/M/N;
%---Figure 5.10(a)--------------------------------------------------
matplot2(X(32:-1:1),Y,ISAR,20);
colormap(1-gray);
colorbar
set(gca,'FontName', 'Arial', 'FontSize',14,'FontWeight',
'Bold');
xlabel('Range [m]');
ylabel('Cross - range [m]');%grid on;
line(-xyout_xout,xyout_yout,'Color','k','LineStyle','.','Marke
rSize',3);
%zero padding;
Enew = Es;
Enew(M*4,N*4) = 0;
XX = X(1):dx/4:X(1)+dx/4*(4*M-1);
YY = Y(1):dy/4:Y(1)+dy/4*(4*N-1);
% ISAR image formatiom
ISARnew = fftshift(fft2(Enew.'));
ISARnew = ISARnew/M/N;
figure;
%---Figure 5.10(b)--------------------------------------------------
matplot2(XX(4*M:-1:1),YY,abs(ISARnew),20);
colormap(1-gray);
colorbar
line(-xyout_xout,xyout_yout,'Color','k','LineStyle','.','Marke
rSize', 3);
set(gca,'FontName', 'Arial', 'FontSize',14,'FontWeight',
'Bold');
xlabel('Range [m]');
ylabel('Cross - range [m]');%grid on;
```

Matlab code 5.3: Matlab file "Figure5-10cd.m"

```
%----------------------------------------------------------------
% This code can be used to generate Figure 5.10 (c-d)
%----------------------------------------------------------------
% This file requires the following files to be present in the
same
% directory:
%
% Esairbus.mat
% airbusteta80_2_xyout.mat
clear all
close all
c=.3; % speed of light
%_____PRE PROCESSING OF ISAR_____
%Find spatial resolutions
```

```
BWx = 80;
BWy = 66;
M = 32;
N = 64;
fc = 4;
phic = 0;
dx = BWx/M;
dy = BWy/N;
% Form spatial vectors
X = -dx*M/2:dx:dx*(M/2-1);
Y = -dy*N/2:dy:dy*(N/2-1);
%Find resoltions in freq and angle
df = c/(2*BWx);
dk = 2*pi*df/c;
kc = 2*pi*fc/c;
dphi = pi/(kc*BWy);
%Form F and PHI vectors
F = fc+[-df*M/2:df:df*(M/2-1)];
PHI = phic+[-dphi*N/2:dphi:dphi*(N/2-1)];
k = 2*pi*F/c;
% Load the backscattered data
load Esairbus
load airbusteta80_2_xyout
%_____ POST PROCESSING OF ISAR_____
ISAR = fftshift(fft2(Es.'));
ISAR = ISAR/M/N;
%---Figure 5.10(c)-----------------------------------------------
matplot2(X(32:-1:1),Y,ISAR,30); colormap(1-gray); colorbar
set(gca,'FontName', 'Arial', 'FontSize',14,'FontWeight',
'Bold');
xlabel('Range [m]'); ylabel('Cross - range [m]');%grid on;
colormap(1-gray);
line(-xyout_xout,xyout_yout,'Color','k','LineStyle','.');
%zero padding with 4 times;
Enew = Es;
Enew(M*4,N*4) = 0;
figure;
% ISAR image formatiom
ISARnew = fftshift(fft2(Enew.'));
ISARnew = ISARnew/M/N;
%ISARnew(1,1)=2.62
load airbusteta80_2_xyout.mat;
%---Figure 5.10(d)-----------------------------------------------
matplot2(X(32:-1:1),Y,ISARnew,30);
colormap(1-gray);
line(-xyout_xout,xyout_yout,'Color','k','LineStyle','.');
set(gca,'FontName', 'Arial', 'FontSize',14,'FontWeight',
'Bold');
xlabel('Range [m]'); ylabel('Cross - range [m]');
```

Matlab code 5.4: Matlab file "Figure5-10ef.m"

```
%----------------------------------------------------------------
% This code can be used to generate Figure 5.10 (e-f)
%----------------------------------------------------------------
% This file requires the following files to be present in the
same
% directory:
%
% ucak.mat
clear all
close all
c=.3; % speed of light
%_____PRE PROCESSING OF ISAR_____
%Find spatial resolutions
BWx = 18;
BWy = 16;
M = 64;
N = 64;
fc = 8;
phic = 0;
% Image resolutions
dx = BWx/M;
dy = BWy/N;
% Form spatial vectors
X = -dx*M/2:dx:dx*(M/2-1);
Y = -dy*N/2:dy:dy*(N/2-1);
%Find resoltions in freq and angle
df = c/(2*BWx);
dk = 2*pi*df/c;
kc = 2*pi*fc/c;
dphi = pi/(kc*BWy);
%Form F and PHI vectors
F = fc+[-df*M/2:df:df*(M/2-1)];
PHI = phic+[-dphi*N/2:dphi:dphi*(N/2-1)];
K = 2*pi*F/c;
%_____ FORM RAW BACKSCATTERED DATA_____
load ucak
l = length(xx);
Es = zeros(M,N);
for m=1:l;
 Es = Es+1.0*exp(j*2*K'*(cos(PHI)*xx(m)+sin(PHI)*yy(m)));
end
%_____ POST PROCESSING OF ISAR (Small BW Small
angle)_____
ISAR=fftshift(fft2(Es.')); ISAR=ISAR/M/N;
%---Figure5-10(e)---------------------------------------------
h=figure;
matplot2(X(M:-1:1),Y,ISAR,25);
colormap(1-gray);
```

```
colorbar
set(gca,'FontName', 'Arial', 'FontSize',14,'FontWeight',
'Bold');
xlabel('Range [m]'); ylabel('Cross - range [m]');%grid on;
colormap(1-gray);%colorbar
%-------------zero padding with 4 times----------
Enew = Es;
Enew(M*4,N*4) = 0;
% ISAR image formatiom
ISARnew = fftshift(fft2(Enew.'));
ISARnew = ISARnew/M/N;
%ISARnew(1,1)=2.62
load airbusteta80_2_xyout.mat;
%---Figure5-10(f)-------------------------------------------------
h=figure;
matplot2(X(M:-1:1),Y,ISARnew,25);
colormap(1-gray);
colorbar
set(gca,'FontName', 'Arial', 'FontSize',14,'FontWeight',
'Bold');
xlabel('Range [m]');
ylabel('Cross - range [m]');%grid on;
```

Matlab code 5.5: Matlab file "Figure5-11.m"

```
%-------------------------------------------------------------
% This code can be used to generate Figure 5-11
%-------------------------------------------------------------
clear all
close all
clc
% Prepare mesh
[X,Y] = meshgrid(-6:.1:6, -6:.1:6);
M = length(X);
N = length(Y) ;
Object=zeros(M,N);
% Set 3 scattering centers
hh = figure;
Object(101,95)=5;
Object(30,96)=2;
Object(100,15)=3;
%---Figure5-11(a)-------------------------------------------------
surf(X,Y,Object);
colormap(1-gray);
axis tight;
set(gca,'FontName', 'Arial', 'FontSize',12,'FontWeight',
'Bold');
xlabel('X [m]');
ylabel('Y [m]');
zlabel('Amplitude')
```

```
view(-45,20)
saveas(hh,'Figure5-11a.png','png');
%Find spatial resolutions
% fc = 10; % center frequency
% phic = 0; % center angle
% c = .3; % speed of light
dx = X(1,2)-X(1,1); % range resolution
dy = dx; % xrange resolution
%Find Bandwidth in spatial frequencies
BWkx = 1/dx;
BWky = 1/dy;
% PSF
h = sinc(BWkx*X/pi).*sinc(BWky*Y/pi);
%---Figure 5-11(b)-----------------------------------------------
hh = figure;
surf(X,Y,abs(h));
axis tight;
colormap(1-gray);
axis([-6 6 -6 6 0 1])
set(gca,'FontName', 'Arial', 'FontSize',12,'FontWeight',
'Bold');
xlabel('X [m]');
ylabel('Y [m]');
zlabel('Amplitude');
view(-45,20)
saveas(hh,'Figure5-11b.png','png');
%Convolution
hh = figure;
ISAR = fft2(fft2(Object).*fft2(h))/M/N;
%---Figure 5-11(c)-----------------------------------------------
surf(X,Y,abs(ISAR));
axis tight;
colormap(1-gray);
set(gca,'FontName', 'Arial', 'FontSize',12,'FontWeight',
'Bold');
xlabel('Range [m]');
ylabel('Cross - range [m]');
zlabel('ISAR ');
view(-45,20)
saveas(hh,'Figure5-11c.png','png');
%---Figure 5-11(c)-----------------------------------------------
hh = figure;
matplot(X(1,1:M),Y(1:N,1),ISAR,30);
colormap(1-gray);
set(gca,'FontName', 'Arial', 'FontSize',12,'FontWeight',
'Bold');
xlabel('Range [m]');
ylabel('Cross - range [m]');
title('ISAR ');
```

```
saveas(hh,'Figure5-11d.png','png');
```

Matlab code 5.6: Matlab file "Figure5-12thru5-18.m"

```
%----------------------------------------------------------------
% This code can be used to generate Figure 5.12 - 5.18
%----------------------------------------------------------------
% Comparison of windowing functions
%--------------------------------
clear all
close all
N = 33;
%---Figure5.12(a)-------------------------------------------------
%---Rectangular window
rect = rectwin(N);
h = figure;
area(rect);
grid;
colormap(gray)
set(gca,'FontName', 'Arial', 'FontSize',12,'FontWeight',
'Bold');
xlabel(' samples ');
ylabel('Amplitude');
title(' Rectangular Window')
axis([-2 N+2 0 2])
%---Figure5.12(b)-------------------------------------------------
rect(16*N)=0;
Frect = fftshift(fft(rect));
Frect = Frect/max(abs(Frect));
h = figure;
plot(mag2db(abs(Frect)),'k','LineWidth',2);
grid
axis tight;
set(gca,'FontName', 'Arial', 'FontSize',12,'FontWeight',
'Bold');
xlabel('samples ');
ylabel('Normalized amplitude[dB]');
title ('Spectrum of Rectangular Window')
axis([1 16*N -120 3])
%---Figure5.13(a)-------------------------------------------------
%---Triangular window
tri = triang(N);
h = figure;
area([0 tri.']);
grid;
colormap(gray)
set(gca,'FontName', 'Arial', 'FontSize',12,'FontWeight',
'Bold');
xlabel(' samples ');
ylabel('Amplitude');
```

```
title (' Triangular Window')
axis([-2 N+4 0 2])
%---Figure5.13(b)-----------------------------------------------
tri(16*N)=0;
Ftri = fftshift(fft(tri));
Ftri = Ftri/max(Ftri);
h = figure;
plot(mag2db(abs(Ftri)),'k','LineWidth',2);
grid;
hold off;
axis tight; set(gca,'FontName', 'Arial', 'FontSize',12,'FontW
eight', 'Bold');
xlabel('samples ');
ylabel('Normalized amplitude [dB]');
title ('Spectrum of Triangular Window')
axis([1 16*N -120 3])
%---Figure5.14(a)-----------------------------------------------
%---Hanning window
han = hanning(N);
h = figure;
area(han);
grid;
colormap(gray);
set(gca,'FontName', 'Arial', 'FontSize',12,'FontWeight',
'Bold');
xlabel(' samples ');
ylabel('Amplitude');
title (' Hanning Window')
axis([-2 N+2 0 2])
%---Figure5.14(b)-----------------------------------------------
han(16*N) = 0;
Fhan = fftshift(fft(han));
Fhan = Fhan/max(Fhan);
h = figure;
plot(mag2db(abs(Fhan)),'k','LineWidth',2);
grid;
hold off;
axis tight; set(gca,'FontName', 'Arial', 'FontSize',12,'FontW
eight', 'Bold');
xlabel('samples ');
ylabel('Normalized amplitude [dB]');
title ('Spectrum of Hanning Window')
axis([1 16*N -120 3])
%---Figure5.15(a)-----------------------------------------------
%---Hamming window
ham = hamming(N);
h = figure;
area(ham);
grid;
```

```
colormap(gray);
set(gca,'FontName', 'Arial', 'FontSize',12,'FontWeight',
'Bold');
xlabel('samples ');
ylabel('Amplitude');
title ('Hamming Window')
axis([-2 N+2 0 2])
%---Figure 5.15(b)-----------------------------------------------
ham(16*N)=0;
Fham = fftshift(fft(ham));
Fham = Fham/max(Fham);
h = figure;
plot(mag2db(abs(Fham)),'k','LineWidth',2);
grid;
hold off;
axis tight; set(gca,'FontName', 'Arial', 'FontSize',12,'FontW
eight', 'Bold');
xlabel('samples ');
ylabel('Normalized amplitude [dB]');
title ('Spectrum of Hamming Window')
axis([1 16*N -120 3])
%---Figure 5.16(a)-----------------------------------------------
%---Kaiser window
ksr = kaiser(N,1.5*pi);
h = figure;
area(ksr);
grid;
colormap(gray);
set(gca,'FontName', 'Arial', 'FontSize',12,'FontWeight',
'Bold');
xlabel('samples ');
ylabel('Amplitude');
title ('Kaiser Window, Beta=1.5*pi')
axis([-2 N+2 0 2])
%---Figure 5.16(b)-----------------------------------------------
ksr(16*N) = 0;
Fksr = fftshift(fft(ksr));
Fksr = Fksr/max(Fksr);
h = figure;
plot(mag2db(abs(Fksr)),'k','LineWidth',2);
grid;
hold off;
axis tight; set(gca,'FontName', 'Arial', 'FontSize',12,'FontW
eight', 'Bold');
xlabel('samples ');
ylabel('Normalized amplitude [dB]');
title ('Spectrum of Kaiser Window, Beta=1.5*pi')
axis([1 16*N -120 3])
%---Figure 5.17(a)-----------------------------------------------
```

```
%---Blackman window
blk = blackman(N);
h = figure;
area(blk);
grid;
colormap(gray);
set(gca,'FontName', 'Arial', 'FontSize',12,'FontWeight',
'Bold');
xlabel('samples ');
ylabel('Amplitude');
title ('Blackman Window')
axis([-2 N+2 0 2])
%---Figure 5.17(b)-------------------------------------------------
blk(16*N) = 0;
Fblk = fftshift(fft(blk));
Fblk = Fblk/max(Fblk);
h = figure;
plot(mag2db(abs(Fblk)),'k','LineWidth',2);
grid;
hold off;
axis tight; set(gca,'FontName', 'Arial', 'FontSize',12,'FontW
eight', 'Bold');
xlabel('samples ');
ylabel('Normalized amplitude [dB]');
title ('Spectrum of Blackman Window')
axis([1 16*N -120 3])
%---Figure 5.18(a)-------------------------------------------------
%---Chebyshev window
cheby = chebwin(N);
h = figure;
area(blk);
grid;
colormap(gray);
set(gca,'FontName', 'Arial', 'FontSize',12,'FontWeight',
'Bold');
xlabel('samples ');
ylabel('Amplitude');
title ('Chebyshev Window')
axis([-2 N+2 0 2])
%---Figure 5.18(b)-------------------------------------------------
cheby(16*N)=0;
Fcheby = fftshift(fft(cheby));
Fcheby = Fcheby/max(Fcheby);
h = figure;
plot(mag2db(abs(Fcheby)),'k','LineWidth',2);
grid;
hold off;
axis tight; set(gca,'FontName', 'Arial', 'FontSize',12,'FontW
eight', 'Bold');
```

```
xlabel('samples ');
ylabel('Normalized amplitude [dB]');
title ('Spectrum of Chebyshev Window')
axis([1 16*N -120 3])
```

Matlab code 5.7: Matlab file "Figure5-19ab.m"
```
%--------------------------------------------------------------
% This code can be used to generate Figure 5.19 (a-b)
%--------------------------------------------------------------
% This file requires the following files to be present in the
same
% directory:
%
% Esplanorteta60.mat
% planorteta60_2_xyout.mat
clear all
close all
c=.3; % speed of light
%_____PRE PROCESSING OF ISAR_____
%Find spatial resolutions
BWx = 12;
BWy = 16;
M = 32;
N = 64;
fc = 6;
phic = 0;
% Image resolutions
dx = BWx/M;
dy = BWy/N;
% Form spatial vectors
X = -dx*M/2:dx:dx*(M/2-1);
Y = -dy*N/2:dy:dy*(N/2-1);
%Find resoltions in freq and angle
df = c/(2*BWx);
dk = 2*pi*df/c;
kc = 2*pi*fc/c;
dphi = pi/(kc*BWy);
%Form F and PHI vectors
F = fc+[-df*M/2:df:df*(M/2-1)];
PHI = phic+[-dphi*N/2:dphi:dphi*(N/2-1)];
% Load the backscattered data
load Esplanorteta60
load planorteta60_2_xyout
%_____ POST PROCESSING OF ISAR_____
ISAR = fftshift(fft2(Es.'));
ISAR = ISAR/M/N;
%---Figure5.19(c)----------------------------------------------
h = figure;
matplot2(X(32:-1:1),Y,ISAR,20);
```

```
colormap(1-gray);
colorbar
set(gca,'FontName', 'Arial', 'FontSize',12,'FontWeight',
'Bold');
xlabel('Range [m]');
ylabel('Cross - range [m]');%grid on;
h = line(-xyout_xout,xyout_yout,'Color','k','LineStyle','.','
MarkerSize',3);
%windowing;
w = hamming(M)*hamming(N).';
Ess = Es.*w;
%zero padding;
Enew = Ess;
Enew(M*4,N*4) = 0;
XX = X(1):dx/4:X(1)+dx/4*(4*M-1);
YY = Y(1):dy/4:Y(1)+dy/4*(4*N-1);
% ISAR image formatiom
ISARnew = fftshift(fft2(Enew.'));
ISARnew = ISARnew/M/N;
%---Figure5.19(d)------------------------------------------------
load planorteta60_2_xyout.mat
h = figure;
matplot2(XX(4*M:-1:1),YY,abs(ISARnew),20);
colormap(1-gray);
colorbar
line(-xyout_xout,xyout_yout,'Color','k','LineStyle','.','Marke
rSize',3);
set(gca,'FontName', 'Arial', 'FontSize',12,'FontWeight',
'Bold');
xlabel('Range [m]');
ylabel('Cross - range [m]');
```

Matlab code 5.8: Matlab file "Figure5-19cd.m"
```
%----------------------------------------------------------------
% This code can be used to generate Figure 5.19 (c-d)
%----------------------------------------------------------------
% This file requires the following files to be present in the same
% directory:
%
% ucak.mat
clear all
close all
c=.3; % speed of light
%_____PRE PROCESSING OF ISAR_____
%Find spatial resolutions
BWx = 18;
BWy = 16;
M = 64;
N = 64;
```

```
fc = 8;
phic = 0;
% Image resolutions
dx = BWx/M;
dy = BWy/N;
% Form spatial vectors
X = -dx*M/2:dx:dx*(M/2-1);
Y = -dy*N/2:dy:dy*(N/2-1);
%Find resoltions in freq and angle
df = c/(2*BWx);
dk = 2*pi*df/c;
kc = 2*pi*fc/c;
dphi = pi/(kc*BWy);
%Form F and PHI vectors
F = fc+[-df*M/2:df:df*(M/2-1)];
PHI = phic+[-dphi*N/2:dphi:dphi*(N/2-1)];
K = 2*pi*F/c;
%_____ FORM RAW BACKSCATTERED DATA_____
load ucak
l = length(xx);
Es = zeros(M,N);
for m=1:l;
 Es=Es+1.0*exp(j*2*K'*(cos(PHI)*xx(m)+sin(PHI)*yy(m)));
end
%_____ POST PROCESSING OF ISAR (Small BW Small
angle)_____
ISAR = fftshift(fft2(Es.'));
ISAR = ISAR/M/N;
h = figure;
matplot2(X(M:-1:1),Y,ISAR,25);
colormap(1-gray);
colorbar
set(gca,'FontName', 'Arial', 'FontSize',12,'FontWeight',
'Bold');
xlabel('Range [m]');
ylabel('Cross - range [m]');%grid on;
colormap(1-gray);
%windowing;
w = hamming(M)*hamming(N).';
Ess = Es.*w;
%-------------zero padding with 4 times----------
Enew = Ess;
Enew(M*4,N*4) = 0;
% ISAR image formatiom
ISARnew = fftshift(fft2(Enew.'));
ISARnew = ISARnew/M/N;
h = figure;
matplot2(X(M:-1:1),Y,ISARnew,25);
colormap(1-gray);
```

```
colorbar
set(gca,'FontName', 'Arial', 'FontSize',12,'FontWeight',
'Bold');
xlabel('Range [m]');
ylabel('Cross - range [m]');%grid on;
```

Matlab code 5.9: Matlab file "Figure5-19ef.m"
```
%----------------------------------------------------------------
% This code can be used to generate Figure 5.19 (e-f)
%----------------------------------------------------------------
% This file requires the following files to be present in the
same
% directory:
%
% Esairbus.mat
% airbusteta80_2_xyout.mat
clear all
close all
c=.3; % speed of light
%_____PRE PROCESSING OF ISAR_____
%Find spatial resolutions
BWx = 80;
BWy = 66;
M = 32;
N = 64;
fc = 4;
phic = 0;
% Image resolutions
dx = BWx/M;
dy = BWy/N;
% Form spatial vectors
X = -dx*M/2:dx:dx*(M/2-1);
Y = -dy*N/2:dy:dy*(N/2-1);
%Find resoltions in freq and angle
df = c/(2*BWx);
dk = 2*pi*df/c;
kc = 2*pi*fc/c;
dphi = pi/(kc*BWy);
%Form F and PHI vectors
F = fc+[-df*M/2:df:df*(M/2-1)];
PHI = phic+[-dphi*N/2:dphi:dphi*(N/2-1)];
load Esairbus
load airbusteta80_2_xyout
%_____ POST PROCESSING OF ISAR_____
ISAR = fftshift(fft2(Es.'));
ISAR = ISAR/M/N;
%---Figure5.19(a)------------------------------------------------
h = figure;
matplot2(X(32:-1:1),Y,ISAR,30);
```

```
colormap(1-gray);
colorbar
set(gca,'FontName', 'Arial', 'FontSize',12,'FontWeight',
'Bold');
xlabel('Range [m]'); ylabel('Cross - range [m]');%grid on;
colormap(1-gray);%colorbar
h=line(-xyout_xout,xyout_yout,'Color','k','LineStyle','.');
%windowing;
w = hamming(M)*hamming(N).';
Ess = Es.*w;
%zero padding with 4 times;
Enew = Ess;
Enew(M*4,N*4) = 0;
% ISAR image formatiom
ISARnew = fftshift(fft2(Enew.'));
ISARnew = ISARnew/M/N;
%---Figure 5.19(b)---------------------------------------------------
h = figure;
matplot2(X(32:-1:1),Y,ISARnew,30);
colormap(1-gray);
colorbar
line(-xyout_xout,xyout_yout,'Color','k','LineStyle','.');
set(gca,'FontName', 'Arial', 'FontSize',12,'FontWeight',
'Bold');
xlabel('Range [m]');
ylabel('Cross - range [m]');
```

REFERENCES

1   E. F. Knott, J. F. Shaeffer, and M. T. Tuley. *Radar cross section*, 2nd ed. Artech House, Norwood, MA, 1993.

2   R. J. Sullivan. *Microwave radar imaging and advanced concepts*. Artech House, Norwood, MA, 2000.

3   R. M. Mersereau and A. V. Oppenheim. Digital reconstruction of multidimensional signals from their projections. *Proc IEEE* 62(10) (1974) 1319–1338.

4   D. A. Ausherman, A. Kozma, J. L. Walker, H. M. Jones, and E. C. Poggio. Developments in radar imaging. *IEEE Trans Aerosp Electron Syst* AES-20(4) (1984) 363–400.

5   C. Ozdemir, R. Bhalla, L. C. Trintinalia, and H. Ling. ASAR—Antenna synthetic aperture radar imaging. *IEEE Trans Antennas Propagat* 46(12) (1998) 1845–1852.

6   M. Abramowitz and I. A. Stegun. *Handbook of mathematical functions*. Dover Publications Inc., New York, 1970.

7   T. Kohler, H. Turbell, and M. Grass. Efficient forward projection through discrete data sets using tri-linear interpolation. IEEE Nuclear Science Symposium 2000, vol. 2, pp. 15/113–15/115, 2000.

8   R. Keys. Cubic convolution interpolation for digital image processing. *IEEE Transactions Acoustics, Speech and Signal Processing* 29 (1981) 1153–1160.

9   F. Lekien and J. Marsden. Tricubic interpolation in three dimensions. *Int J Numer Methods Eng* 63 (2005) 455–471.

10   R. B. Blackman and J. W. Tukey. *"Particular pairs of windows." The measurement of power spectra, from the point of view of communications engineering.* Dover, New York, 1959, pp. 95–101.

11   L. D. Enochson and R. K. Otnes. Programming and analysis for digital time series data, U.S. Deptartment of Defense, Shock and Vibration Information Center, pp. 142, 1968.

12   J. F. Kaiser. Nonrecursive digital filter design using the I0- sinh window function. Proc. 1974 IEEE Symp. Circuits and Systems, pp. 20–23, 1974.

13   A. V. Oppenheim and R. W. Schafer. *Discrete-time signal processing.* Prentice-Hall, Upper Saddle River, NJ, 1999, pp. 468–471.

14   F. J. Harris. *Multirate signal processing for communication systems.* Prentice Hall PTR, Upper Saddle River, NJ, 2004, pp. 60–64.

CHAPTER SIX

Range-Doppler Inverse Synthetic Aperture Radar Processing

In Chapter 4, the base algorithm for inverse synthetic aperture radar (ISAR) imaging is provided. This algorithm is based on the assumption that the target is stationary and the data are collected over a finite number of stepped look angles. In real scenarios, however, the target is usually in motion and therefore, the aspect diverse data can only be collected if the target's motion allows different look angles to the radar during the *coherent processing time* of the radar. The radar usually sends chirp (linear frequency modulated [LFM]) pulses or stepped frequency pulses to catch different look angles of the target. After the radar receiver collects the echoed pulses from the target, the ISAR image can only be formed in the two-dimensional (2D) range-Doppler space since the radar line of sight (RLOS) angle values with respect to target axis are unknown to the radar. This phenomenon will be explained in the forthcoming subsections.

In this chapter, we will examine the ISAR imaging techniques for real-world scenarios when the target is not stationary with respect to radar and the Doppler frequency shift-induced backscattered data are collected by the radar. In particular, commonly used ISAR waveforms, namely the *chirp* (LFM) and the *stepped frequency continuous wave* (SFCW) pulse waveforms, are utilized. The 2D range-Doppler ISAR imaging algorithms that employ these waveforms are presented.

Inverse Synthetic Aperture Radar Imaging with MATLAB Algorithms, First Edition.
Caner Özdemir.
© 2012 John Wiley & Sons, Inc. Published 2012 by John Wiley & Sons, Inc.

231

6.1   SCENARIOS FOR ISAR

As mentioned in the previous chapters, the ISAR provides an electromagnetic (EM) image of the target that is moving with respect to radar. The backscattered signal at the radar receiver is processed such that this signal is transformed to time (or range) and Doppler frequency (or cross range). The time (or range) processing is accomplished by utilizing the frequency bandwidth of the radar pulse such that the points in the range (or RLOS) direction can be resolved. The movement of the target with respect to radar provides Doppler frequency shifts as the target moves and therefore the radar can collect the scattering from the target. The Doppler frequency analysis makes it possible to resolve the points along the cross-range axis, which is defined as the axis perpendicular to the RLOS direction.

In real-world applications, the target can be aerial, such as an airplane or helicopter, or ground or sea based, such as a ship or a tank. In most scenarios, aerial targets are usually imaged with the help of a ground-based radar, whereas ground/sea-based platforms are usually imaged via an airborne radar.

6.1.1   Imaging Aerial Targets via Ground-Based Radar

As illustrated in Figure 6.1, this case represents when the radar is stationary on the earth's surface and the target is an aerial one that has a general motion with respect to radar. As mentioned in previous chapters, the range or the line-of-sight resolution is achieved by using an adequate frequency bandwidth. If the target is rotating, the angular diversity of the target can be readily constituted between the received pulses. When the target is not rotating and is moving straight, as illustrated in Figure 6.1, its motion can be devised into radial translation motion, the motion along the RLOS axis, and tangential

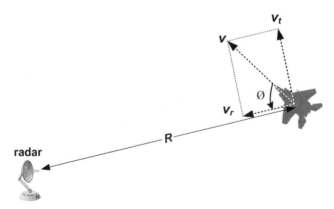

FIGURE 6.1   For the ISAR operation, the aspect diversity is constituted target's rotational and/or tangential motion with respect to radar.

motion, the motion along the axis that is perpendicular to the RLOS. If this is the case, angular diversity of the target is realized for a longer time as the target's tangential motion produces Doppler frequency shifts slower than the case of a rotating target. If the target's distance from the radar is R and moving with a speed of v as depicted in Figure 6.1, it has a tangential speed of $v_t = v \cdot sin\phi$. Therefore, the corresponding angular rotational speed becomes equal to

$$\omega = \frac{v_t}{R}. \tag{6.1}$$

If the coherent integration time, also called the *coherent processing time*, the *image frame time*, or the *dwell time* of radar is T, the total angular width seen by the radar is

$$\Omega = \omega T. \tag{6.2}$$

How this angular width produces the required frequency Doppler shift will be presented in Section 6.3 together with the associated signal processing to resolve the points in the cross-range dimension.

The aerial targets usually move on a straight path and rarely make rotational movements. Therefore, the necessary angular diversity required for a possible ISAR image can be obtained by the target's tangential motion with respect to radar as depicted in Figure 6.1.

Table 6.1 lists some values of angular width for different range distances and different tangential speed values for an aerial target. Noticing the fact that most jet fighters have a typical speed of about 800–900 km/h, Case #1 in Table 6.1 corresponds to the case where an airplane is flying mostly in the tangential direction. If the radar is 13 km away, the radar observes a rotational angular speed of 0.73°/second according to Equation 6.1. Considering a nominal

TABLE 6.1 Angular Speed and the Total Look-Angle Width Observed by Radar for a Target that Has a Motion Component in the Translational Direction

Case ID	Target's Distance (R)	Target's Tangential Speed (v_t)	Corresponding Angular Rotational Speed (ω)	Integration Time (T)	Total Angular Width Seen by Radar (Ω)
1	13 km	600 km/h	0.73°/second	3 seconds	2.2°
2	13 km	40 km/h	0.05°/second	3 seconds	0.15°
3	4 km	600 km/h	2.39°/second	1 second	2.39°
4	4 km	40 km/h	0.80°/second	4 seconds	3.2°
5	4 km	600 km/h	2.39°/second	4 seconds	9.56°

coherent integration time of 3 seconds, the radar look-angle width of target becomes $2.2°$, which is good enough to form a good ISAR image. As listed in Case #2, however, the tangential speed of the target can be relatively smaller in some cases such that the look-angle width happens to be much lower than $1°$. For this scenario, a good quality ISAR image may not be possible since the scattering centers on the target along the cross-range direction may not be resolved with such a small look-angle width. If the target is much closer, as listed in Cases #3 and #4, the integration time should be chosen accordingly to have a logical value for the angular width to be seen by the radar. Angular widths between $2°$ and $7°$ are practical to get a fast ISAR image. If the integration time is not taken into account, as in Case #5, where the total angular width seen by radar becomes as big as $10°$, the resultant ISAR image may have unwanted motion effects such as blurring and defocusing due to the fact that the scattering centers on the target may occupy several range bins during the integration time of radar. Furthermore, the use of fast Fourier transform (FFT) and therefore fast formation of the ISAR image will not be possible since the small-angle approximation will not be valid for this case.

It is also important to note that the target may maneuver during the integration time of the radar such that it may yaw, roll, or pitch while progressing at the same time. If this is the case, the target's rotational motion with respect to radar would be mostly governed by the target's rotational motion in its own axis. For such cases, the target's look-angle width will be much wider when compared with the case shown in Figure 6.1. Therefore, very small integration time values will be sufficient to construct a good quality ISAR image for such scenarios.

In most real-world applications, the target's motion parameters such as translational velocity, translational acceleration, rotational velocity, and rotational acceleration are unknown to the radar. Furthermore, a target's initial angular position with respect to RLOS is also an unidentified quantity. Therefore, the appropriate cross-range indexing in meters could not be possible in most cases. For this reason, the resultant ISAR image may not be displayed on the *range–cross-range* plane, but on the *time-Doppler* or *range-Doppler* plane.

6.1.2   Imaging Ground/Sea Targets via Aerial Radar

When it comes to the aerial radar, the main application of ISAR imaging is to identify and/or classify sea or ground platforms such as tanks, ships, and vessels. Since both the target and the radar are in motion in this case, the analysis and the processing become more complex due to the fact that the radar's motion with respect to the target provides additional Doppler shift in the phase of the received signal. The problem of directing the radar antenna's beam toward the target, that is, tracking, is another issue that needs to be controlled, which is not an easy task most of the time. Therefore, effective target tracking systems are essential on the radar site.

It is also worthwhile to mention that the propagation characteristics of the air medium also play an important role in collecting a reliable received signal. The propagation characteristics of EM waves in different weather conditions (foggy, rainy, stormy, snowy, etc.) differ from the ideal case of a calm air situation. The atmospheric noise, background noise, and the radar platform's electronic noise itself may influence the quality of the received signal.

While the main source of Doppler frequency shift is due to the target's rotational and/or tangential motion with respect to radar for aerial targets, the Doppler shift caused by the tangential motion of ground/sea targets is generally quite small since the speeds of these platforms are relatively much smaller. The major source of Doppler frequency shift for such targets is due to the targets' rotational motion about its own axis, that is, yawing, rolling, and pitching. As demonstrated in Figure 6.2, yaw, roll, and pitch motions of the platform can produce the required angular variation during the coherent integration time (or illumination period) of the radar.

A ship's yawing, rolling, or pitching motions are usually caused by the wave motions for different sea-state conditions. In oceanography, the term *sea state* is commonly used to describe the general condition of the free surface on a large body of water with respect to wind waves and swell at a certain location and moment [1]. The sea-state code index ranges from 0 (calm) to 9 (severe or phenomenal) depending on the wave height. Sea-state values of 3 to 4 correspond to slight to moderate sea conditions and represent wave heights from 0.5 m to 2.5 m.

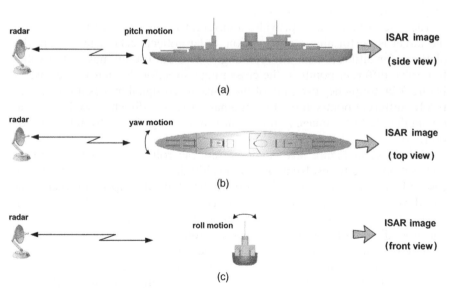

FIGURE 6.2   Resulting 2D ISAR images for (a) pitching, (b) yawing (turning), and (c) rolling platform.

Yaw is the rotational motion of the ship about the vertical (up-down) axis and is caused by temporary bearing changes. Maximum yaw angle for a ship can be as large as tens of degrees. The period of yaw motion is generally equal to wave period [2].

Roll is the rotational motion of the ship about the longitudinal (front-back) axis. The sea-state condition, frequency of the hitting wave, and the ship's righting arm curve are the most important parameters that influence the maximum value of the roll angle. A typical value for the maximum roll angle is a few degrees for sea-state codes of 3 to 4. However, this value can jump to tens of degrees for a sea-state code of 8.

Pitch is the rotational motion of the ship about the transverse (side-to-side) axis. Pitch motion primarily depends on the sea-state condition and the length between the perpendiculars [2]. Longer ships tend to allow smaller pitch angles. Typical pitch angles are in the range of $1°–2°$ for a sea-state code of 4 and can be as large as $5°–11°$ for a sea-state code of 8 [2].

While the range resolution is achieved by multifrequency sampling of the platform's backscattered echo, the cross-range (or the Doppler frequency shift) resolution is achieved by multiangle sampling of the received signal. The corresponding 2D ISAR images for a pitching, yawing, and rolling platform are illustrated in Figure 6.2a–c, respectively: In Figure 6.2a, the platform is performing a pitch motion so that the radar collects the backscattered returns from the platform for different look angles of elevation. This elevation angle diversity provides the spatial resolution in the altitudinal direction. Also, utilizing the frequency bandwidth of the received pulses provides range resolution along the longitudinal (or the range) axis. As a result, a 2D ISAR image that shows the side view of the platform can be obtained.

The case in Figure 6.2b provides a different ISAR image of the target. As the target is performing a yaw motion, radar collects echo signals from the platform for different azimuth look angles. This data setup makes it possible to resolve different points in the cross-range direction. Similar to the case in Figure 6.2a, frequency diversity of the transmitted signal makes it possible to resolve different points in the longitudinal (or range) direction so that we can obtain the 2D ISAR image as if the platform is being viewed from the top (or bottom).

If the case in Figure 6.2c is considered, the roll motion of the platform makes it possible to resolve points in the altitudinal direction as similar to the case in Figure 6.2a. Since the platform is rotated by $90°$ in azimuth when compared with the first case, the frequency diversity provides range resolution in the beam direction of the ship platform as seen in Figure 6.2c. Therefore, the corresponding 2D ISAR image shows the front (or back) view of the target.

In most real-world applications, the target's position with respect to radar and the target's axial motion with respect to radar are random, as shown in Figure 6.3. If this is the situation, the target's ISAR image is displayed on the 2D projection plane where the range axis is the RLOS axis, and the cross-range axis is the same direction as the target's maneuver (pitch, yaw, or roll) axis,

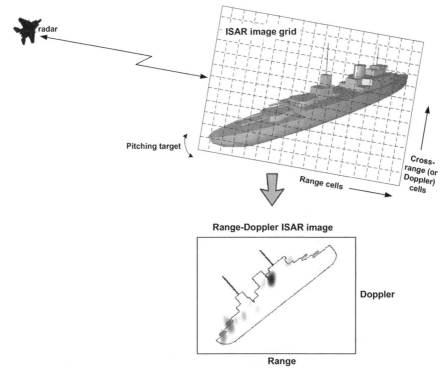

FIGURE 6.3   The formation of the ISAR grid for a maneuvering platform.

which should also be perpendicular to the range axis as demonstrated in Figure 6.3.

6.2   ISAR WAVEFORMS FOR RANGE-DOPPLER PROCESSING

In real scenarios of ISAR, images of various platforms including airplanes, helicopters, ships, and tanks are usually formed by collecting the multifrequency multiaspect received signals from these targets with one of the following popular waveforms:

1. Stretch or LFM or *chirp* pulse train
2. SFCW pulse train

These waveforms were already listed and studied in Chapter 2, Sections 2.6.5 and 2.7. Next, we are going to revisit chirp pulse train and stepped-frequency pulse train waveforms and their uses in ISAR range-Doppler processing.

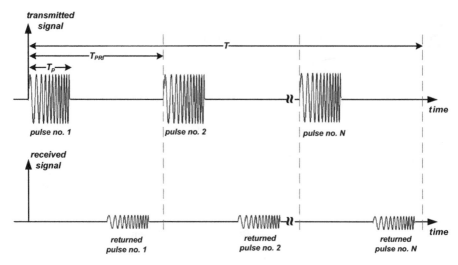

FIGURE 6.4   The chirp pulse train is utilized in range-Doppler processing of ISAR.

6.2.1 Chirp Pulse Train

A common radar pulse train that consists of a total of N chirp pulse waveforms is shown in Figure 6.4. Here T_p is the pulse duration or pulse length, T_{PRI} is the pulse repetition interval (PRI), and T is the *dwell time*, often called the *coherent integration time*, and is determined by

$$\begin{aligned} T &= N \cdot T_{PRI} \\ &= N/PRF, \end{aligned} \tag{6.3}$$

where PRF is the pulse repetition frequency. To avoid ambiguity in range, every returned pulse should arrive at the radar before the next pulse is transmitted. If the target is at R distant from the radar, therefore, the minimum value of the PRI should be as the following to avoid the ambiguity in range determination:

$$T_{PRI_{min}} = \frac{c}{2R}, \tag{6.4}$$

which means that the PRF should be always less than the maximum value:

$$PRF_{max} = \frac{2R}{c}. \tag{6.5}$$

If the target is at the range of 30 km, for example, the PRF should be less than 200 μs to avoid ambiguity in the range.

The variation of the frequency within a chirp provides the necessary frequency bandwidth to resolve the points along the range dimensions. The bandwidth of a chirp pulse is selected according to the required range resolution as

$$B = \frac{c}{2 \cdot \Delta r},$$ (6.6)

where Δr is the desired range resolution. The instantaneous frequency of a single chirp pulse waveform is given by

$$f_i = f_o + K \cdot t; \quad 0 \le t \le T_p,$$ (6.7)

where f_o is the starting frequency of the chirp and K is the *chirp rate*. Therefore, the chirp rate of the LFM pulse should be selected in the following way to provide the necessary bandwidth for range processing:

$$K = \frac{B}{T_p}.$$ (6.8)

Coherent integration time for the chirp pulse radar is as given in Equation 6.3.

6.2.2 Stepped Frequency Pulse Train

SFCW pulse is one of the most frequently used radar waveforms in ISAR imaging. A detailed explanation of the stepped frequency waveform is given in Chapter 2, Section 2.6.3 and its usage in range–cross range ISAR imaging is demonstrated in Chapter 4. Here, we will show the usage of SFCW pulse train in range-Doppler ISAR imaging.

In the SFCW pulse train operation, a total of M identical bursts of N pulses are generated to be transmitted toward the target as demonstrated in Figure 6.5. Each pulse in each burst is composed of a single frequency sinusoidal wave. The frequency of the first pulse is f_L, and the frequencies of the subsequent pulses are incremented by Δf such that the nth pulse in any burst is given by

$$f_n = f_L + (n-1) \cdot \Delta f.$$ (6.9)

Therefore, the frequency of the Nth pulse in any burst is

$$\begin{aligned} f_N &\triangleq f_H \\ &= f_L + (N-1) \cdot \Delta f. \end{aligned}$$ (6.10)

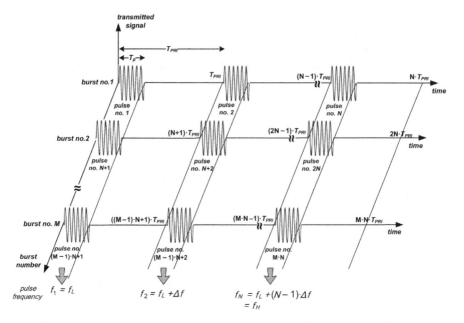

FIGURE 6.5 Representation of stepped frequency transmitted signal of M bursts each having N stepped frequency waveforms.

Therefore, the frequency bandwidth of the stepped frequency pulse train is then equal to

$$B = N \cdot \Delta f. \tag{6.11}$$

The total time passed for one burst is

$$
\begin{aligned}
T_{burst} &= N \cdot T_{PRI} \\
&= \frac{N}{PRF},
\end{aligned} \tag{6.12}
$$

and the coherent integration time or the dwell time for the M burst is equal to

$$
\begin{aligned}
T &= M \cdot T_{burst} \\
&= M \cdot N \cdot T_{PRI} \\
&= \frac{M \cdot N}{PRF}.
\end{aligned} \tag{6.13}
$$

To avoid ambiguity in range determination, the pulse response should arrive before the next pulse is transmitted. Therefore, the maximum PRF value is the same as in the case of chirp pulse illumination and given as in Equation 6.5.

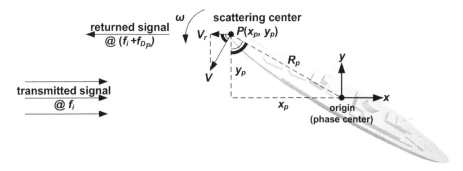

FIGURE 6.6 Target's rotational motion causes Doppler shift in the frequency of the received pulses for a point on the target.

6.3 DOPPLER SHIFT'S RELATION TO CROSS RANGE

Let us start with the analysis of Doppler shift caused by target motion by assuming that the target only has a rotational motion component as illustrated in Figure 6.6. The target has a rotational motion with an angular velocity of ω. The goal here is to find the Doppler shift at the received signal caused by the rotational movement of a point scatterer, P, on the target. As illustrated in Figure 6.6, the point $P(x_p, y_p)$ is located at R_p away from the rotation axis of the target. Then the tangential velocity of the point P is equal to

$$v = R_p \cdot \omega. \tag{6.14}$$

The radial velocity of this scattering center along the radar's line of sight direction is represented as v_r and can be found by using the two similar triangles in the figure as

$$v_r = \frac{y_p}{R_p} \cdot v. \tag{6.15}$$

Substituting Equation 6.15 into Equation 6.16 will yield

$$v_r = \frac{y_p}{R_p} \cdot (R_p \cdot \omega)$$
$$= y_p \cdot \omega, \tag{6.16}$$

which clearly states that the "radial velocity" of a point scatterer on the target is directly related to its "cross-range value." Now, we can easily calculate the Doppler frequency shift caused by radial velocity of the point scatterer as

$$f_{Dp} = \frac{2v_r}{c} \cdot f_i$$

$$= \frac{2\omega}{\lambda_i} \cdot y_p,$$

(6.17)

where f_i and λ_i are the instantaneous frequency and the corresponding wavelength of the chirp pulse waveform, respectively. Therefore, the EM wave returned from point P will have a Doppler shift with an amount of f_{Dp} as calculated in Equation 6.18. This result clearly demonstrates that the Doppler frequency shift caused by the motion of the point scatterer is directly proportional to its cross-range position, y_p. Therefore, if the returned signal from all the scattering centers is plotted in the Doppler shift domain, the resulting plot is proportional to the target's cross-range profile. If the angular speed of the target is correctly estimated, then the cross-range profile of the target can be correctly labeled.

6.3.1   Doppler Frequency Shift Resolution

For the general case, the Doppler frequency shift of a point at (x, y) on the target is equal to

$$f_D = \frac{2\omega}{\lambda_i} \cdot y.$$

(6.18)

Then, the resolution for the Doppler frequency shift, Δf_D, can be readily found as

$$\Delta f_D = \frac{2\omega}{\lambda_i} \cdot \Delta y.$$

(6.19)

As found earlier in Chapter 4 (Eq. 4.45), the cross-range resolution is given by $\Delta y = (\lambda/2)/\Omega$ where Ω is the total angular width or the total viewing angle of the target by the radar. Ω can be easily related to angular velocity, ω, and the total viewing time of the target (or the dwell time), T, as

$$\Omega = \omega \cdot T.$$

(6.20)

Then the cross-range resolution becomes equal to

$$\Delta y = \frac{\lambda_i/2}{(\omega T)}.$$

(6.21)

Finally, Δf_D can be determined in terms of T by substituting Equation 6.21 into Equation 6.19 as

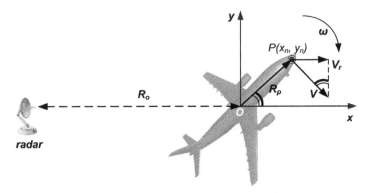

FIGURE 6.7 Geometry for Doppler processing of a rotating target.

$$\Delta f_D = \frac{2\omega}{\lambda_i} \cdot \frac{\lambda_i}{2\omega T}$$

$$= \frac{1}{T}.$$

(6.22)

Therefore, the resolution for the Doppler frequency shift is the inverse of the total viewing time which is an expected result according to the Fourier theory.

6.3.2   Resolving Doppler Shift and Cross Range

Let us consider the geometry seen in Figure 6.7 where the target is rotating with a rotational velocity of ω. The origin of the geometry is R_o away from the radar.

Assuming that the target is at the far field of the radar, the phase of the returned signal from the nth scattering center located at $P(x_n, y_n)$ on the target will have the form of

$$\varphi = e^{-j2kR_n(t)},$$

(6.23)

where $R_n(t)$ is the radial distance of the point scatterer from the radar and can be written as

$$R_n(t) = (R_o + x_n) + v_{r_n} \cdot t.$$

(6.24)

Here, v_{r_n} corresponds to the radial velocity of the nth point scatterer. Then the phase in Equation 6.24 can be rewritten as the following:

$$\varphi = e^{-j2k(R_o + x_n)} \cdot e^{-j2kv_{r_n} \cdot t}.$$

(6.25)

Note that the first term is constant with respect to time and only the second term is time varying. Substituting $v_{r_n} = \lambda \cdot f_{D_n}/2$ into Equation 6.26 will yield

$$\varphi = e^{-j2k(R_0+x_n)} \cdot e^{-j2\left(\frac{2\pi}{\lambda}\right)\left(\frac{\lambda \cdot f_{Dn}}{2}\right) \cdot t}$$

$$= e^{-j2k(R_0+x_n)} \cdot e^{-j2\pi f_{Dn}t}, \tag{6.26}$$

where f_{D_n} is the Doppler frequency shift for the nth point scatterer. It is clear that there exists a Fourier transform relationship between time variable, t, and the Doppler frequency shift variable, f_{D_n}. Therefore, taking the inverse Fourier transform (IFT) of the received signal with respect to time, the Doppler frequency shift for the nth point scatterer can be easily resolved. As listed in Equation 6.18, cross-range is proportional to the Doppler frequency shift. If the rotational velocity ω is predicted, the cross-range dimension y_n can also be correctly labeled.

6.4 FORMING THE RANGE-DOPPLER IMAGE

Let us assume that the target in Figure 6.7 is modeled as if it contains N point scatterers located at (x_n, y_n) where n runs from 1 to N. Then, the received signal can be approximated as the following:

$$E^s(k,t) \cong \sum_{n=1}^{N} A_n \cdot e^{-j2k(R_0+x_n)} \cdot e^{-j2\pi f_{Dn}t}, \tag{6.27}$$

where A_n is the complex magnitude of the nth scattering center. Taking the origin of the target as the phase center of the geometry, the phase term e^{-j2kR_0} can be suppressed to get

$$E^s(f,t) \cong \sum_{n=1}^{N} A_n \cdot e^{-j2\pi\left(\frac{2f}{c}\right)x_n} \cdot e^{-j2\pi f_{Dn}t}. \tag{6.28}$$

Taking the 2D IFT of the backscattered signal with respect to $(2f/c)$ and (t), we get

$$F_2^{-1}\{E^s(f,t)\} \cong \sum_{n=1}^{N} A_n \cdot F_1^{-1}\left\{e^{-j2\pi\left(\frac{2f}{c}\right)x_n}\right\} \cdot F_1^{-1}\{e^{-j2\pi f_{Dn}t}\}$$

$$= \sum_{n=1}^{N} A_n \cdot \delta(x - x_n) \cdot \delta(f_D - f_{Dn}) \tag{6.29}$$

$$\triangleq ISAR(x, f_D).$$

This result clearly shows that the resulting 2D image data are on the range-Doppler frequency plane. As is obvious from the above analysis, the range components of the scattering centers, x_ns, are easily resolved as the same way that we did in the conventional ISAR imaging. The other dimension is the Doppler frequency axis that is proportional to the cross-range axis. Provided that the target's rotational velocity is known or estimated correctly, the transformation from Doppler frequency space to cross-range space can be performed by applying the following transformation formula:

$$y = \frac{\lambda_c}{2\omega} \cdot f_D, \tag{6.30}$$

where λ_c is the wavelength corresponding to the center frequency. After this transformation, the cross-range components, y_ns, are also resolved, and we can form the ISAR image in range cross-range plane.

6.5 ISAR RECEIVER

Most ISAR systems are designed for either chirp or SFCW pulse train waveforms. Some systems utilize other *stretch* waveforms as well [3]. The ISAR receiver circuitry is, therefore, designed according to the type of illumination waveform.

6.5.1 ISAR Receiver for Chirp Pulse Radar

LFM pulse train or chirp pulse train waveform is widely used in SAR and ISAR applications, thanks to its easy applicability. The general block diagram for chirp pulse ISAR receiver is shown in Figure 6.8. The receiver processes the received signal pulse-by-pulse such that range profile corresponding to each pulse is obtained. Doppler frequency shifts for each range bin are determined with the help of Fourier transform operation so that the final 2D range-Doppler image of the target is obtained.

Let us analyze the ISAR receiver in Figure 6.8 in more detail. First, the chirp pulse return from the target is collected and fed to the intermediate frequency (IF) amplifier such that the signal level is amplified at the IF stage for further processing. Then, the matching filtering is applied to compress each of the incoming pulses. As demonstrated in Chapter 3, Section 3.4.1, the output of the matched filtering (or the pulse compressor) is the compressed version of the received pulse. The result is nothing but the one-dimensional (1D) range profile of the target for that particular pulse. At this point, N range profiles corresponding to N pulse returns are produced. Then quadrature detection (QD) follows to detect the amplitude and the phase information of the returned signal at the baseband frequencies. The details of QD will be given in Section 6.6.

The entire signal processing scheme up to this point is analog. As the next step, the I and Q pairs at the output of the QD are sampled and digitized by using samplers and analog-to-digital (A/D) converters such that each range

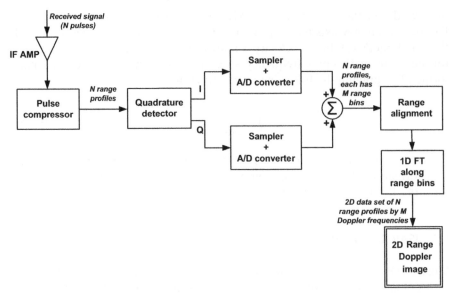

FIGURE 6.8   ISAR receiver block diagram for chirp pulse illumination.

profile is digitized to M range cells (or range bins). Then, the digitized range profiles of length M are put side-by-side to align the range positions such that each range cell has to correspond to the same respective range positions along the target. Otherwise, image blurring occurs due to this *range walk*, the movement of range positions from profile to profile. The process of range alignment will be explained in Section 6.7.

After compensating the range walk in the 2D data set, 1D discrete Fourier transform (or DFT) can be applied along azimuthal time instants to transform the returns to Doppler frequency space. The resulting 2D matrix is the *N-by-M range-Doppler ISAR image* of the target. In Section 6.8, we will present the detailed processes and the algorithm for range-Doppler ISAR imaging.

6.5.2   ISAR Receiver for SFCW Radar

SFCW signal is also one of the commonly used waveforms in radar imaging because it can provide digital data for fast processing for a reliable SAR/ISAR image. The SFCW transmitter sends out M repeated bursts of stepped frequency waveforms. In each burst, a total of N stepped frequency waveforms are transmitted. A common block diagram for a SFCW-based ISAR receiver is illustrated in Figure 6.9.

The receiver collects the total scattered field data that are composed of M bursts with N stepped frequency pulses. This received signal is fed to an IF amplifier such that the signal level is amplified at the IF stage for further

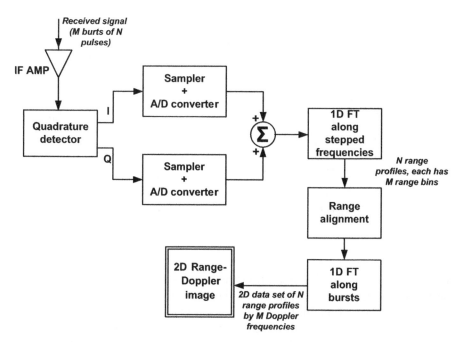

FIGURE 6.9 ISAR receiver block diagram for stepped frequency radar illumination.

processing. Afterward, QD is used to gather the amplitude and the phase information of the returned signal around the baseband frequencies. Then, the *I* and *Q* pairs at the output of the QD are sampled and digitized using samplers and A/D converters such that a matrix of *M*-by-*N* is constituted for *M* bursts that corresponds to *M* azimuthal time instants and *N* stepped frequencies. Taking 1D IFT along the stepped frequencies will yield a total of *M* different range profiles, each having *N* range bins. If the radial velocity of the target is not small and/or the PRF is not high enough (so the dwell time is long), the range profiles may not line up and alignment of range profiles may be required before processing in the azimuth direction, as will be clarified in Section 6.7. If the range arrangement is not made, the ISAR image becomes blurred due to this movement of range positions from profile to profile. For this case, range cells should be aligned for the whole 2D data set. Then, 1D IFT along bursts (or azimuthal time instants) will transform the data into the Doppler frequency shift domain. The resulting 2D matrix is the ISAR image of the target in range-Doppler frequency domain.

6.6 QUADRADURE DETECTION

The process of QD is commonly used in radar systems to acquire received signal phase information relative to the transmitted signal carrier. QD can also

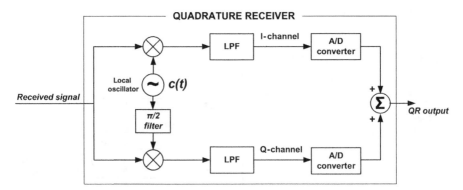

FIGURE 6.10   Block diagram for quadrature detection.

be regarded as the mixing operation that carries the received signal to the baseband to obtain the amplitude and the phase information of the received signal in the form of quadrature, that is, I and Q components.

The block diagram of QD is shown in Figure 6.10. The QD receiver is usually applied after the pulse compression filter in SAR/ISAR imaging as shown in Figure 6.8. The input signal to the QD receiver is fed to the *inphase* (I) *channel* and 90° delayed with respect to the reference signal at the local oscillator (LO) or the *quadrature* (Q) *channel*.

Let us assume that the transmitted signal has the form of a simple sinusoid,

$$s(t) = A_i \cdot \cos(2\pi f_i t), \tag{6.31}$$

where f_i is the instantaneous frequency within the bandwidth of the transmitted signal. Then, the received signal from a point scatterer that is R_o away from the radar has the following form:

$$E(t) = A_i \cdot \cos\left(2\pi f_i \left(t - \frac{2R_o}{c}\right)\right), \tag{6.32}$$

where A_i is the backscattered amplitude. The processing in I- and Q-channels are given in detail below.

6.6.1   I-Channel Processing

The received signal is multiplied with the pilot signal, $c(t)$, that is generated by the LO:

$$c(t) = B_i \cdot \cos(j2\pi f_i t). \tag{6.33}$$

The multiplier output in the I-channel yields the following:

$$E(t) \cdot c(t) = A_i B_i \cdot \cos\left(2\pi f_i\left(t - \frac{2R_o}{c}\right)\right) \cdot \cos(j2\pi f_i t)$$

$$= \frac{A_i B_i}{2} \cdot \left[\cos\left(4\pi f_i t - 2\pi f_i \frac{2R_o}{c}\right) + \cos\left(2\pi f_i \frac{2R_o}{c}\right)\right].$$

(6.34)

After low pass filtering, the first term at $(2f_i)$ frequency is filtered out, and the second term, that is, the time invariant (or DC) component, will stay as

$$s^{[I]} = C_i \cdot \cos\left(2\pi f_i \frac{2R_o}{c}\right).$$

(6.35)

6.6.2 Q-Channel Processing

The Hilbert (or $-\pi/2$) filter puts a $\pi/2$ radian (or 90°) delay to the pilot signal as

$$\hat{c}(t) = B_i \cdot \cos\left(j2\pi f_i t - \frac{\pi}{2}\right).$$

(6.36)

The multiplier in the Q-channel produces the following output:

$$E(t) \cdot \hat{c}(t) = A_i B_i \cdot \cos\left(2\pi f_i\left(t - \frac{2R_o}{c}\right)\right) \cdot \cos\left(j2\pi f_i t - \frac{\pi}{2}\right)$$

$$= \frac{A_i B_i}{2} \cdot \left[\cos\left(4\pi f_i t - 2\pi f_i \frac{2R_o}{c} - \frac{\pi}{2}\right) + \cos\left(-2\pi f_i \frac{2R_o}{c} + \frac{\pi}{2}\right)\right]$$

(6.37)

$$= \frac{A_i B_i}{2} \cdot \left[\cos\left(4\pi f_i t - 2\pi f_i \frac{2R_o}{c} - \frac{\pi}{2}\right) + \sin\left(-2\pi f_i \frac{2R_o}{c}\right)\right].$$

The low pass filtering opreration filters out the first term and keeps the second DC term as

$$s^{[Q]} = C_i \cdot \sin\left(-2\pi f_i \frac{2R_o}{c}\right)$$

$$= -C_i \cdot \sin\left(2\pi f_i \frac{2R_o}{c}\right).$$

(6.38)

Both channels are processed with an A/D converter such that $s^{[I]}$ and $s^{[Q]}$ signals are digitized for M different frequencies (see Fig. 6.10). Afterward, baseband I and Q signals are summed to give the final output as

$$S_{out}[f_i] = s^{[I]} + s^{[Q]}$$

$$= C_i \cdot \cos\left(2\pi f_i \frac{2R_o}{c}\right) - jC_i \cdot \sin\left(2\pi f_i \frac{2R_o}{c}\right)$$

(6.39)

$$= C_i \cdot \exp\left(-j2\pi f_i\left(\frac{2R_o}{c}\right)\right).$$

The phase of this output signal has the delay of $(2R_o/c)$ compared to the transmitted signal which obviously shows the location of the scatterer. The amplitude of this output signal is directly related to the backscattering field amplitude of the scatterer.

To be able to employ digital processing of the received signal, the data should be sampled and digitized with the help of an AD converter. The range resolution has already been defined as

$$\Delta r = \frac{c}{2B}. \tag{6.40}$$

If the frequency bandwidth is to be digitized to a total of M discrete frequencies, then

$$\Delta f = \frac{B}{M}, \tag{6.41}$$

which is also equal to

$$\Delta f = \frac{c}{2R_{max}}, \tag{6.42}$$

where $R_{max} = M \cdot \Delta r$ is the maximum range or unambiguous range extent seen by the radar. Therefore, the frequency variable f_i can be replaced by the following discrete variable:

$$f_i = f_o + i \cdot \Delta f; \qquad i = 0, 1, 2, \ldots, M - 1. \tag{6.43}$$

Here, f_o denotes the initial or the start-up frequency. The output of the AD converter is the digitized version of the N range profiles, each having a total of M range cells (or range bins). Since the target is in motion in general, normally, Doppler shifts occur between the received pulses. The target may have translational motion and/or rotational motion with respect to radar. In any case, any cross-range point on the image will produce Doppler shifts along the received pulses.

After digitizing the whole received signal, the data can be represented in a 2D form such that the time response of each chirp (or the range profile) is plotted in a column for every pulse received (see Fig. 6.11a).

6.7   RANGE ALIGNMENT

At the end of QD and before applying the azimuth compression to each bin of the range profiles, alignment of range is necessary in most cases. This alignment is applied to compensate for the phenomenon called *range walk*. In the

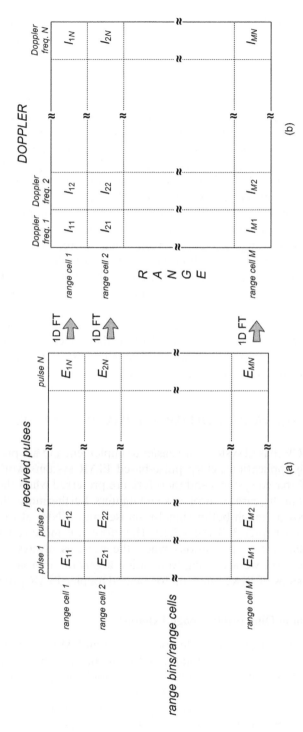

FIGURE 6.11 Formation of the range-Doppler ISAR image via digital processing of received signal.

251

typical ISAR setup, range walk is mainly induced by the target's radial translational motion with respect to radar. The change of range value from profile to profile results in one scatterer to "walk" among the range bins. For constant-velocity targets, the translational velocity of the target can be estimated and the alignment of range profiles can be performed accordingly. In real scenarios, however, the target's motion can be more complex such that it may contain both radial and the tangential components of higher orders. If this is the case, the range alignment is not simple; many motion compensation algorithms have been developed to solve this problem [4–10]. For instance, finding a prominent scatterer in a range profile and tracking it among the other range profiles can sometimes be effective [9, 10]. Various motion compensating algorithms, including the prominent scattering technique, will be covered in Chapter 8.

In some cases when the target's velocity is low and the dwell time is short, the change in the range may stay within the range resolution such that the "walk" settles within a range cell. Therefore, no range correction is needed. In some other cases, when the integration time is sufficiently short such that the target's motion can be approximated to a constant radial velocity, then an effective range alignment method can be applied to attain walk-free range profiles.

When all the range profiles are aligned, the Doppler processing can then be reliably applied by inverse Fourier transforming the collected data for every range cell. This operation provides the final ISAR matrix of the target in the 2D range-Doppler frequency plane as illustrated in Figure 6.11b.

6.8   DEFINING THE RANGE-DOPPLER ISAR IMAGING PARAMETERS

Although the SFCW-based systems are easier to implement and are preferable in ISAR imaging applications, chirp pulse-based ISAR systems work much faster than SFCW-based systems and therefore are preferred when the target is moving fast, as in the case of airplanes and fighters. Furthermore, the chirp pulse systems provide much better signal-to-noise ratio (SNR) at the image output as given in Chapter 3, Section 3.4.1. Therefore, it is more reliable and applicable for real-world applications when the noise is always available and unavoidable. Next, we are going to mention a general approach for the implementation steps of ISAR imaging for range-Doppler ISAR processing.

6.8.1   Image Frame Dimension (Image Extends)

For an ISAR application, the ultimate goal is to get an EM reflectivity of the target; therefore, the frame of the image, that is, the dimensions of the image in the range and cross-range (or Doppler) plane, should be specified to cover the whole target (see Fig. 6.12). It is always safe to select the size of the image frame to be at least two to three times larger than the actual size of target's projection on the range and cross-range frame of $X_p \cdot Y_p$ to avoid aliasing.

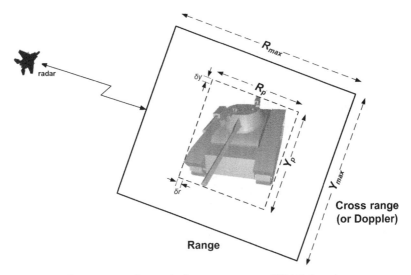

FIGURE 6.12   Some design parameters of ISAR imaging.

6.8.2   Range–Cross-Range Resolution

The range resolution should be selected according to the target's size in the range direction. If the range resolution is selected as δr, the number of range points along the target will have a maximum value of

$$N = \text{floor}\left(\frac{R_{max}}{\delta r}\right), \tag{6.44}$$

where the "floor" function rounds its argument to the lower nearest integer. Similarly, if the cross-range resolution is selected as Δy, the number of cross-range points along the target will have a maximum value of

$$M = \text{floor}\left(\frac{Y_{max}}{\Delta y}\right). \tag{6.45}$$

Therefore, the values of N and M should be selected such that the target features are clearly distinguished with the selected range and cross-range resolutions.

6.8.3   Frequency Bandwidth and the Center Frequency

Based on the decided value for the range resolution, the required frequency bandwidth should be

$$B = \frac{c}{2 \cdot \delta r}. \tag{6.46}$$

For fast processing of the ISAR data, the center frequency can be conveniently selected as at least 10 times the frequency bandwidth, that is, $f_c > 10B$. Otherwise, direct integration of the ISAR integral should be employed or a polar reformatting scheme should be applied, as explained in Chapter 4, Section 4.6.

6.8.4 Doppler Frequency Bandwidth

As listed in Equation 6.20, the Doppler frequency shift resolution of a target that is rotating with an angular velocity of ω with respect to radar is

$$\Delta f_D = \frac{2f_c}{c} \cdot \omega \Delta y. \tag{6.47}$$

Therefore, the total Doppler frequency bandwidth can be found as

$$\begin{aligned} BW_{f_D} &= \Delta f_D \cdot M \\ &= \frac{2f_c}{c} \cdot \omega(\Delta y \cdot M) \\ &= \frac{2f_c}{c} \cdot \omega Y_{max}. \end{aligned} \tag{6.48}$$

6.8.5 PRF

The Doppler frequency bandwidth shown above covers all the cross-range scatterers extending over the cross-range width of Y_{max}. When the chirp pulse radar operation is considered, therefore, the required *PRF* in order to unambiguously sample the scattered field over the cross-range width of Y_{max} is

$$\begin{aligned} PRF &= \frac{1}{T_2} \\ &\geq \frac{2f_c}{c} \cdot \omega Y_{max}. \end{aligned} \tag{6.49}$$

Therefore, the minimum value of this PRF is equal to

$$PRF_{min} = \frac{2f_c}{c} \cdot \omega Y_{max}. \tag{6.50}$$

or the maximum value for the *PRI*

$$PRI_{max} = \frac{c}{2f_c \omega Y_{max}}. \tag{6.51}$$

On the other hand, the minimum value for the PRI is dictated by the target's distance from the radar's position, as the return of the pulse should arrive before the next pulse leaves the transmitter:

$$PRI_{min} = \frac{2R_o}{c}.$$
(6.52)

Therefore, the PRI should be selected between these two limit values to avoid ambiguity in the cross-range dimension.

When the SFCW operation is considered, a total of N pulses are to be sent for a single burst. If the pulse width of a single frequency pulse is T_P,

$$\frac{PRF}{N} = \frac{1}{N \cdot T_P}$$
$$\geq \frac{2f_c}{c} \cdot \omega Y_{max}.$$
(6.53)

Therefore, minimum PRF for stepped frequency radar should be

$$PRF_{min} = N \cdot \frac{2f_c}{c} \cdot \omega Y_{max}.$$
(6.54)

6.8.6 Coherent Integration (Dwell) Time

To achieve the selected cross-range resolution, the value for the total viewing time (or the dwell time) should be equal to the following for chirp pulse illumination:

$$T = N \cdot PRI$$
$$= N \cdot \frac{c}{2f_c \omega Y_{max}}$$
$$= \frac{c}{2f_c \omega \Delta y},$$
(6.55)

which in turn provides a total viewing angle width of

$$\Omega = \omega T.$$
(6.56)

When the SFCW is considered, coherent integration time becomes equal to

$$T = M \cdot N \cdot PRI$$
$$= M \cdot N \cdot \frac{c}{2M \cdot f_c \omega Y_{max}}$$
$$= \frac{c}{2f_c \omega \delta y},$$
(6.57)

which is identical to the result in Equation 6.55. Therefore, SFCW systems should be times N faster than the chirp-pulse systems to have the same dwell time value.

6.8.7   Pulse Width

For the unmodulated (or the single frequency) pulse, the requirement for the minimum value for the duration or the width of the transmitted pulse is

$$T_{pmin} = \frac{R_{max}}{c}. \tag{6.58}$$

On the other hand, the linear frequency modulation within the chirp pulse makes it possible to have a longer pulse width, as explained in Chapter 3, Section 3.4. Therefore, the real pulse width value, T_p, is decided by considering the amount of power to be induced on the pulse. Once T_p is decided, the compression ratio, D, can be chosen according to the following condition:

$$D \leq \frac{T_p}{T_{pmin}}. \tag{6.59}$$

After the compression ratio is found, the chirp rate, K, can be calculated using Equation 3.47, in Chapter 3, as

$$K = \frac{D}{T_p^2} \tag{6.60}$$

so that the chirp pulse

$$s_{tx}(t) \sim \exp\left(j2\pi\left(f_c t + K\frac{t^2}{2} \right) \right), |t| \leq T_p/2 \tag{6.61}$$

is formed to be transmitted toward the target.

6.9   EXAMPLE OF CHIRP PULSE-BASED RANGE-DOPPLER ISAR IMAGING

This example will demonstrate the range-Doppler ISAR image of a target that is illuminated by the chirp pulse radar, as in the scenario illustrated in Figure 6.13. The radar's location is taken as the origin and the target's center is assumed to be located at some (x_o, y_o) point on the 2D coordinate system. The target reflectivity is assumed to be characterized by a number of point scatterers of equal magnitude as their locations in the 2D Cartesian coordinate system are plotted in Figure 6.14. The target platform has a constant velocity of v_x along the x direction. A supposed scenario with the corresponding parameters listed in Table 6.2 is chosen for this example.

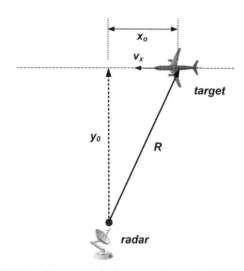

FIGURE 6.13 The scenario for range-Doppler ISAR imaging.

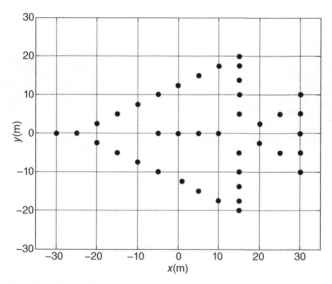

FIGURE 6.14 Fictitious fighter consists of perfect point scatterers of equal reflectivity.

The radar parameters of chirp pulse waveform are listed in Table 6.2. Therefore, the transmitter sends out the following chirp pulses:

$$
s_{tx}(t) \sim \begin{cases} \exp\left[j2\pi(f_o t + K \dfrac{t^2}{2}) \right] & (m-1)\cdot T_2 \leq t \leq (m-1)\cdot T_2 + T_p \\ 0 & elsewhere, \end{cases} \tag{6.62}
$$

TABLE 6.2 Target Simulation Parameters for Chirp Pulse Illumination

Parameter Name	Symbol	Value
Target parameters		
Target's initial position in x	x_o	0 m
Target's initial position in y	y_o	24 km
Target's velocity along x	v_x	120 m/s
Radar parameters		
Center frequency of chirp	f_c	10 GHz
Frequency bandwidth of chirp	B	2.5 GHz
Pulse duration of single pulse	T_P	0.4 μs
Chirp rate	$K \triangleq \dfrac{T_P}{B}$	6.25 e15 s/Hz

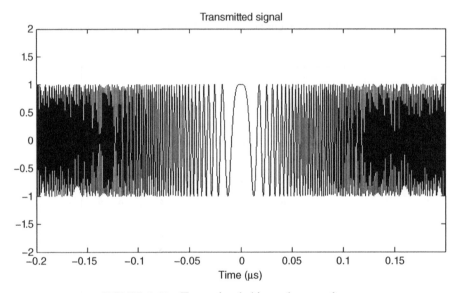

FIGURE 6.15 Transmitted chirp pulse waveform.

where

$$m = 1: N_p: \text{pulse number variable}$$

$$T_2 = 1/PRF \triangleq PRI: \text{pulse repetition interval.}$$

The transmitted chirp pulse waveform is plotted in Figure 6.15. The received signal from a point scatterer at R_o away from the radar will be in the form of the following:

$$s_{rx}(t) \sim \begin{cases} \exp\left[j2\pi(f_o\left(t - \dfrac{2R}{c}\right) + K\dfrac{\left(t - \dfrac{2R}{c}\right)^2}{2}) \right] & \begin{aligned} & (m-1)\cdot T_2 + \dfrac{2R}{c} \leq t \\ & \leq (m-1)\cdot T_2 + \dfrac{2R}{c} + T_p \end{aligned} \\ 0 & \textit{elsewhere,} \end{cases} \quad (6.63)$$

where

$$R = \left((x_o - v_x \cdot t)^2 + y_o^2\right)^{1/2}. \quad (6.64)$$

As is obvious from Equation 6.63, the path of the EM wave, R, is changing with time because of the motion of the target. Therefore, frequency of the received signal, s_{rx}, will have some Doppler shift components as expected. It is also important to note that there will be some additive noise due to the clutter from the scene, atmospheric effects, and from the electronic circuitry of the radar. Therefore, the total received signal at the receiver, $g_{rx}(t)$, will be the summation of the received signal plus the additive noise as

$$g_{rx}(t) = s_{rx}(t) + n(t). \quad (6.65)$$

It is preferable to have as high an SNR as possible; however, this cannot be achieved most of the time. Therefore, in this example, a very bad SNR value of 0.0013 (or -28.93 dB) is considered. The additive noise signal is chosen to be white Gaussian.

To be able to process the received signal digitally, the received pulses should be sampled in time. Since the bandwidth of the transmitted signal is B, this corresponds to a minimum sampling time interval of

$$t_s = \frac{1}{B}. \quad (6.66)$$

Therefore, each pulse is sampled by N_{sample} points of

$$N_{sample} \cong \frac{T_P}{t_s}. \quad (6.67)$$

After this sampling process, the received data can be written as a 2D matrix of M_p times N_{sample}. The range compression now can be applied for each digitized pulse by applying a pulse compression procedure that includes matched filtering of each pulse return with the replica of the original pulse, as explained in detail in Chapter 3, Section 3.5. The matched filter response of

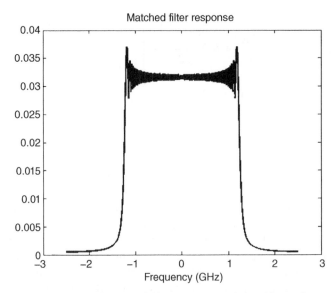

FIGURE 6.16   Matched filter response of the chirp pulse.

the transmitted chirp pulse is depicted in Figure 6.16. This signal is used as the frequency-domain replica of the transmitted pulse to be used in matched filtering process. After applying the matched filter operation, the resultant range compressed data are plotted in Figure 6.17 where the range profile for each azimuth time instant can be easily observed at different range bins. The effect of additive noise can also be observed from the image as the noise shows up as a clutter all around the image. The range profiles for different azimuth time instants are well aligned, thanks to high PRF rate of 3000. Therefore, no range alignment procedure is needed before proceeding to the azimuth compression. If the PRF was lower, dwell time, $T = M_p/PRF$, will be longer, and the range profiles for different received pulses may not be lined up; therefore, range alignment procedure should be employed before applying the processing in the azimuth dimension.

Finally, an IFT operation is performed along the pulse index so that the points in the cross-range dimension can be resolved as they appeared in the different Doppler frequency shift values. Therefore, the final ISAR image is obtained on the range-Doppler plane as depicted in Figure 6.18 where the point scatterers on the target are resolved well in range direction and fairly well in the cross-range direction due to some finite velocity of the target along the azimuth direction. The noise in the receiver is highly suppressed, thanks to matched filtering. Although the received noise energy was about 29 dB higher than that of the received signal energy, the noise level of the image is at least

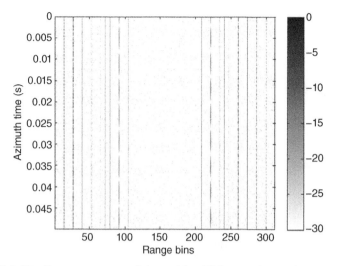

FIGURE 6.17   Range compressed data with additive random noise are present.

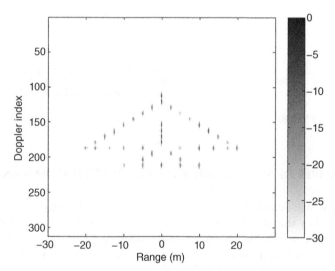

FIGURE 6.18   Range-Doppler ISAR image of the target with additive random noise is present.

25 dB lower than the target image points; this outcome is easily observed from Figure 6.18. The range cross–range ISAR image in Figure 6.19 can only be constructed provided that the angular velocity of the target is known or esti- mated. Then, the transformation from Doppler frequency shift axis to cross- range axis is accomplished by the formula given in Equation 6.31.

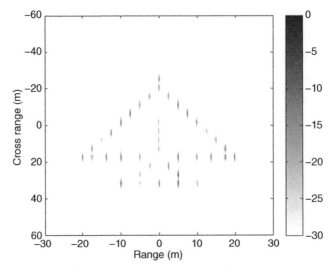

FIGURE 6.19 Range cross–range ISAR image of the target with additive random noise is present.

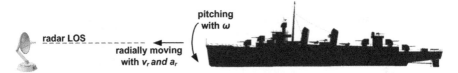

FIGURE 6.20 A geometry for ISAR imaging scenario.

6.10 EXAMPLE OF SFCW-BASED RANGE-DOPPLER ISAR IMAGING

This example will demonstrate the range-Doppler ISAR image of a target that is illuminated by the SFCW-based radar, as in the scenario illustrated in Figure 6.20. For this scenario, the target has both radial and rotational motion components. The target is assumed to have a radial translational velocity component of v_r and a radial translational acceleration of a_r. Furthermore, the target is also assumed to have a rotational velocity component of ω. Both the target and the radar parameters that are used in the simulation is listed in Table 6.3.

The target is assumed to be composed of perfect point scatterers of equal magnitude as the locations of these scattering centers are shown in Figure 6.21. A Matlab code (see Matlab code 6.2 at the end of this chapter) is used to calculate the theoretical backscattered electric field from the target. White Gaussian noise was also added to the calculated field to include the effect of the noise originated by any cause. An SNR value of 3.55 (or 5.50 dB) was assumed during the simulation.

TABLE 6.3   Target Simulation Parameters for SFCW Illumination

Parameter Name	Symbol	Value
Target parameters		
Target's initial position in range	R_0	4 km
Target's radial velocity	v_r	5 m/s
Target's radial acceleration	a_r	0.04 m/s^2
Target's rotational velocity	ω	1.2°/second
Radar parameters		
Starting frequency	f_o	9 GHz
Frequency bandwidth	B	125 MHz
Pulse repetition frequency	PRF	35 KHz
Number of pulses	N_{pulse}	128
Number of bursts	M_{burst}	128

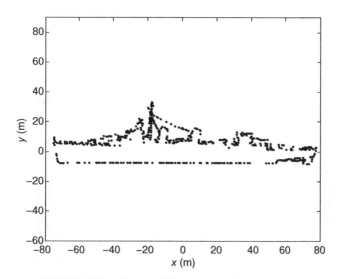

FIGURE 6.21   Target with perfect point scatterers.

First, the range profiles from the target are obtained by applying the 1D IFT operation along the frequency-diverse data. The resultant range profiles for different burst indexes are plotted in Figure 6.22. Since the range profiles seem to be aligned from burst to burst, no range alignment procedure is applied. Then, the IFT operation along the bursts makes it possible to resolve the points in the cross-range direction such that they are lined up in the Doppler frequency shift axis. Therefore, the resultant image is nothing but the 2D ISAR image in the range-Doppler plane as plotted in Figure 6.23.

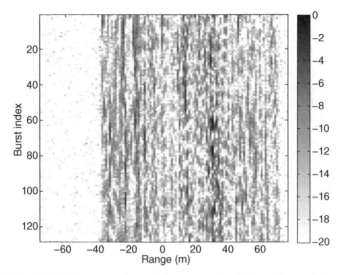

FIGURE 6.22 Range profiles of the target for different burst indexes.

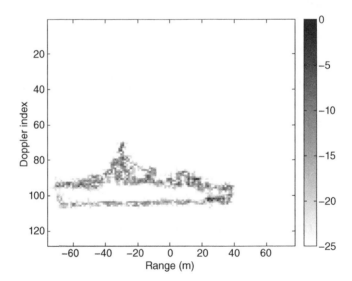

FIGURE 6.23 Range-Doppler ISAR image of the target.

6.11 MATLAB CODES

Below are the Matlab source codes that were used to generate all of the Matlab-produced figures in Chapter 6. The codes are also provided in the CD that accompanies this book.

Matlab code 6.1: Matlab file "Figure6-14thru19.m"

```
%----------------------------------------------------------------
% This code can be used to generate Figure 6.14 thru 19
%----------------------------------------------------------------
% This file requires the following files to be present in the
same
% directory:
%
% fighter.mat
clear all
close all


%---Radar parameters---------------------------------------------
c = 3e8; % speed of EM wave [m/s]
fc = 10e9; % Center frequency of chirp [Hz]
BWf = 2.5e9; % Frequency bandwidth of chirp [Hz]
T1 =.4e-6; % Pulse duration of single chirp [s]

%---target parameters---------------------------------------------
Vx = 120; % radial translational velocity of target [m/s]
Xo = 0e3; % target's initial x coordinate wrt radar
Yo = 24e3; % target's initial y coordinate wrt radar
Xsize = 180; % target size in cross-range [m]
Ysize = 60; % target size in range [m]

%----set parameters for ISAR imaging-----------------------------
% range processing
Ro = sqrt(Xo^2+Yo^2); % starting range distance [m]
dr = c/(2*BWf); % range resolution [m]
fs = 2*BWf; % sampling frequency [Hz]
M = round(T1*fs); % range samples
Rmax = M*dr; % max. range extend [m]
RR = -M/2*dr:dr:dr*(M/2-1); % range vector [m]
Xmax = 1*Xsize; % range window in ISAR [m]
Ymax = 1*Ysize; % cross-range window in ISAR [m]

% Chirp processing
U = Vx/Ro; % rotational velocity [rad/s]
BWdop = 2*U*Ysize*fc/c; % target Doppler bandwith [Hz]
PRFmin = BWdop; % min. PRF
PRFmax = c/(2*Ro); % max. PRF
N = floor(PRFmax/BWdop);% # of pulses
PRF = N*BWdop; % Pulse repetition frequency [Hz]
T2 = 1/PRF; % Pulse repetition interval
T = T2*N; % Dwell time [s] (also = N/PRF)

% cross- range processing
dfdop = BWdop/N; % doppler resolution
lmdc = c/fc; % wavelength at fc
```

```
drc = lmdc*dfdop/2/U; % cross-range resolution
RC = -N/2*drc:drc:(N/2-1)*drc; % cross-range vector

%---load the coordinates of the scattering centers on the fighter------
load fighter
%---Figure 6.14-----------------------------------------------------------
h = figure;
plot(-Xc,Yc,'o', 'MarkerSize',8,'MarkerFaceColor', [0, 0,
1]);grid;
set(gca,'FontName', 'Arial', 'FontSize',12,'FontWeight',
'Bold');
axis([-35 35 -30 30])
xlabel('X [m]'); ylabel('Y [m]');
%--- sampling & time parameters ---------------------------
dt = 1/fs; % sampling time interval
t = -M/2*dt:dt:dt*(M/2-1); % time vector along chirp pulse
XX = -Xmax/2:Xmax/(M-1):Xmax/2;
F = -fs/2:fs/(length(t)-1):fs/2; % frequency vector

slow_t = -M/2*dt:T2: -M/2*dt+(N-1)*T2;

%--- transmitted signal -----------------------------------
Kchirp = BWf/T1; % chirp pulse parameter
s = exp(j*2*pi*(fc*t+Kchirp/2*(t.^2)));% original signal
sr =exp(j*2*pi*(fc*t+Kchirp/2*(t.^2)));% replica
H = conj(fft(sr)/M); % matched filter transfer function

%---Figure 6.15-----------------------------------------------------------
h = figure;
plot(t*1e6, s, 'k','LineWidth',0.5)
set(gca,'FontName', 'Arial', 'FontSize',12,'FontWeight',
'Bold');
title('transmitted signal');
xlabel(' Time [\mus]')
axis([min(t)*1e6 max(t)*1e6 -2 2 ]);

%---Figure 6.16-----------------------------------------------------------
h = figure;plot(F*1e-9,abs(fftshift(H)), 'k','LineWidth',2)
set(gca,'FontName', 'Arial', 'FontSize',12,'FontWeight',
'Bold');
title('Matched filter response');
xlabel(' Frequency [GHz]')

%--- Received Signal --------------------------------------
for n=1: N
 Es(n,1:M) =zeros(1,M);
 for m =1: length(Xc);
 x = Xo+Xc(m)-Vx*T2*(n-1);
 R = sqrt((Yo+Yc(m))^2+x^2);
```

```
 Es(n,1:M) =
Es(n,1:M)+exp(j*2*pi*(fc*(t-2*R/c)+Kchirp/2*((t-2*R/c).^2)));
 end
 % define noise
 noise=5*randn(1,M);
 NS(n,1:M)=noise;

 % Matched filtering
 EsF(n,1:M) = fft(Es(n,1:M)+noise)/M;
 ESS(n,1:M) = EsF(n,1:M).*H;
 ESS(n,1:M) = ifft(ESS(n,1:M));
end;

E_signal = sum(sum(abs(EsF.^2)));
E_noise = sum(sum(abs(NS.^2)));
SNR = E_signal/E_noise;
SNR_db = 10*log10(SNR);

%---Figure 6.17---------------------------------------------------------
rd=30; % dynamic range of display
h=figure;
matplot2(1:N,slow_t,(ESS),rd);
colormap(1-gray);
colorbar
set(gca,'FontName', 'Arial', 'FontSize',12,'FontWeight',
'Bold');
xlabel('Range bins');
ylabel('Azimuth time [s]');

%---Figure 6.18---------------------------------------------------------
h=figure;
matplot2(RC,1:N,fftshift(fft(ESS.*win)),rd);
colormap(1-gray);
colorbar
set(gca,'FontName', 'Arial', 'FontSize',12,'FontWeight',
'Bold');
xlabel('Range [m]');
ylabel('Doppler index');

%---Figure 6.19---------------------------------------------------------
win = hanning(N)*ones(1,M); % prepare window in cross-range
direction
h=figure;
matplot2(RC,XX,fftshift(fft(ESS.*win)),rd);
colormap(1-gray);
colorbar
set(gca,'FontName', 'Arial', 'FontSize',12,'FontWeight',
'Bold');
xlabel('Range [m]');
```

```
ylabel('Cross-Range [m]');
axis([-30 30 -60 60])
```

Matlab code 6.2: Matlab file "Figure6-21thru23.m"

```
%------------------------------------------------------------
% This code can be used to generate Figure 6.21 thru 6.23
%------------------------------------------------------------
% This file requires the following files to be present in the
same
% directory:
%
% CoutUssFletcher.mat
clear all
close all
clc

%---Radar parameters-----------------------------------------
pulses = 128; % # no of pulses
burst = 128; % # no of bursts
c = 3.0e8; % speed of EM wave [m/s]
f0 = 9e9; % Starting frequency of SFR radar system [Hz]
bw = 125e6; % Frequency bandwidth [Hz]
T1 = (pulses-1)/bw; % Pulse duration [s]
PRF = 35e3; % Pulse repetition frequency [Hz]
T2 = 1/PRF; % Pulse repetition interval [s]

%---target parameters-----------------------------------------
theta0 = 0; % Initial angle of target's wrt target [degree]
w = 1.2; % Angular velocity [degree/s]
Vr = 5.0; % radial velocity of EM wave [m/s]
ar = 0.04; % radial accelation of EM wave [m/s^2]
R0 = 4e3; % target's initial distance from radar [m]
dr = c/(2*bw); % range resolution [m]
W = w*pi/180; % Angular velocity [rad/s]

%---load the coordinates of the scattering centers on the fighter------
load CoutUssFletcher

%---Figure 6.21-----------------------------------------------
n = 10;
Xc =(xind(1:n:6142)-93.25)/1.2;
Yc =-zind(1:n:6142)/1.2;
h = figure;
plot(Xc,-Yc,'o','MarkerSize',3,'MarkerFaceColor', [0, 0, 1]);
set(gca,'FontName', 'Arial', 'FontSize',12,'FontWeight',
'Bold');
axis([-80 80 -60 90])
xlabel('X [m]');
ylabel('Y [m]');
```

```
%---Scattering centers in cylindirical coordinates-----------------------
[theta,r] = cart2pol(Xc,Yc);
Theta = theta+theta0*0.017455329; %add initial angle
i = 1:pulses*burst;
T = T1/2+2*R0/c+(i-1)*T2;%calculate time vector
Rvr = Vr*T+(0.5*ar)*(T.^2);%Range Displacement due to radial
vel. & acc.
Tetw = W*T;% Rotational Displacement due to angular vel.


i = 1:pulses;
df = (i-1)*1/T1; % Frequency incrementation between pulses
k = (4*pi*(f0+df))/c;
k_fac = ones(burst,1)*k;


%------Calculate backscattered E-field------------------------------------
 Es(burst,pulses)=0.0;
 for scat=1:1:length(Xc);
 arg = (Tetw - theta(scat) );
 rngterm = R0 + Rvr - r(scat)*sin(arg);
 range = reshape(rngterm,pulses,burst);
 range = range.';
 phase = k_fac.* range;
 Ess = exp(j*phase);
 Es = Es+Ess;
 end
 Es = Es.';

% define noise
noise=10*randn(burst,pulses);

E_signal = sum(sum(abs(Es.^2)));
E_noise = sum(sum(abs(noise.^2)));
SNR = E_signal/E_noise
SNR_db = 10*log10(SNR)

Es = Es+noise.';

%---Figure 6.22----------------------------------------------------------
% Check out the range profiles
X = -dr*((pulses)/2-1):dr:dr*pulses/2;Y=X/2;
RP = fft((Es.'));
RP = fftshift(RP,1);
h = figure;
matplot2(X,1:burst,RP.',20);
colormap(1-gray);
colorbar;
set(gca,'FontName', 'Arial', 'FontSize',12,'FontWeight',
'Bold');
xlabel('Range [m]');
ylabel('Burst index');
```

```
%Form ISAR Image (no compansation)
%---Figure 6.23-----------------------------------------------
ISAR = abs(fftshift(fft2((Es))));
h = figure;
matplot2(X,1:burst,ISAR(:,pulses:-1:1),25);
colormap(1-gray);
colorbar;
set(gca,'FontName', 'Arial', 'FontSize',12,'FontWeight',
'Bold');
xlabel('Range [m]');
ylabel('Doppler index'); _____
```

REFERENCES

1 http://en.wikipedia.org/wiki/Sea_state

2 A. W. Doerry. Ship dynamics for maritime ISAR imaging, Technical Report, Sandia National Laboratories, SAND2008-1020, February. 2008.

3 J. C. Curlander and R. N. McDonough. *Synthetic aperture radar systems and signal processing.* John Wiley and Sons, New York, 1991.

4 J. C. Kirk. Motion compensation for synthetic aperture radar. *IEEE Trans Aerosp Electron Syst* 11 (1975) 338–348.

5 H. Wu, et al. Translational motion compensation in ISAR image processing. *IEEE Trans Image Process* 14(11) (1995) 1561–1571.

6 C. C. Chen and H. C. Andrews. Target-motion-induced radar imaging. *IEEE Trans Aerosp Electron Syst* 16(1) (1980) 2–14.

7 T. Itoh, H. Sueda, and Y. Watanabe. Motion compensation for ISAR via centroid tracking. *IEEE Trans Aerosp Electron Syst* 32(3) (1996) 1191–1197.

8 X. Li, G. Liu, and J. Ni. Autofocusing of ISAR images based on entropy minimization. *IEEE Trans Aerosp Electron Syst* 35(4) (1999) 1240–1251.

9 T. Küçükkılıç. Isar imaging and motion compensation, MS thesis, Middle East Technical University, 2006.

10 Y. Wang, H. Ling, and V. C. Chen. ISAR motion compensation via adaptive joint time-frequency technique. *IEEE Trans Aerosp Electron Syst* 34(2) (1998) 670–677.

CHAPTER SEVEN

Scattering Center Representation of Inverse Synthetic Aperture Radar

In radar imaging, the scattering center concept provides various advantages, especially when dealing with radar cross sections of objects and synthetic aperture radar/inverse synthetic aperture radar (SAR/ISAR) imaging. After the object is illuminated by an electromagnetic (EM) wave, some locations on the object provide localized radiation/scattering energy toward the observation point. The locations of these concentrated sources of scattering energy are called *scattering centers*. The use of the scattering center model provides a very sparse representation of ISAR imagery such that EM scattering from complex bodies can be modeled as if it is emitting from a discrete set of points on the target [1–5] as illustrated in Figure 7.1. Such representation is so powerful that it may provide many advantages, including the following:

1. A simple and sparse representation of the EM scattering and/or SAR/ISAR imagery can be obtained.
2. Since the model is sparse, the scattering data and/or SAR/ISAR image can be compressed with high data compression ratios.
3. Because of the fact that the new data set is much smaller than the original data set, the reconstruction of the scattering data and/or SAR/ISAR image can be obtained quickly.
4. The model can be used to interpolate the scattering data and/or SAR/ISAR image with infinite resolution.

Inverse Synthetic Aperture Radar Imaging with MATLAB Algorithms, First Edition.
Caner Özdemir.
© 2012 John Wiley & Sons, Inc. Published 2012 by John Wiley & Sons, Inc.

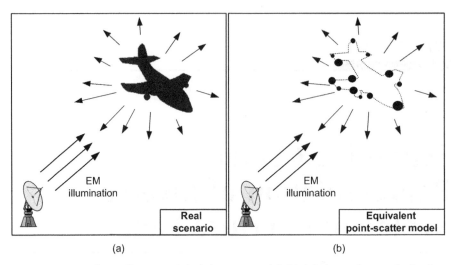

(a)   (b)

FIGURE 7.1   Point-radiator model of the scattered field: (a) the real scenario for far-field scattering from a target, (b) equivalent point-scatterer model based on scattering centers.

5. The model can be used to extrapolate the scattering data and/or SAR/ISAR image within some finite bandwidths of frequencies and angles.
6. The model provides insights to the scattering mechanisms that may lead to understanding the cause-and-effect relationship of the scattering phenomena from the target.

The scattering center representation is usually implemented through a parameterization scheme based on the point-scatterer model. Therefore, the model is first defined and the model parameters are extracted from the scattered field data and/or SAR/ISAR image afterward by applying an extraction algorithm. Once the model is properly extracted, the data can be reconstructed and/or interpolated/extrapolated by applying a reconstruction routine. All these steps will be explained throughout this chapter.

7.1   SCATTERING/RADIATION CENTER MODEL

This model is motivated by the observation that an ISAR image exhibits strong point-scatterer-like behavior as can easily be seen from the ISAR images in Chapters 4–6. It is well known that the backscattered signature can be modeled by a very sparse set of point radiators called *scattering centers* on the target [1–9]. Since these scattering centers can also be regarded as the secondary point radiators, they are also called *radiation centers* [8, 9] in some applications.

When the incident field impinges on the object, surface currents flow on the surface of the object. At some regions of the target, these surface currents interfere constructively (i.e., in-phase), and the resultant scattering wave amplitude becomes greater. These dominant sources of radiation or scattering are called *hot spots* or *scattering centers* in the radar literature. Corner reflector-type structures and planar surfaces are capable of providing strong specular scattering that can be regarded as the scattering centers. In some parts of the target, however, surface currents interfere destructively (i.e., out-of-phase), and the overall scattering amplitude is decreased. Planar regions on the target (if they do not constitute any specular scattering) provide very little scattering. Those parts of target are sometimes called *cold regions* or *ghost regions*.

The main idea in the scattering center representation comes from the fact that the scattered field from a target can be approximated as if it is coming from a finite number of point scatterers on the target (Fig. 7.1). These point scatterers happen to be around the hot spots on the target. With this model, these scattering centers are responsible for all the scattering energy as the scattered field can be parameterized by a finite set of discrete point scatterers as

$$
\begin{aligned}
E^s(k, \emptyset) &\cong \sum_{n=1}^{N} A_n \cdot e^{-j2\vec{k}\cdot\vec{r}_n} \\
&= \sum_{n=1}^{N} A_n \cdot e^{-j2(k_x \cdot x_n + k_y \cdot y_n)} \\
&= \sum_{n=1}^{N} A_n \cdot e^{-j2k(cos\emptyset \cdot x_n + sin\emptyset \cdot y_n)}.
\end{aligned}
\tag{7.1}
$$

Here, $E^s(k, \emptyset)$ represents the backscattered electric field collected at different frequencies and angles. A_n is the complex amplitude of the nth scattering center and $\vec{r}_n = x_n \cdot \hat{x} + y_n \cdot \hat{y}$ is the displacement vector of the location of the nth scattering center. Equation 7.1 suggests that the total scattered field can be written as the sum of N different point radiators at different (x_n, y_n) locations.

This point radiator or scattering center model is more conveniently applied in the image domain rather than in the Fourier domain. Therefore, the employment of the scattering center model is more meaningful in the image domain since the ISAR image itself consists of finite number of point scatterers together with their point spread functions (PSFs). Therefore, an ISAR image is conveniently parameterized via a finite set of point radiators as

$$
ISAR(x, y) \cong \sum_{n=1}^{N} A_n \cdot h(x - x_n, y - y_n),
\tag{7.2}
$$

where A_n is the complex amplitude, (x_n, y_n) is the location of the nth scattering center, and $h(x, y)$ is called the PSF or *ray spread function* whose formula has already been derived in Chapter 5, Equation 5.26 as

$$h(x, y) = \left(e^{j2k_{xc} \cdot x} \frac{BW_{k_x}}{\pi} \operatorname{sinc}\left(\frac{BW_{k_x}}{\pi} x \right) \right) \cdot \left(e^{j2k_{yc} \cdot y} \frac{BW_{k_y}}{\pi} \operatorname{sinc}\left(\frac{BW_{k_y}}{\pi} y \right) \right), \quad (7.3)$$

where BW_{k_x} and BW_{k_y} are the finite bandwidths in k_x and k_y, respectively. Here, k_{xc} and k_{yc} are the center spatial frequencies in x and y directions. When the small bandwidth and the small angle approximation holds true, the above PSF can be rewritten in terms of the frequency and the angle variables as

$$h(x, y) = \left(e^{j \frac{4\pi f_c}{c}(x + \emptyset_c y)} \cdot \frac{4 f_c \cdot B \cdot \Omega}{c^2} \right) \cdot \operatorname{sinc}\left(\frac{2B}{c} x \right) \cdot \operatorname{sinc}\left(\frac{2 f_c \cdot \Omega}{c} y \right). \quad (7.4)$$

In the above equation, B and Ω are the frequency bandwidth and the look-angle width, respectively. Also, f_c and $\emptyset_c$ are the center values for the frequency bandwidth and aspect width, respectively.

7.2 EXTRACTION OF SCATTERING CENTERS

Once the model is defined, these scattering centers can now be extracted from the image together with their corresponding PSFs. Although it is more convenient and practical to extract these point radiators or scattering/radiation centers directly in the image domain, it is also possible to convey the extraction process in the Fourier (or frequency-aspect) domain. Because the ISAR image itself is the display of these point radiators, it is easier to implement the extraction procedure in the image domain. As we shall see next, it is computationally more desirable as well.

7.2.1 Image Domain Formulation

7.2.1.1 Extraction in the Image Domain   There are a few methods used to extract the scattering centers from the ISAR image [4]. Among those, the common image processing algorithm used for the extraction of scattering/radiation centers is the well-known CLEAN algorithm. The CLEAN algorithm was first introduced to perform deconvolution on images created in radio astronomy [10, 11]. It has also been successfully utilized for scattering/radiation center extraction [9, 12, 13] in radar imaging applications. CLEAN is a robust, iterative procedure that successively picks out the highest point in the image, assumes it is a point radiator (or a scattering center) with the corresponding amplitude, and removes its point spread response from the image. At the nth iteration of CLEAN, therefore, if A_n is the strength of the highest point in the image with the location of (x_n, y_n), the two-dimensional (2D) residual image can be written as

$$\begin{bmatrix} 2D\ residual \\ image \end{bmatrix}^n = \begin{bmatrix} 2D\ residual \\ image \end{bmatrix}^{n-1} - A_n \cdot h(x - x_n, y - y_n), \quad (7.5)$$

that is to say, "2D residual image at the nth step is equal to 2D residual image at the $(n-1)$th step minus the highest scattering center strength with its corresponding PSF in the residual image for the nth step." The extraction process is iteratively continued until the maximum strength in the residual image reaches a user-defined threshold value. Typically, the energy in the residual image decreases quickly during the initial stages of the iteration and tapers off after reaching the noise floor.

An example for the extraction of scattering centers is demonstrated through Figures 7.2–7.4. The scattering center extraction procedure in Equation 7.5 is applied to the ISAR image shown in Figure 7.2a. The CLEAN algorithm is utilized to extract a total of 250 scattering centers by using the PSF described in Equation 7.3. The locations of extracted scattering centers are plotted in Figure 7.2b. Their sizes are shown with respect to their relative amplitudes. As seen from Figure 7.2, most of the extracted scattering centers happen to be located on the target. However, we also notice that some scattering centers turn out to be out of the outline of the target. This is mostly because of the fact that multibounce mechanisms in ISAR are delayed in the range and dislocated in the cross range, as already discussed in Chapter 4, Section 4.5.3. Also, as the CLEAN algorithm successively extracts the scattering center with its PSF, the sidelobes of the sinc-type PSF may constitute some imaginary scattering centers around the true scattering centers. These scattering centers may be out of the outline of the target, as can be seen from Figure 7.2. However, CLEAN will work succesfully even if imaginary scattering centers are created during the extraction process.

The strengths of the extracted scattering centers are plotted in Figure 7.3 versus the scattering center number in the order of extraction. As seen from the figure, the CLEAN algorithm converges rapidly as the energy in the residual image reduces significantly at each iteration of the extraction procedure. After extracting a sufficient number of scattering centers (about 250 in this example), the energy in the residual image starts to decrease gradually. This situation occurs when the maximum strength in the residual image reaches the noise floor of the original image. Therefore, when this level is reached, the CLEAN process should be terminated. Otherwise, the extraction algorithm would try to clean the noise by the selected point-radiator model, which may continue infinitely without converging. This is because of the fact that the noise (either numerical or measurement) is not composed of point radiators. For that reason, the iterative search in CLEAN is usually halted after reaching a user-defined threshold value (which is usually the noise floor level of the image).

As we have already stated, the scattering center model provides a sparse and simple representation of scattered field patterns and/or ISAR images. This feature will be demonstrated with the help of the above numerical example. The original image size of Figure 7.2a is 128×256 with complex entries. Therefore, this image occupies $128 \times 256 \times 2 \times 8$ bit $= 524,288$ bytes of disk space. However, when the scattering center model is used, one complex entry for the amplitude and two real entries for the location of each scattering center

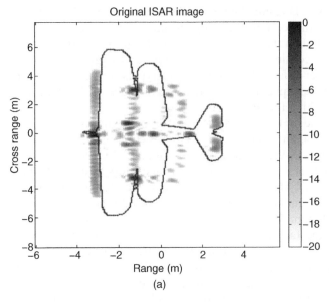

(a)

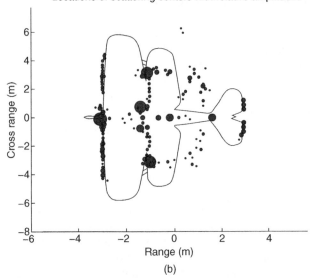

(b)

FIGURE 7.2 (a) Original ISAR image, (b) location of extracted scattering centers displayed with respect to their relative strengths.

are needed. For our example of 250 scattering centers, therefore, a disk space of $250 \times (2 + 2) \times 8$ bit $= 8000$ bytes is required to store all the scattering centers. So, a data compression ratio of approximately 66-to-1 is achieved.

To summarize, it has been shown that a sparse model based on scattering center representation can be constructed. Once available, this model can be

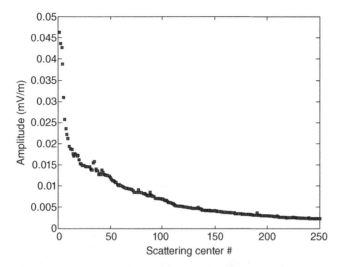

FIGURE 7.3   The amplitudes of the extracted scattering centers.

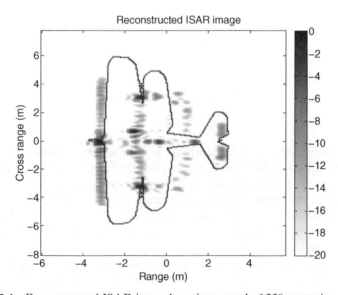

FIGURE 7.4   Reconstructed ISAR image by using a total of 250 scattering centers.

used to reconstruct the scattered field and the ISAR image with a very fine resolution and good fidelity, as will be shown next.

7.2.1.2   Reconstruction in the Image Domain   Once the scattering centers are extracted, the reverse process, that is, reconstruction of the ISAR image, can be readily done in real time by means of the extracted scattering centers.

Furthermore, the Fourier domain (or frequency-aspect) data can also be reconstructed very easily with good fidelity, as we shall see next.

Image Reconstruction   The image can be reconstructed by putting all the extracted scattering centers back into the ISAR image, together with their PSFs, as

$$ISAR_r(x, y) = \sum_{n=1}^{N} A_n \cdot h(x - x_n, y - y_n). \tag{7.6}$$

This reconstruction formula is the same as the extraction formula given in Equation 7.2. Since the parameters (A_n, x_n, y_n) are already obtained after the extraction process, the reconstructed ISAR image can be obtained quickly. An example of a reconstructed ISAR image is shown in Figure 7.4 where a total of 250 extracted scattering centers are used to reconstruct the original ISAR image in Figure 7.2a. Comparison of the original image in Figure 7.2a with the reconstructed image in Figure 7.4 clearly validates that almost perfect reconstruction can be attained with the help of scattering centers. This example, of course, shows the effectiveness of the scattering center model in representing the ISAR image.

Field Reconstruction   The scattered electric field in frequencies and angles can be reconstructed easily by using the extracted scattering centers with the following formula:

$$E_r^s(k, \emptyset) = \sum_{n=1}^{N} A_n \cdot e^{-j2(k_x \cdot x_n + k_y \cdot y_n)}$$
$$= \sum_{n=1}^{N} A_n \cdot e^{-j2k(\cos\emptyset \cdot x_n + \sin\emptyset \cdot y_n)}. \tag{7.7}$$

One will easily notice that this formula is obtained by exploding the point-radiator model given in Equation 7.1. Therefore, the scattering centers are being used in its literal meaning such that extracted scattering centers are being used as the point radiators. Furthermore, this formula is essentially nothing but the 2D Fourier transform of Equation 7.6, as expected. Therefore, the field can be reconstructed directly using the extracted scattering centers and with the help of Equation 7.7.

An example of field reconstruction is demonstrated with the help of Figures 7.5–7.10. In Figure 7.5a, original backscattered field data for the original ISAR image given in Figure 7.2a are shown. These original data are displayed in 2D multifrequency multiaspect plane. Then, by applying the expression in Equation 7.7, the reconstructed field pattern for the same frequency-angle variables is presented in Figure 7.5b. For the reconstruction of the backscattered field pattern, all of the 250 extracted scattering centers were used. Comparing both the original and the reconstructed images visually, agreement between both sets of data can be easily seen. Since the scattered field pattern is reconstructed by the analytical formula in Equation 7.7, the angular or spectral resolution can be selected at any value that offers infinite resolution. In practice, we have

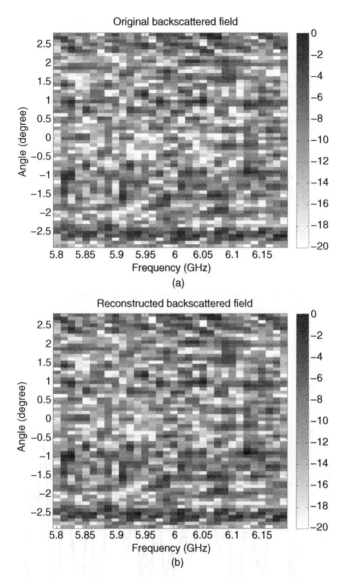

FIGURE 7.5   (a) Original backscattered field for the ISAR example in Figure 7.2, (b) reconstructed field patterns by using a total of 250 scattering centers.

infinite resolution in reconstructing the scattered field values. In Figure 7.6, for example, the same scattering centers were used to reconstruct the backscattered field by sampling both the frequency and aspect variables 10 times more. Image size, therefore, was increased 100 times compared to the original image. As a result, the resolution of this new image is much finer than that of the original image.

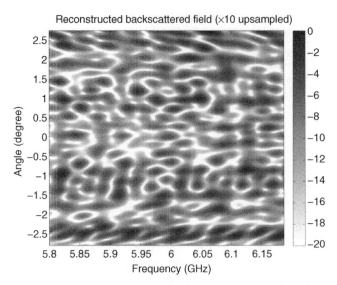

FIGURE 7.6 Reconstructed backscattered field patterns sampled 10 times more compared to the original image.

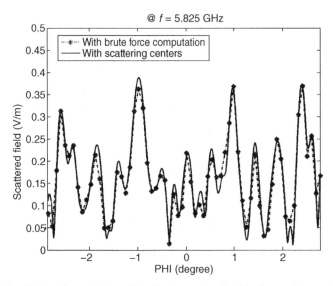

FIGURE 7.7 Comparison of the original pattern obtained by brute-force computation to the reconstructed pattern by scattering center in the angle domain.

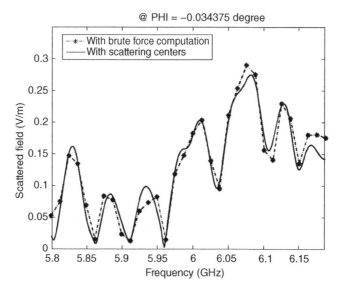

FIGURE 7.8 Comparison of the original pattern obtained by brute-force computation to the reconstructed pattern by scattering center in the frequency domain.

To be able to observe the success of field reconstruction using scattering centers, we will show the agreement between the original and the reconstructed fields in the one-dimensional frequency domain or aspect domain. To demonstrate the results for the same ISAR example, the reconstructed field patterns are compared with the original field patterns in Figures 7.7 and 7.8. First, the aspect field pattern comparison is done for azimuth angles varying from −2.78° to 2.78° at the frequency of 5.825 GHz. The original field data obtained by brute-force computation have 64 distinct points. After using scattering center presentation of ISAR imaging and using the formula in Equation 7.7, the field reconstruction is accomplished for the same frequency and the angle range. Since there is no restriction in choosing the angular granularity in the process of reconstructing, we use 10 times more data points such that there exist a total of 640 data points in the new reconstructed field pattern, as depicted in Figure 7.7. The comparison of the original angular field data computed by brute-force calculation and the reconstructed angular field data computed by using the scattering centers clearly shows the success of the scattering center model in angular field reconstruction. This image clearly demonstrates that the scattering center model can be used to interpolate the original field data with high accuracy. This, of course, provides tremendous computation time-savings for RCS calculation or ISAR image formation with fine resolutions.

The result of another comparison is plotted in Figure 7.8, where original and reconstructed field patterns in the frequency domain are shown in the

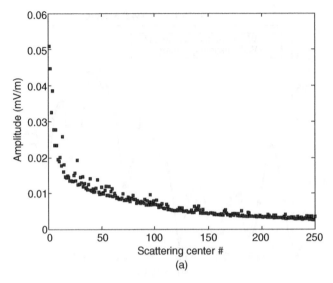

(a)

Locations of scattering centers with relative amplitudes

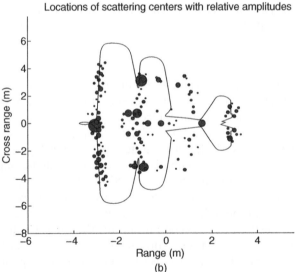

(b)

FIGURE 7.9 (a) Amplitudes of the extracted scattering centers, (b) location of extracted scattering centers displayed with respect to their relative strengths (extraction was done in frequency-aspect domain).

same plot. This time, the radar look angle is kept constant at the azimuth angle of −0.034375° and the frequency is altered from 5.80 GHz to 7.1875 GHz. Again, the reconstruction formula in Equation 7.7 is used to estimate the frequency field pattern that is 10 times more sampled than the original field pattern. The visual comparison of original and reconstructed frequency-domain electric field patterns clearly shows the fidelity of the scattering centers

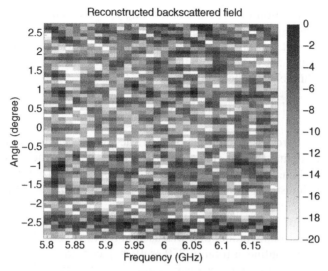

FIGURE 7.10   Reconstructed field patterns by using a total of 250 scattering centers.

in estimating the field pattern. Again, the data between the original data points were successfully interpolated in a short time, thanks to the scattering center model.

7.2.2   Fourier Domain Formulation

7.2.2.1   Extraction in the Fourier Domain   The scattering center representation can also be employed in the the Fourier domain (or frequency-aspect domain). For this case, the scattering centers are to be extracted from the frequency-aspect data. This process is conceptually possible, but requires more computer resources as will be explained next.

The extraction in the Fourier domain can be achieved by the famous method known as "matching pursuit" [14]. In the extraction process, the frequency-aspect field projection of each scattering center is iteratively extracted from the 2D Fourier domain electric field data. During each iteration process, the scattering center location that gives the largest value of the following inner product is recorded:

$$A_n = max_{(x_n, y_n)} \left\{ <E^s(k, \emptyset), \varphi(x, y)> \right\}. \tag{7.8}$$

Here, A_n is the highest value of the inner product between the Fourier domain scattered electric field and the phase term $\varphi(x, y)$ is equal to $e^{-j2(k_x \cdot x_n + k_y \cdot y_n)}$. The inner product in the above equation is defined as the following:

$$<E^s(k, \emptyset), \varphi(x, y)> = \int_{k_L}^{k_H} \int_{\emptyset_L}^{\emptyset_H} E^s(k, \emptyset) \cdot \varphi^*(x, y) d\emptyset dk. \tag{7.9}$$

So, at the nth iteration of the search, the (x_n, y_n) pair that gives the largest value of A_n corresponds to the location of corresponding scattering center. Once the strongest scattering center at the nth iteration is determined, that is, (A_n, x_n, y_n) values are determined, its associated scattered field is subtracted from the residual electric field as

$$\left[E^s(k,\emptyset)\right]_n = \left[E^s(k,\emptyset)\right]_{n-1} - A_n \cdot \varphi(x_n, y_n). \tag{7.10}$$

This extraction process is continued until the energy in the residual field reaches a user-defined threshold value. Usually, this value is the noise floor of the collected data. The main problem in this extraction routine is that it requires an exhaustive search over 2D data for every iteration of the search process.

An example of Fourier domain extraction is demonstrated via the same data set for the airplane model that was used in Section 7.2.1. By applying the matching pursuit routine, a total of 250 scattering centers are extracted from the multifrequency, multiaspect backscattered electric field data. In Figure 7.9a, amplitudes of the extracted scattering centers are given. During the Fourier domain extraction, the original data grid is four times more sampled to better estimate the locations of the scattering centers. However, the iterative search takes more time. If the data grid is sampled more densely, the amplitudes of the scattering centers converge faster. This is in turn at the price of increased computation time of the extraction algorithm. The locations of extracted scattered centers are displayed in Figure 7.9b with respect to their relative amplitudes.

7.2.2.2 *Reconstruction in the Fourier Domain*   Once the scattering centers are extracted, the reconstruction either in the Fourier domain or in the image domain can be done using (A_n, x_n, y_n) values for each scattering centers. Both reconstruction routines will be given next.

Field Reconstruction   The scattered electric field in frequencies and angles can be reconstructed by putting back the field projections of all extracted scattering centers by using the following formula:

$$E_r^s(k,\emptyset) = \sum_{n=1}^{N} A_n \cdot e^{-j2k(cos\emptyset \cdot x_n + sin\emptyset \cdot y_n)}. \tag{7.11}$$

Provided that the extraction has been employed such that most of the energy has been extracted from the original field data, the reconstruction field should be almost equal to the original field, that is, $E_r^s(k,\emptyset) \cong E^s(k,\emptyset)$. An example of reconstruction field pattern is demonstrated in Figure 7.10 where the 2D frequency-aspect backscattered field is reconstructed using the 250 scattering centers for the same airplane model. The comparison of the reconstructed field in Figure 7.10 with the original field in Figure 7.5a clearly demonstrates the

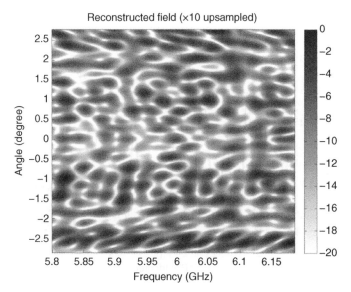

FIGURE 7.11 Reconstructed field pattern (10 times more sampled) gives more detailed estimate of the field pattern in between the original field data points.

success of the reconstruction process. Since there is no limit in reconstructing the field at a desired granularity, the reconstruction can be employed at very fine resolutions. In Figure 7.11, the frequency-aspect backscattered data for the airplane model are reconstructed by the same 250 scattering centers with 10 times more sampled data points in each domain. Therefore, the field value in between the original data points can be estimated (or interpolated) with good fidelity. The demonstration of this phenomenon is given in Figure 7.12. The field data comparison over angles between the original field data and the reconstructed field data (10 times more sampled) is shown in Figure 7.12a. Similarly, a comparison of field patterns with the original and the reconstructed (10 times more sampled) over frequencies is plotted in Figure 7.12b. As seen from both figures, the model successfully predicts the data in between points.

Image Reconstruction   A fast and convenient way of reconstructing the ISAR image by using scattering centers extracted in the Fourier domain is by taking the 2D inverse Fourier transform (IFT) of the reconstructed field pattern. Therefore, the reconstructed ISAR image can be readily formed by

$$ISAR_r(x, y) = \mathcal{F}_2^{-1}\left\{E_r^s(k, \emptyset)\right\}. \qquad (7.12)$$

An example of the reconstructed ISAR image using the scattering centers extracted in the Fourier domain is shown in Figure 7.13. In this figure, a 2D reconstructed ISAR image of the same airplane model is obtained by applying

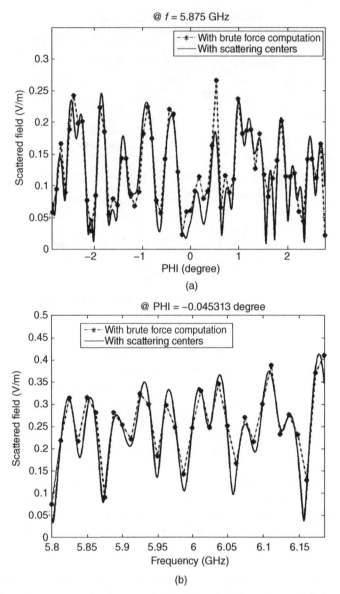

FIGURE 7.12 Comparison of the original pattern obtained by brute-force computation to the reconstructed pattern obtained by scattering center extraction in the Fourier domain: (a) comparison in the azimuth angle domain, (b) comparison in the frequency domain.

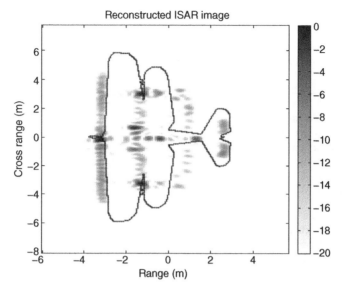

FIGURE 7.13   Reconstructed ISAR image by using a total of 250 scattering centers.

the formula in Equation 7.12. If this reconstructed image is compared with the original ISAR image in Figure 7.2a, one can see that an almost perfect match between the two, with only small discrepancies, is achieved.

7.3   MATLAB CODES

Below are the Matlab source codes that were used to generate all of the Matlab-produced Figures in Chapter 7. The codes are also provided in the CD that accompanies this book.

Matlab code 7.1: Matlab file "Figure7-2thru7-8.m"
```
%------------------------------------------------------------------
% This code can be used to generate Figure 7.2 thru 7.8
%------------------------------------------------------------------
% This file requires the following files to be present in the
same
% directory:
%
% Es60.mat
% planorteta60_2_xyout.mat
clear all
close all
```

```
c = .3; % speed of light
fc = 6; % center frequency
phic = 0*pi/180; % center of azimuth look angles
thc = 90*pi/180; % center of elevation look angles

%_____PRE PROCESSING OF ISAR_____
BWx = 12; % range extend
M = 32; % range sampling
BWy = 16; % xrange extend
N = 64; % xrange sampling

dx = BWx/M; % range resolution
dy = BWy/N; % xrange resolution

% Form spatial vectors
X = -dx*M/2:dx:dx*(M/2-1);
Y = -dy*N/2:dy:dy*(N/2-1);
XX =-dx*M/2:dx/4:-dx*M/2+dx/4*(4*M-1);
YY =-dy*N/2:dy/4:-dy*N/2+dy/4*(4*N-1);

%Find resoltions in freq and angle
df = c/(2*BWx); % frequency resolution
dk = 2*pi*df/c; % wavenumber resolution
kc = 2*pi*fc/c;
dphi = pi/(kc*BWy);% azimuth resolution

%Form F and PHI vectors
F = fc+[-df*M/2:df:df*(M/2-1)]; % frequency vector
PHI = phic+[-dphi*N/2:dphi:dphi*(N/2-1)];% azimuth vector
K = 2*pi*F/c; % wavenumber vector
dk = K(2)-K(1); % wavenumber resolution

%_____GET THE DATA_____
load Es60 % load E-scattered
load planorteta60_2_xyout.mat % load plane outline
% ISAR
ISAR = fftshift(fft2(Es));
ISAR = ISAR/M/N; % the image

% ISAR 4x UPSAMPLED-------------------
Enew = Es;
Enew(M*4,N*4) = 0;
ISARnew = fftshift(fft2(Enew));
ISARnew = ISARnew/M/N;

%_____2D-CLEAN_____
% prepare 2D sinc functions
 sincx = ones(1,M);
 sincx(1,M+1:M*4) = 0;
 hsncF = fft(sincx)/M;
```

```matlab
sincy=ones(1,N);
sincy(1,N+1:N*4) = 0;
hsncPHI = fft(sincy)/N;

%initilize
hh = zeros(4*M,4*N);

ISARres = ISARnew.';
Amax = max(max(ISARnew));
ISARbuilt = zeros(N*4,M*4);

% loop for CLEAN
for nn=1:250,
[A,ix] = max(max(ISARres));
[dum,iy] = max(max(ISARres.'));
hsincX = shft(hsncF,ix);
hsincY = shft(hsncPHI,iy);
hhsinc = hsincX.'*hsincY;
ISARres = ISARres-A*hhsinc.';
SSs(nn,1:3) = [A XX(ix) YY(iy)];
II=ISARres;
II(1,1) = Amax;
% Image Reconstruction
ISARbuilt = ISARbuilt-A*hhsinc.';
end

%_____IMAGE COMPARISON_____
%---Figure 7.2(a)---------------------------------------------
h = figure;
matplot(X,Y,abs(ISARnew(4*M:-1:1,:).'),20);
colorbar;
colormap(1-gray);
set(gca,'FontName', 'Arial', 'FontSize',14,'FontWeight',
'Bold');
line( -xyout_xout,xyout_yout,'Color','k','LineStyle','.');
xlabel('Range [m]');
ylabel('X-Range [m]');
title('Original ISAR image')

%---Figure 7.4-----------------------------------------------
h = figure;
matplot(X,Y,abs(ISARbuilt(:,4*M:-1:1)),20);
colorbar;
colormap(1-gray);
set(gca,'FontName', 'Arial', 'FontSize',14,'FontWeight',
'Bold');
line( -xyout_xout,xyout_yout,'Color','k','LineStyle','.');
xlabel('Range [m]');
ylabel('X-Range [m]');
```

```
title('Reconstructed ISAR image')
%_____SCATTERING CENTER INFO
DISPLAY_____

%---Figure 7.3------------------------------------------------
h = figure;
plot(abs(SSs(:,1)),'square', 'MarkerSize',4,'MarkerFaceColor',
[0, 0, 1]);
set(gca,'FontName', 'Arial', 'FontSize',14,'FontWeight',
'Bold');
xlabel('Scattering Center #');
ylabel('Amplitude [mV/m]');
%---Figure 7.2(b)---------------------------------------------
h = figure;
hold
for m = 1:150
  t = round(abs(SSs(m,1))*20/abs(SSs(1,1)))+1
  plot(-SSs(m,2),SSs(m,3),'o', 'MarkerSize',t,'MarkerFaceColor
', [0, 0, 1]);
end
hold
line(-xyout_xout,xyout_yout,'Color','k','LineStyle','.');
axis([min(X) max(X) min(Y) max(Y)])
set(gca,'FontName', 'Arial', 'FontSize',14,'FontWeight',
'Bold');
xlabel('Range [m]');
ylabel('X-Range [m]');
title('Locations of scattering centers with relative
amplitudes ')

%_____RECONSTRUCT THE FIELD
PATTERN_____
ESR = zeros(320,640);
ESr = zeros(32,64);
k = K;
kk = k(1):(k(32)-k(1))/319:k(32);
pp = PHI(1):(PHI(64)-PHI(1))/639:PHI(64);
for nn = 1:250;
  An = SSs(nn,1);
  xn = SSs(nn,2);
  yn = SSs(nn,3);
  ESR = ESR+An*exp(j*2*xn*(kk-k(1)).')*exp(j*2*kc*yn*(pp-PHI
(1)));
  ESr = ESr+An*exp(j*2*xn*(k-k(1)).')*exp(j*2*kc*yn*(PHI-PHI
(1)));
end

%---Figure 7.5(a)---------------------------------------------
```

```
h = figure;
matplot(F,PHI*180/pi,abs((Es.')),20);
colorbar;
colormap(1-gray)
set(gca,'FontName', 'Arial', 'FontSize',14,'FontWeight',
'Bold');
ylabel('Angle [Degree]');
xlabel('Frequency [GHz]');
title('Original back-scattered field')
%---Figure 7.5(b)-------------------------------------------------
h = figure;
matplot(F,PHI*180/pi,abs((ESr.')),20);
colorbar;
colormap(1-gray)
set(gca,'FontName', 'Arial', 'FontSize',14,'FontWeight',
'Bold');
ylabel('Angle [Degree]');
xlabel('Frequency [GHz]');
title('Reconstructed back-scattered field')
%---Figure 7.6---------------------------------------------------
h = figure;
matplot(F,PHI*180/pi,abs((ESR.')),20);
colorbar;
colormap(1-gray)
set(gca,'FontName', 'Arial', 'FontSize',14,'FontWeight',
'Bold');
ylabel('Angle [Degree]');
xlabel('Frequency [GHz]');
title('Reconstructed back-scattered field (x10 upsampled)')
%---Figure 7.7---------------------------------------------------
nn = 3;
h = figure;
plot(PHI*180/pi,abs(Es(nn,:)),'k-.*','MarkerSize',8,'LineWi
dth',2);
hold;
plot(pp*180/pi,abs(ESR(10*(nn-1)+1,:)),'k-','LineWidth',2);
hold;
set(gca,'FontName', 'Arial', 'FontSize',14,'FontWeight',
'Bold');
xlabel('PHI [Degree]');
ylabel('Scat. field [V/m]');
tt=num2str(F(nn));
ZZ=['@ f = ' tt ' GHz'];
axis([PHI(1)*180/pi PHI(64)*180/pi 0 0.5]);
title(ZZ);
drawnow;
legend('with brute force computation','with scattering
centers')
```

```
%---Figure 7.8----------------------------------------------------
nn = 11;
h = figure;
plot(F,abs(Es(:,nn)),'k-.*','MarkerSize',8,'LineWidth',2);
hold;
plot(kk*c/2/pi,abs(ESR(:,10*(nn-1)+1)),'k-','LineWidth',2);
hold;
set(gca,'FontName', 'Arial', 'FontSize',14,'FontWeight',
'Bold');
xlabel('Frequency [GHz]');
ylabel('Scat. field [V/m]');
tt=num2str(PHI(nn));
ZZ=['@ PHI = ' tt ' Deg.'];
title(ZZ);drawnow; axis([F(1) F(32) 0 0.35])
legend('with brute force computation','with scattering
centers')
```

Matlab code 7.2: Matlab file 'Figure7-9thru7-13.m'

```
%------------------------------------------------------------------
% This code can be used to generate Figure 7.9 thru 7.13
%------------------------------------------------------------------
% This file requires the following files to be present in the
same
% directory:
%
% Es60.mat
% planorteta60_2_xyout.mat
clear all
close all

c = .3; % speed of light
fc = 6; % center frequency
phic = 0*pi/180; % center of azimuth look angles
thc = 90*pi/180; % center of elevation look angles

%_____PRE PROCESSING OF ISAR_____
BWx = 12; % range extend
M = 32; % range sampling
BWy = 16; % xrange extend
N = 64; % xrange sampling

dx = BWx/M; % range resolution
dy = BWy/N; % xrange resolution

% Form spatial vectors
X = -dx*M/2:dx:dx*(M/2-1);
Y = -dy*N/2:dy:dy*(N/2-1);
XX = -dx*M/2:dx/4:-dx*M/2+dx/4*(4*M-1);
YY = -dy*N/2:dy/4:-dy*N/2+dy/4*(4*N-1);
```

```
%Find resoltions in freq and angle
df = c/(2*BWx); % frequency resolution
dk = 2*pi*df/c; % wavenumber resolution
kc = 2*pi*fc/c;
dphi = pi/(kc*BWy);% azimuth resolution


%Form F and PHI vectors
F = fc+[-df*M/2:df:df*(M/2-1)]; % frequency vector
PHI = phic+[-dphi*N/2:dphi:dphi*(N/2-1)];% azimuth vector
K = 2*pi*F/c; % wavenumber vector
dk = K(2)-K(1); % wavenumber resolution


%_____GET THE DATA_____
load Es60 % load E-scattered
load planorteta60_2_xyout.mat % load plane outline
%_____MATCHING PURSUIT_____
collectedData = zeros(200,3); %initilize scattering center
info
ES = Es;
Power1 = sum(sum(Es).^2); % initial power of the data
axisX = X(1):dx/4:X(32);
axisY = Y(1):dy/4:Y(64);
cosPhi = cos(PHI);
sinPhi = sin(PHI);


for N = 1:250; % extract 250 scattering centers
 Amax = 0;
 p1Max = zeros(size(ES));
 for Xn = axisX
 for Yn = axisY
 p1 = exp(-j*2*K.'*(cosPhi.*Xn+sinPhi.*Yn));
 A = sum(sum(ES.*p1))/(size(ES,1)*size(ES,2));
 if A > Amax
 Amax = A;
 collectedData(N,1:3) = [A Xn Yn];
 p1Max = conj(p1);
 end
 end
 end
 ES = ES-(Amax.*p1Max);
end


%-------Field Reconsctruction----------
Esr = zeros(32,64);
for N = 1:250
 A = collectedData(N,1);
```

```
  x1 = collectedData(N,2);
  y1 = collectedData(N,3);
  Esr = Esr+A*exp(j*2*K.'*(cosPhi.*x1+sinPhi.*y1));
end


%---Figure 7.9(a)---------------------------------------------
%---SCATTERING CENTER INFO DISPLAY----------------------------
load planorteta60_2_xyout.mat
SSs = collectedData;
h = figure;
plot(abs(SSs(1:250,1)),'square','MarkerSize',4,'MarkerFaceCo
lor', [0, 0, 1]);
set(gca,'FontName', 'Arial', 'FontSize',14,'FontWeight',
'Bold');
xlabel('Scattering Center #');
ylabel('Amplitude [mV/m]');
%---Figure 7.9(b)---------------------------------------------
h = figure;
hold
for m=1:150
  t = round(abs(SSs(m,1))*20/abs(SSs(1,1)))+1
  plot(-SSs(m,2),SSs(m,3),'o','MarkerSize',t,'MarkerFaceColor',
[0, 0, 1]);
end
hold
line( -xyout_xout,xyout_yout,'Color','k','LineStyle','.');
axis([min(X) max(X) min(Y) max(Y)])
set(gca,'FontName', 'Arial', 'FontSize',14,'FontWeight',
'Bold');
xlabel('Range [m]');
ylabel('X-Range [m]');
title('Locations of scattering centers with relative
amplitudes ')


%-ISAR IMAGE COMPARISON---------------------------------------
Enew = Es;
Enew(M*4,N*4) = 0;
ISARorig = fftshift(fft2(Enew));
ISARorig = ISARorig/M/N;

h = figure;
matplot(X,Y,abs(ISARorig(4*M:-1:1,:).'),20);
colorbar;
colormap(1-gray);
set(gca,'FontName', 'Arial', 'FontSize',14,'FontWeight',
'Bold');
line(-xyout_xout,xyout_yout,'Color','k','LineStyle','.');
xlabel('Range [m]');
ylabel('X-Range [m]');
```

```
title('Original ISAR image')
Enew = Esr;
Enew(M*4,N*4) = 0;
ISARrec = fftshift(fft2(Enew));
ISARrec = ISARrec.'/M/N; % reconstructed ISAR image

%---Figure 7.13 ------------------------------------------------
h = figure;
matplot(X,Y,abs(ISARrec(:,4*M:-1:1)),20);
colorbar;
colormap(1-gray);
set(gca,'FontName', 'Arial', 'FontSize',14,'FontWeight',
'Bold');
line(-xyout_xout,xyout_yout,'Color','k','LineStyle','.');
xlabel('Range [m]');
ylabel('X-Range [m]');
title('Reconstructed ISAR image')
%-------FIELD COMPARISON--------------------
%------------------------------------------
h = figure;
matplot(F,PHI*180/pi,abs((Es.')),20);
colorbar;
colormap(1-gray)
set(gca,'FontName', 'Arial', 'FontSize',14,'FontWeight',
'Bold');
ylabel('Angle [Degree]');
xlabel('Frequency [GHz]');
title('Original back-scattered field')
%---Figure 7.10 ------------------------------------------------
h = figure;
matplot(F,PHI*180/pi,abs((Esr.')),20);
colorbar;
colormap(1-gray)
set(gca,'FontName', 'Arial', 'FontSize',14,'FontWeight',
'Bold');
ylabel('Angle [Degree]');
xlabel('Frequency [GHz]');
title('Reconstructed back-scattered field')
%-------RECONSTRUCT THE FIELD PATTERN x10-------------
Esr = zeros(320,640);
k = K;
kk = k(1):(k(32)-k(1))/319:k(32);
pp = PHI(1):(PHI(64)-PHI(1))/639:PHI(64);
csP = cos(pp);
snP = sin(pp);
for N = 1:250
 A = collectedData(N,1);
 x1 = collectedData(N,2);
 y1 = collectedData(N,3);
```

```
  Esr = Esr+A*exp(j*2*kk.'*(csP.*x1+snP.*y1));
end


%---Figure 7.11 ------------------------------------------------------
h = figure;
matplot(F,PHI*180/pi,abs((Esr.')),20);
colorbar;
colormap(1-gray)
set(gca,'FontName', 'Arial', 'FontSize',14,'FontWeight',
'Bold');
ylabel('Angle [Degree]');
xlabel('Frequency [GHz]');
title('Reconstructed field (x10 upsampled)')
%---Figure 7.12(a)---------------------------------------------------
nn = 7;
h = figure;
plot(PHI*180/pi,abs(Es(nn,:)),'k-.*','MarkerSize',8,'LineWi
dth',2);
hold;
plot(pp*180/pi,abs(Esr(10*(nn-1)+1,:)),'k-','LineWidth',2);
hold;
set(gca,'FontName', 'Arial', 'FontSize',14,'FontWeight',
'Bold');
xlabel('PHI [Degree]'); ylabel('Scat. field [V/m]');
tt = num2str(F(nn));
ZZ = ['@ f = ' tt ' GHz'];
axis([PHI(1)*180/pi PHI(64)*180/pi 0 0.35])
title(ZZ);
drawnow;
legend('with brute force computation','with scattering
centers')
%---Figure 7.12(b)---------------------------------------------------
nn = 4;
h = figure;
plot(F,abs(Es(:,nn)),'k-.*','MarkerSize',8,'LineWidth',2);
hold;
plot(kk*c/2/pi,abs(Esr(:,10*(nn-1)+1)),'k-','LineWidth',2);
hold;
set(gca,'FontName', 'Arial', 'FontSize',14,'FontWeight',
'Bold');
xlabel('Frequency [GHz]'); ylabel('Scat. field [V/m]');
tt = num2str(PHI(nn));
ZZ = ['@ PHI = ' tt ' Deg.'];
title(ZZ);
drawnow;
axis([F(1) F(32) 0 0.5])
legend('with brute force computation','with scattering
centers')
```

REFERENCES

1   W. P. Yu, L. G. To, and K. Oii. N-Point scatterer model RCS/Glint reconstruction from high-resolution ISAR target imaging. Proc. End Game Measurement and Modeling Conference, Point Mugu, CA, January 1991, pp. 197–212.

2   M. P. Hurst and R. Mittra. Scattering center analysis via Prony's method. *IEEE Trans Antennas Propagat* 35 (1987) 986–988.

3   R. Carriere and R. L. Moses. High-resolution radar target modeling using a modi-fied Prony estimator. *IEEE Trans Antennas Propagat* 40 (1992) 13–18.

4   R. Bhalla and H. Ling. 3-D scattering center extraction using the shooting and bouncing ray technique. *IEEE Trans Antennas Propagat* AP-44 (1996) 1445–1453.

5   V. C. Chen and H. Ling. *Time-frequency transforms for radar imaging and signal analysis*. Artech House, Boston, MA, 2002.

6   N. Y. Tseng and W. D. Burnside. A very efficient RCS data compression and recon-struction technique, Technical Report No. 722780-4, Electroscience Lab, Ohio State University, November. 1992.

7   S. Y. Wang and S. K. Jeng. Generation of point scatterer models using PTD/SBR technique. Antennas and Propagation Society International Symposium, Newport Beach, CA, June 1995, pp. 1914–1917.

8   C. Özdemir, R. Bhalla, H. Ling, and A. Radiation. Center representation of antenna radiation patterns on a complex platform. *IEEE Trans Antennas Propagat* AP-48 (2000) 992–1000.

9   C. Özdemir, R. Bhalla, and H. Ling. Radiation center representation of antenna synthetic aperture radar (ASAR) images. IEEE Antennas and Propagat. Society Int. Symp., Atlanta, Vol. I, 338–341, IEEE, Atlanta, 1998.

10 J. A. Högbom. Aperture synthesis with a non-regular distribution of interferometer baselines. *Astron Astrophys Suppl* 15 (1974) 417–426.

11 A. Selalovitz and B. D. Frieden. A "CLEAN"-type deconvolution algorithm. *Astron Astrophys Suppl* 70 (1978) 335–343.

12 J. Tsao and B. D. Steinberg. Reduction of sidelobe and speckle artifacts in micro-wave imaging: The CLEAN technique. *IEEE Trans Antennas Propagat* 36 (1988) 543–556.

13 R. Bhalla, J. Moore, and H. Ling. A global scattering center representation of complex targets using the shooting and bouncing ray technique. *IEEE Trans Antennas Propagat* 45 (1997) 1850–1856.

14 S. G. Mallat and Z. Zhang. Matching pursuit with time-frequency dictionaries. *IEEE Trans Signal Process* 41 (1993) 3397–3415.

CHAPTER EIGHT

Motion Compensation for Inverse Synthetic Aperture Radar

For the operational inverse synthetic aperture radar (ISAR) situation, the target's relative movement with respect to the radar sensor provides the angular diversity required for range-Doppler ISAR imagery as given in Chapter 6. For the ground-based ISAR systems, for example, collecting back-scattered energy from an aerial target that is moving with a constant velocity for a sufficiently long period of time can provide the necessary angular extend to form a successful ISAR image. On the other hand, real targets such as planes, ships, helicopters, and tanks do have usually complicated motion components while maneuvering. These may include translational and rotational (yaw, roll, and pitch) motion parameters such as velocity, acceleration, and jerk. Moreover, all these parameters are unknown to the radar engineer, which adds further complexities to the problem. Therefore, trying to estimate these motion parameters and also trying to invert the undesired effects of motion on the ISAR image is often called *motion compensation* (MOCOMP).

Since the motion parameters are unknown to the radar sensor, the MOCOMP process can be regarded as a blind process and is also assumed to be one of the most challenging tasks in ISAR imaging research. The MOCOMP procedure has to be employed in all SAR and ISAR applications to obtain a clear and focused image of the scene or the target. In SAR applications, for example, the information gathered from the radar platform's inertial measurement system, gyro, and/or global positioning system (GPS) is generally used to correct the motion effects on the phase of the received signal [1, 2]. However,

Inverse Synthetic Aperture Radar Imaging with MATLAB Algorithms, First Edition.
Caner Özdemir.
© 2012 John Wiley & Sons, Inc. Published 2012 by John Wiley & Sons, Inc.

the situation is very different in ISAR applications such that all motion parameters, including velocity, acceleration, jerk, and the type of maneuver (straight motion, yaw, roll, and pitch), are not known by the radar. Therefore, these parameters must somehow be estimated and then eliminated to have a successful ISAR image of the target.

If an efficient compensation routine is not applied, the resultant ISAR image is defocused and blurred in slant range and cross-range dimensions. Various methods from many researchers have been suggested to mitigate or eliminate these unwanted motion effects in ISAR imaging [3–12]. The single scattering referencing algorithm [3], the multiple-scatterer method [4], the centroid tracking algorithm [5, 6], the entropy minimization method [7, 8], the phase gradient autofocusing technique [9, 10], the cross-correlation method [11, 12], and the joint time-frequency (JTF) methods [13–16] are popular ones among numerous ISAR MOCOMP techniques.

8.1 DOPPLER EFFECT DUE TO TARGET MOTION

When the scatterer on the target is moving, the Doppler shift posed by the scatterer's line of-sight velocity sets "inaccurate" distance information about the position of the scatterer to the phase of the received electromagnetic (EM) wave. While the scatterer is moving fast, it may occupy several pixels in the image during the integration interval of ISAR. Therefore, the phase of the backscattered wave is altered such that the resultant ISAR image is mislocated in cross range and defocused in both range and cross-range domains. If the scatterer is not moving fast, the ISAR image may not be blurred. However, the location of the scatterer will still not be true due to the Doppler shift imposed by the target's movement.

The effect of the target's motion to the phase of the backscattered wave and/or to the ISAR image is investigated based on the geometry illustrated in Figure 8.1. In the general case, the target may have both radial and rotational motion during the illumination period of radar. For this reason, the point scat-

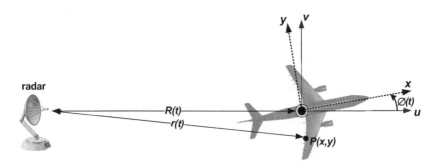

FIGURE 8.1   Geometry for a moving target with respect to radar.

terer at $P(x, y)$ on the target is assumed to have both radial and rotational motion components. According to the practical convention of radar imaging, the phase center is selected in the middle of the target and is assumed to be the origin. We would like to estimate the phase error induced due to target motion. If the target is situated at the far field of the radar, the distance of point P from the radar can be approximated as [14]

$$r(t) \cong R(t) + x \cdot cos\emptyset(t) - y \cdot sin\emptyset(t). \tag{8.1}$$

Here, $R(t)$ is the target's translational range distance from the radar and $\emptyset(t)$ represents the rotational angle of the target with respect to the radar line of sight (RLOS) axis, u. Expanding $R(t)$ and $\emptyset(t)$ into their Taylor series, they can be represented to yield

$$R(t) = R_o + v_t t + \frac{1}{2}a_t t^2 + \cdots$$
$$\emptyset(t) = \emptyset_o + \omega_r t + \frac{1}{2}\alpha_r t^2 + \cdots. \tag{8.2}$$

Here, R_o is the initial range of the target, and v_t and a_t are the target's translational velocity and the acceleration, respectively. Higher order terms starting from the target's translational jerk follow these first three terms. Similarly, $\emptyset_o$ is the initial angle of the target with respect to the RLOS axis. ω_r and α_r are the angular velocity and the angular acceleration of the target, respectively.

The phase of the backscattered signal from point P can be written as

$$\Phi(t) = -2k \cdot r(t)$$
$$= -2\pi f \frac{2r(t)}{c}. \tag{8.3}$$

Therefore, the Doppler frequency shift due to motion can be calculated by taking the time derivative of this phase as

$$f_D = \frac{1}{2\pi} \frac{\partial}{\partial t} \Phi(t)$$
$$= -\frac{2f}{c} \frac{\partial}{\partial t} r(t)$$
$$= -\frac{2f}{c} (v_t + a_t t + \cdots)$$
$$+ \frac{2f}{c} (\omega_r + \alpha_r t + \cdots) \cdot (x \cdot sin\emptyset(t) + y \cdot cos\emptyset(t)). \tag{8.4}$$

Here, the first and second terms represent the radial (or translational) and the rotational Doppler frequency shifts, respectively. If the motion of the target is to be approximately described with v_t and ω_r, then the translational and the rotational Doppler frequency shifts are reduced to the following terms:

$$f_D^{trans} \cong -\frac{2f}{c} v_t$$

$$f_D^{rot} \cong \frac{2f}{c} \omega_r \cdot (x \cdot \sin(\emptyset_o + \omega_r t) + y \cdot \cos(\emptyset_o + \omega_r t)). \tag{8.5}$$

Therefore, the translational Doppler shift becomes directly related to the target's translational velocity, v_t. On the other hand, the rotational Doppler shift posed by the target is more complex as it depends on many parameters that can be deduced from Equation 8.5.

8.2   STANDARD MOCOMP PROCEDURES

Where monostatic ISAR imaging is concerned, the received signal can be theoretically approximated as the integration of the backscattered echoes from all the scatterers inside the radar beam:

$$s(t) = \iint_{-\infty}^{\infty} A(x, y) \cdot \exp\left(-j4\pi f \frac{r(t)}{c}\right) \cdot dy dx. \tag{8.6}$$

Here $A(x, y)$ is the backscattered signal intensity from any point scatterer at (x, y), and f corresponds to the frequency of the radar waveform. Substituting the range equation in Equation 8.1 into Equation 8.6, the received signal is deduced to be equal to

$$s(t) = \exp\left(-j4\pi f_o \frac{R(t)}{c}\right)$$

$$\cdot \iint_{-\infty}^{\infty} A(x, y) \cdot \exp\left(-j4\pi \frac{f}{c}(x \cdot \cos\emptyset(t) - y \cdot \sin\emptyset(t))\right) \cdot dy dx. \tag{8.7}$$

If the target's initial range, R_o, and the linear translational velocity, v_t, are known, the phase term prior to the above integral can be removed by multiplying Equation 8.7 by the term of "$\exp(j4\pi f R(t)/c)$." This is called *range tracking* or the *coarse MOCOMP*, and it is the standard procedure for compensating the translational motion effects. After obtaining the phase-compensated backscattered signal, a Fourier transform operation can then be applied to image the backscattered signal intensity function, $A(x, y)$ [16]. If the scatterers pass different range cells during the coherent integration time, the resultant phase-compensated image will still be defocused. Therefore, a finer compensating technique called *Doppler tracking* that attempts to make the Doppler shifts constant is required [17–19]. This procedure is also called the *fine MOCOMP*.

The signal processing tools used to reduce motion errors, that is, MOCOMP techniques, are usually treated in two steps: First, the effects due to translational motion components are solved. Then, the errors associated with the rotational movement of the target are dealt with.

8.2.1   Translational MOCOMP

The target's *radial or translational motion* is defined as the movement of the target along the range axis (or RLOS axis) of the radar. The target's translational motion is one of the most significant components that affects image quality in the ISAR image. The main effect of the target's translational motion is shifting the positions of the scatterers on the target along the the range axis. This is because of the fact that target's radial distance changes for consecutive radar pulses as they are sent at different time instants while the target is moving. Therefore, the phase of the collected electric field data is misaligned along the pulses so that the Doppler frequencies that are used to estimate the exact locations of the target are spread out over a finite number of range cells.

When the Fourier transform is directly applied to the collected data that contain translational motion, the location of the point scatterer is poorly estimated since the scatterer is visible for all of that finite number of range cells. Therefore, the scatterers look as if they "walk" over the range cells. The *range walk* phenomena can negatively affect the range resolution, range accuracy, and signal-to-noise ratio of the resulting ISAR image. Therefore, the target's resultant image before the compensation may be smeared in the cross-range direction and defocused in range and cross-range directions. The amount of smearing, of course, depends on the amount of target's radial motion (or the radial velocity). Although there may be little or no smearing effects for small radial velocity values as in the case of slowly moving ship targets, the image smearing can be drastic for fast moving targets such as fighter airplanes. Usually, an algorithm is applied to overcome the range walk issue by trying to align the range bins. The common name for keeping the scatterers in their range cells is *range tracking*.

8.2.1.1   Range Tracking   There are different range tracking methods that have been employed by various researchers [3, 5, 6, 11, 12, 15, 20–25]. The *cross-correlation method* [3, 11, 12], for example, calculates the correlation coefficients between the adjacent range profiles and tries to estimate the range walk between the range profiles.

Another widely applied range tracking method is called *target centroid tracking* [5, 6]. The main idea in this method is to estimate the radial motion of the target centroid point and to compensate in such a way that the range and the Doppler shift of the target centroid are kept constant.

Another famous scheme is called *prominent point processing* (PPP) [11, 15, 21–25]. The first step in the PPP algorithm is to select prominent points on the target. Assuming that the phase components of these dominant point scatterers are known, the translational motion error can be mitigated by unwrapping the higher order phase component of the first prominent point as the second step of the algorithm. In the third step, the rotational motion error is eliminated by extracting the relationship between the rotation angle and the dwell time from the phase of the second prominent points. Finally, the rotational velocity can be estimated by measuring the phase of the third prominent points [11, 15].

8.2.1.2 Doppler Tracking While the range tracking procedures are capable of aligning all the scatterers at their correct range cells, the Doppler frequency shifts of these scatterers may be still varying with time in the phase of the received signal. In other words, the Doppler frequency shifts may not be constant between the scatterers. These Doppler frequency shifts can be caused by the target's movement along the range direction and changes in the instantaneous radial velocity of the target with respect to radar [12]. The procedure that tries to make the Doppler frequency shifts constant among the range cells is called *Doppler tracking* [5, 11, 12, 14].

Some of the popular Doppler tracking algorithms that have been widely applied are listed as follows: dominant scatterer algorithm [12, 26], sub-aperture approach [9], cross-range centroid tracking algorithm [5], phase gradient autofocus technique [10], and multiple PPP technique [27, 28].

8.2.2 Rotational MOCOMP

The target's *rotational motion* is defined as the movement that causes aspect change of the target from the RLOS. As thoroughly presented in Chapter 6, a small value of rotation of the target with respect to the RLOS is adaquate to form the range-Doppler ISAR image. However, when the rotation of the target is not small for the coherent integration time of the radar, the Fourier-based ISAR imaging algorithm produces degradations such as blurring and smearing in the resultant ISAR image.

In most ISAR scenarios, the target's motion contains not only rotational motion components but also translational motion components. Usually, the unfavorable effect caused by the translational motion is more severe than that caused by rotational motion. Therefore, the rotational MOCOMP is applied after the translational MOCOMP in the general treatment of ISAR MOCOMP.

Let us consider the scenario in Figure 8.1, wherein the target has a general motion of both translational and rotational motion. Substituting Equation 8.1 into Equation 8.3, we get the following:

$$\Phi(t) = -\frac{4\pi f}{c}(R(t) + x \cdot cos\emptyset(t) - y \cdot sin\emptyset(t)). \tag{8.8}$$

To have rotationally motion-compensated ISAR image, the phase of the received signal should only be a linear function of angular velocity as demonstrated in Chapter 6. With this construct, a Fourier-based ISAR imaging procedure will be able to resolve the points in the cross-range directions. However, the phase in Equation 8.3 is a nonlinear function of angular velocity and is quite complex. Let us simplify this phase by assuming that the target has only radial velocity and angular velocity such that

$$R(t) = R_o + v_t t$$
$$\emptyset(t) = \emptyset_o + \omega_r t.$$
(8.9)

Then, the phase of the received signal becomes

$$\Phi(t) = -\frac{4\pi f}{c}[R_o + v_t t] - \frac{4\pi f}{c}[x \cdot \cos(\emptyset_o + \omega_r t) - y \cdot \sin(\emptyset_o + \omega_r t)].$$
(8.10)

The first term is related to the translational motion, and the second term is responsible for the rotational motion. Without loss of generality, we can set $\emptyset_o = 0$ to have

$$\Phi_{rot}(t) = -\frac{4\pi f}{c}[x \cdot \cos(\omega_r t) - y \cdot \sin(\omega_r t)].$$
(8.11)

For sufficiently small values of angular velocity or the coherent integration time of ISAR or both, that is, the argument "$\omega_r t$" is small, then $\cos(\omega_r t) \approx 1$ and $\sin(\omega_r t) \approx \omega_r t$. Therefore,

$$\Phi_{rot}(t) \cong -\left(\frac{4\pi f}{c}\right) \cdot x - \left(\frac{4\pi f}{c}\omega_r t\right) \cdot y.$$
(8.12)

Noting that the Doppler frequency shift is equal to $f_D = 2\omega_r y/\lambda$, this equation can then be rewritten as

$$\Phi_{rot}(t) \cong -\left(\frac{4\pi f}{c}\right) \cdot x - (2\pi f_D) \cdot t.$$
(8.13)

This result gives exactly the same phase value as in Equation 6.26 (Chapter 6) that is necessary for an ideal range-Doppler processing. When the "$\omega_r t$" is not small, this rotational motion has to be compensated for by applying a compensating procedure to have a distortion-free ISAR image.

JTF-based methods have proven to be very effective for removing the rotational motion effects from the ISAR image [13–16]. An example based on JTF processing for compensating the rotational motion will be given in Section 8.3.3.3. Another popular method for rotational MOCOMP is PPP [11, 15, 21–25].

8.3 POPULAR MOCOMP TECHNIQUES IN ISAR

As the names of MOCOMP methods have been listed in the above paragraphs, there are many algorithms studied by numerous researchers. In this section, we will present the most famous and popular MOCOMP algorithms associated with some numerical examples in Matlab.

8.3.1 Cross-Correlation Method

The cross-correlation method is one of the basic and most applied range tracking algorithms. The presented algorithm here relies on the stepped frequency continuous wave (SFCW) radar configuration. Let us assume that radar system sends out the stepped frequency waveform of M bursts each having N pulses toward the target. The target's translational velocity, v_t, is assumed to be constant. Therefore, radar collects the two-dimensional (2D) backscattered electric field data, $E^s[m, n]$, of size $M \cdot N$. Then, the phase of the mth burst and the nth pulse can be written in terms of v_t as [11]

$$\varphi\{E^s[m, n]\} = -\frac{4\pi f_n}{c}(R_o - T_{PRI} \cdot v_t \cdot (n-1+N \cdot (m-1))), \quad \begin{array}{l} m = 1, 2, \ldots, M \\ n = 1, 2, \ldots, N \end{array}, \quad (8.14)$$

where f_n is the stepped frequency value for the nth transmitted pulse, R_o is the initial radial location of the target from radar, and T_{PRI} is the time lag between adjacent pulses or simply the *pulse repetition interval* (PRI). Similarly, the phase of the $(m + 1)$th burst and nth pulse is equal to

$$\varphi\{E^s[(m+1), n]\} = -\frac{4\pi f_n}{c}(R_o - T_{PRI} \cdot v_t \cdot (n-1+N \cdot m)). \quad (8.15)$$

Therefore, the phase difference between the adjacent bursts can be calculated as

$$\begin{aligned} \Delta \varphi_{burst-to-burst} &= \varphi\{E^s[(m+1), n]\} - \varphi\{E^s[m, n]\} \\ &= \frac{4\pi f_n}{c} v_t \cdot T_{PRI} \cdot N \\ &= \frac{4\pi f_n}{c} \Delta R_{burst-to-burst}, \end{aligned} \quad (8.16)$$

where $\Delta R_{burst-to-burst} = v_t \cdot (T_{PRI} \cdot N)$ is the so-called "range walk" between the adjacent bursts. This range shift can be compensated by applying the following steps:

1. First, one-dimensional (1D) fast Fourier transform (FFT) is applied along the pulses such that a total of M range profile vector, RP_m of length N is obtained.

2. One of the range profiles is taken as the reference. In practice, the first one, RP_1, is usually chosen due to the fact that its phase is usually either in advance or in lag when compared to phases of all the others.

3. The cross-correlations of the magnitudes of other $(M-1)$ range profiles to that of the reference range profile are calculated via computing the following cross-correlation factor:

$$CCR_m = \left| IFFT\left(FFT\left(|RP_{ref}|\right)\right) \cdot FFT\left(|RP_m|\right)^* \right|, m = 1, 2, \ldots, M-1. \quad (8.17)$$

Notice that each CCR_m vector is also of length N.

4. The locations of the peak value for the calculated cross correlations indicate the range shifts (or time delays) that are required to align each RP with respect to the reference range profile of RP_{ref}:

$$K_m = index\left[\max(CCR_m)\right], m = 1, 2, \ldots, M-1. \quad (8.18)$$

5. The resultant index vector is usually smoothed by fitting to a lower order polynomial so that the gradual change between the index vector K is almost constant:

$$S_m = smooth\left[K_m\right], m = 1, 2, \ldots, M-1. \quad (8.19)$$

6. Therefore, the range walk between the nth range profile, RP_m, and the reference range profile, RP_{ref}, can then be approximated as

$$\Delta R_{n-to-ref} \cong S_m \cdot \Delta r, \quad (8.20)$$

where Δr is the range resolution and is given by

$$\Delta r = c/(2B), \quad (8.21)$$

where B is the total frequency bandwidth.

7. As the last step, the compensating phase vector for the range profile RP_m equals

$$\Delta \varphi_{m-to-ref} = \frac{4\pi f_n}{c} \Delta R_{m-to-ref}, \quad n = 1, 2, \ldots, N. \quad (8.22)$$

Then, the motion compensated range profile can be obtained by using this correcting phase as

$$RP'_m = FFT\left\{\Delta \varphi_{m-to-ref} \cdot IFFT(RP_m)\right\}. \quad (8.23)$$

Once all M range profiles are corrected, a motion-compensated ISAR image can then be generated using conventional ISAR imaging routines.

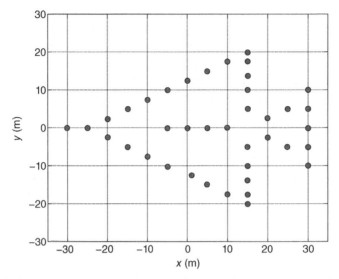

FIGURE 8.2   A fighter target composed of perfect point scatterers.

8.3.1.1 Example for the Cross-Correlation Method   We will demonstrate
the concept of range tracking by applying the cross-correlation method over
a numerical example. A hypothetical fighter, shown in Figure 8.2, composed
of perfect point scatterers of equal magnitudes is chosen for this example. The
target, at an initial radial distance of $R_o = 16$ km, is moving toward the radar
with a radial velocity of $v_t = 70$ m/s. The target has a radial acceleration value
of $a_t = 0.1$ m/s^2. We also assume that target is rotating slowly with an angular
velocity of $\varphi_r = 0.03$ rad/s.

The radar sends 128 bursts, each having 128 modulated pulses. The frequency of the first pulse is $f_o = 10$ GHz and the total frequency bandwidth is
$B = 128$ MHz. Pulse repetition frequency (PRF) of the radar system is chosen
as 20 KHz.

First, we obtained the conventional range-Doppler ISAR image of the
fighter by employing traditional ISAR imaging procedures without applying
any compensation for the motion of the target. The resultant raw range-
Doppler ISAR is depicted in Figure 8.3. As can be clearly seen from the figure,
the effect of target's motion is severe in the resultant ISAR image such that
the image is broadly blurred in the range and Doppler domains, and the true
locations of the target's scattering centers cannot be retrieved from the image.

Next, the cross-correlation method is applied to track the range and compensate for the motion of the target. First, the range profiles of the target are
obtained by applying 1D inverse Fourier transform (IFT) operation to the
backscattered electric field along the frequencies. After taking the first range
profile, PR_1, as the reference, the cross-correlation between the reference

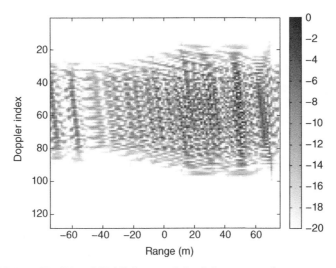

FIGURE 8.3 Traditional ISAR image of the fighter target (no compensation).

range profile and the others are calculated by using the formula in Equation 8.17. As explained in the algorithm, the index for the maximum value of the correlations indicates the time shift required to align the range profiles. After finding these indices and multiplying them with the calculated range resolution value of $\Delta r = c/(2B)$ (1.17 m for this example), we get the range shifts of the range profiles with respect to PR_1. These range walks and their smoothed versions with respect to the range profile index are plotted in Figure 8.4 as solid and dashed lines, respectively. While smoothing the range profile shifts to a lower order polynomial (a line for this example), *Robust Lowess* method is utilized [29].

Furthermore, the difference between the consecutive range walks is plotted in Figure 8.5a where these differences are almost constant. Then, the target's radial translational speed can be found by dividing these range walk differences by the time differences between each burst. This calculated speed versus range profile index is plotted in Figure 8.5b. From this figure, we see that the speed is almost constant, around 70 m/s. If we take the average of this speed vector, we get an estimated average value of $v_t^{est} = 70.81$ m/s for the target's radial translational speed, which is very close to the actual speed of $v_t = 70$ m/s.

At the last step of the algorithm, the phase contribution caused by the target's motion can be compensated for by multiplying the scattered field data with the below phase term as

$$E_{comp}^s[m, n] = E^s[m, n] \cdot \exp\left(j4\pi \frac{f_n}{c} \cdot v_t^{est} \cdot (n - 1 + N \cdot (m - 1)) \right), \begin{array}{l} m = 1, 2, \dots, M \\ n = 1, 2, \dots, N. \end{array}$$

$$(8.24)$$

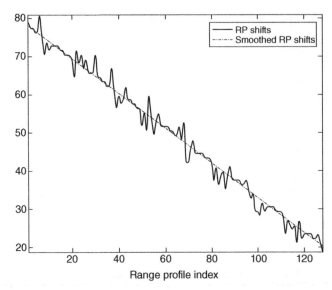

FIGURE 8.4 Range profiles shifts and their smoothened versions versus range profile index.

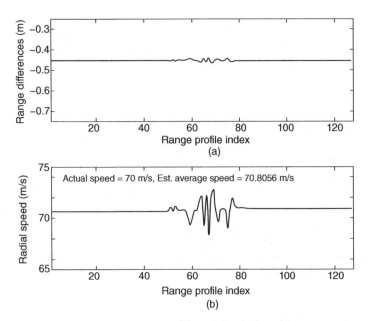

FIGURE 8.5 (a) Range differences and (b) radial velocity with respect range profile index.

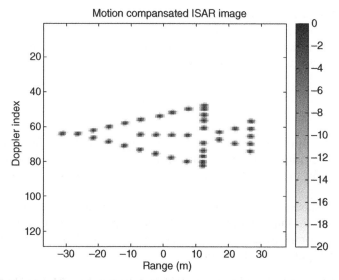

FIGURE 8.6 Motion-compensated ISAR image of the fighter target.

Once the phase of the collected scattered field is compensated, the ISAR image can then be obtained by applying the regular ISAR imaging procedures. Figure 8.6 shows the resultant motion-compensated ISAR image after applying the whole process explained above. The dominant motion effects of translational velocity $v_t = 70$ m/s are successfully eliminated, and the image of the fictitious fighter is almost perfectly focused.

We also notice that radial translation acceleration of $a_t = 0.1$ m/s^2 and angular speed $\varphi_r = 0.03$ rad/s have little effect on the resultant motion-compensated image as observed from Figure 8.6. When these values are taken to be sufficiently greater, however, the image distortion/blurring is unavoidable if only the range tracking procedure is used as the compensation tool.

8.3.2 Minimum Entropy Method

Another popular tool that has been used as a MOCOMP technique for removing translational motion effects in ISAR images is called the *minimum entropy method*. In fact, the concept of entropy is commonly utilized in engineering to measure the disorder of any system [30].

In SAR/ISAR imagery, the entropy phenomenon is used in a similar manner to estimate the disorder in the image [7, 8, 31]. In ISAR imaging, the minimum entropy method tries to estimate the motion parameters (such as velocity and acceleration) of the target. This task is accomplished by calculating the entropy of the energy in the image and minimizing this parameter by iteratively trying out the possible values of motion parameters. The details of this method will be presented next.

8.3.2.1   Definition of Entropy in ISAR Images   We assume that the target has translational motion parameters, both the translational velocity of v_t and a radial acceleration of a_t. With this construct, the phase of the backscattered signal can be written in the following form:

$$\varphi\{E^s\} = -\frac{4\pi f}{c}\left(R_o + \left(v_t t + \frac{1}{2}a_t t^2\right)\right). \tag{8.25}$$

Here, R_o is the initial radial distance of the target from the radar. The sign of v_t can be either plus or minus for an approaching or retreating target, respectively. Similarly, the sign of a_t can be either plus or minus for an accelerating or decelerating target, respectively.

The first phase, $-4\pi f R_o/c$, is constant for all time values and therefore can be suppressed for imaging purposes. With this convention, the effect of motion can then be compensated if the scattered electric field is multiplied by the following compensating phase term:

$$S = \exp\left(j\frac{4\pi f}{c}\left(v_t t + \frac{1}{2}a_t t^2\right)\right). \tag{8.26}$$

Therefore, the goal of the algorithm is to estimate the motion parameters of v_t and a_t to be able to successfully remove their effects from the phase of the received signal. If the ISAR image matrix is I and has M columns and N rows, then the Shannon entropy, $\tilde{E}$, is defined as [31]

$$\tilde{E}(I) = -\sum_{m=1}^{M}\sum_{n=1}^{N} I'[m, n]\cdot log_{10}(I'[m, n]), \tag{8.27}$$

where

$$I'[m, n] = \frac{I[m, n]}{\sum_{m=1}^{M}\sum_{n=1}^{N} I[m, n]}. \tag{8.28}$$

Here I' is the normalized version of the ISAR image. The normalization is accomplished by dividing the image pixels by the total energy in the image. Once the entropy is defined for the ISAR image itself, the goal is to find the corresponding compensation vector (so the motion parameters) such that the new ISAR image has the minimum entropy (or the disorder). The process of searching for the correct values of motion parameters can be done iteratively as will be demonstrated with a numerical example.

8.3.2.2   Example for the Minimum Entropy Method   In this example, we will demonstrate the use of the minimum entropy method for compensating the motion effects in an ISAR image. First, we use a target composed of discrete perfect point scatterers that have equal scattering amplitudes as shown in

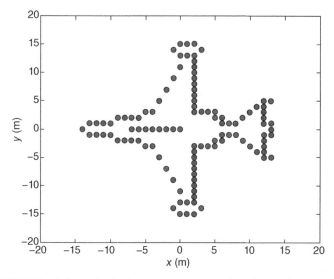

FIGURE 8.7   A hypothetical target composed of perfect point scatterers.

Figure 8.7. The target is moving away from the radar with a radial translational velocity of $v_t = 4$ m/s and with a radial translational acceleration of $a_t = 0.6$ m/s². The target's angular velocity is chosen as $\varphi_r = 0.06$ rad/s. The target's initial radial distance from the radar is taken as $R_o = 500$ m.

The radar sends 128 bursts, each having 128 modulated pulses. The frequency of the starting pulse is $f_o = 8$ GHz, and the total frequency bandwidth is $B = 384$ MHz. The PRF is chosen as 14.5 KHz. Without applying any compensation, the range-Doppler ISAR image of the target is constructed with the help of conventional ISAR imaging techniques, and the corresponding range-Doppler ISAR image is obtained as shown in Figure 8.8. As is obvious from the image, the uncompensated ISAR image is highly distorted and blurred due to both the translational and the rotational motion of the target.

In Figure 8.9, the spectrogram of received time pulses is plotted. This figure clearly demonstrates the progressive shift in the frequency (or in the phase) of the consecutive received time pulses. This shift occurs due to the change of target's range distance from the radar during the integration time of the ISAR process. If a successful MOCOMP practice is applied, there is expected to be no (or minimal) range shift between consecutive time pulses.

Then, the minimum entropy methodology is applied to the ISAR image data in Figure 8.8. The algorithm iteratively searches for the values of v_t and a_t by minimizing the entropy of the compensated ISAR image of

$$I' = \mathcal{F}_2^{-1}\{S \cdot E^s\}, \tag{8.29}$$

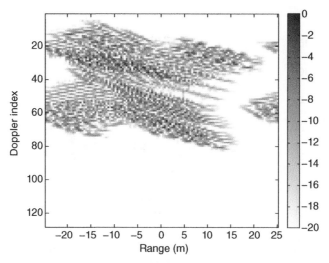

FIGURE 8.8   Conventional ISAR image of the airplane target (no compensation).

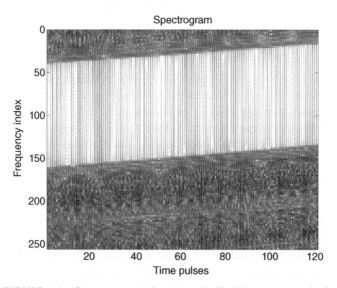

FIGURE 8.9   Spectrogram of range cells (before compensation).

where S is defined as in Equation 8.26 for different values of v_t and a_t. The estimated values for v_t and a_t are found by iteratively searching the minimum value of the entropy as defined in Equation 8.27. Figure 8.10 shows the graph of the entropy value for different values of translational velocity and acceleration. The 2D search space makes a minimum where v_t^{est} parameter becomes equal to 4 m/s and a_t^{est} parameter equals to 0.6 m/s^2 as demonstrated in Figure 8.10. Therefore, the algorithm successfully estimates the correct values of v_t and a_t.

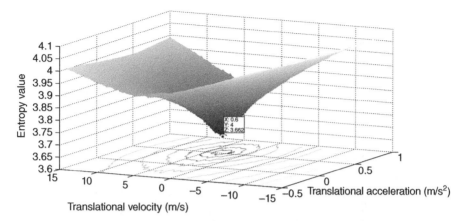

FIGURE 8.10 Entropy plot for translational radial velocity translational radial acceleration search space.

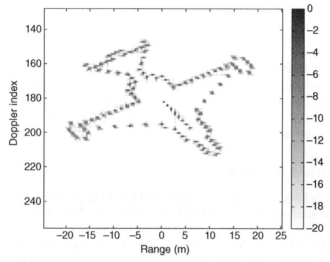

FIGURE 8.11 ISAR image of the airplane after applying minimum entropy compensation.

The effect of motion in the scattered field can then be mitigated by multiplying it with the compensating phase term as given in Equation 8.29. Consequently, the motion-compensated ISAR image is obtained as shown in Figure 8.11 by applying the regular FFT-based ISAR imaging technique. The compensated ISAR image clearly demonstrates that the unwanted effects due to target's motion are eliminated after applying the minimum entropy methodology. The target's scattering centers are very well displayed with good

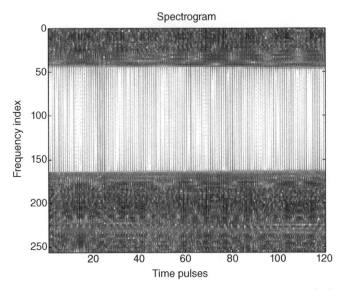

FIGURE 8.12   Spectrogram of range cells (after compensation).

resolution. A further check is done by looking at the spectrogram of the motion-compensated received time pulses as illustrated in Figure 8.12. As is obvious from this spectrogram, the range delays between the time pulses are eliminated such that all frequency (or the phase) values of the returned pulses are aligned successfully.

8.3.3   JTF-Based MOCOMP

The JTF tools have been extensively utilized in EM applications ranging from analysis and interpretation of EM signals [32–34] to radar signature and target classification [35, 36]. JTF representations, including short-time Fourier transform (STFT), Wigner-Ville distributions (WVD), continuous wavelet transform (CWT), adaptive wavelet transform (AWT), and Gabor wavelet transform (GWT), have been shown to be quite effective in analyzing the ISAR image of the complicated moving targets as well [16, 37]. In this section we will present the use of JTF representation in compensating the motion effects in ISAR imaging.

8.3.3.1   Received Signal from a Moving Target   In real-world scenarios, the target's maneuver can be so complex that Doppler frequency shifts in the received signal may vary with time. If the target has complex motion such as yawing, pitching, rolling, or more generally maneuvering, regular Fourier-based MOCOMP techniques may not be sufficient to model the behavior of the motion. Therefore, the use of JTF tools may provide insights in understanding and characterizing the Doppler frequency variations such that translational and rotational motion parameters such as velocity, acceleration, and jerk can be estimated with good fidelity.

Let us assume that the target has a complex maneuver that can be written as a linear combination of both the translational and rotational motion components. If $R(t)$ is the target's translational range distance from the radar, and $\emptyset(t)$ is the rotational angle of the target with respect to RLOS axis as illustrated in Figure 8.1, expanding $R(t)$ and $\emptyset(t)$ into Taylor series yields the formulation listed in Equation 8.2. We first assume that the target is modeled on point scatterers, such that there exists a total number of K point-scatterers on the target. The time-domain backscattered signal at the radar receiver can then be represented as the following sum from each scattering centers on the target as

$$g(t) = \sum_{k=1}^{K} A_k(x_k, y_k) \cdot \exp\left(-j4\pi \frac{f_o}{c}(R(t) + x_k \cdot \cos\emptyset(t) - y_k \cdot \sin\emptyset(t))\right). \quad (8.30)$$

Here, $A_k(x_k, y_k)$ is the backscattered field amplitude from the kth point scatterer. When the range profiles are concerned only, the time-domain backscattered signal at a selected range cell, x, can be written in a similar manner as follows:

$$g(x, t) = \sum_{k=1}^{K} A_k(x, y_k) \cdot \exp\left(-j4\pi \frac{f_o}{c}(R(t) + x \cdot \cos\emptyset(t) - y_k \cdot \sin\emptyset(t))\right). \quad (8.31)$$

Here, x-axis corresponds to the range direction, and t is the coherent processing interval that can also be regarded as the pulse number. Substituting $R(t)$ and $\emptyset(t)$ as listed in Equation 8.2 into Equation 8.31 and displaying only the leading phase terms, one can get [14]

$$g(x, t) = \sum_{k=1}^{K} A_k(x, y_k) \cdot \exp\left\{-j4\pi \frac{f_o}{c}[(R_o + x) + (v_t + \omega_r y_k)t \right.$$
$$\left. + \frac{1}{2}(a_t - \omega_r^2 x + \alpha_r y_k)t^2 + \cdots \right\}. \quad (8.32)$$

The first term in the phase is constant and can be ignored for the imaging process. To have a motion-free range-Doppler image of the target, $R(t)$ should be fixed at R_o, and $\emptyset(t)$ should linearly vary with time as $\emptyset(t) = \omega_r t$. If these ideal conditions are met, the Fourier transform operation will successfully focus the cross-range points (i.e., y_ks) onto their correct locations. Therefore, the MOCOMP procedure should be applied to other phase terms starting from the second order in aiming to suppress them in the phase of the received signal.

8.3.3.2 An Algorithm for JTF-Based Rotational MOCOMP One effective solution, suggested by Chen [14, 16], is to apply JTF tools to extract the instantaneous Doppler frequency information of the time-varying range-Doppler data such that time snapshots of the time-varying ISAR image can be constructed. The JTF-based schematic algorithm that can take the time-snapshot ISAR images of a rotating target is illustrated in Figure 8.13. The methodology can be separated into the following steps:

1. Pulsed radar (either linear frequency modulated [LFM] or SFCW) collects the scattered field from the target for the coherent integration time. After the received signal is digitized, let us suppose that we have a matrix size of $M \cdot N$. For the SFCW radar operation, the matrix is obtained from M bursts, each having N pulses.
2. In the second step, 1D IFT operation is applied among bursts to get the 1D range profiles for N pulses.
3. Then, multiple JTF transforms are employed to the pulses for every range cell value. Each JTF transformation operation at the single range cell yields a time-Doppler matrix that has a dimension of $M \cdot P$. If the target's rotational velocity ω_r is known, the Doppler axis can be readily replaced with a cross-range axis by using the following relationship:

$$y = \frac{f_D \cdot \lambda_c}{2\omega_r}, \tag{8.33}$$

where y is the cross-range variable, f_D is the instantaneous Doppler frequency shift, and λ_c is the wavelength of center frequency.

4. After the JTF transformation operations are employed to all of the range cells, a three-dimensional (3D) time-range-Doppler (or time-range-cross-range) cube of size $M \cdot N \cdot P$ is constructed. This cube has the property of providing the range-Doppler (or range-cross-range) image at a selected time instant.
5. As the final step, a total of N range-Doppler (or range-cross-range) ISAR images of the target can be obtained by taking different slices of the time-range-Doppler (or time-range-cross-range) cube as illustrated in Figure 8.13. The resultant 2D ISAR images correspond to the time snapshots of the target while rotating.

8.3.3.3 Example for JTF-Based Rotational MOCOMP An example for the above algorithm is demonstrated for the airplane model in Figure 8.14a. The model consists of ideal point scatterers that imitate a fighter aircraft. The simulation of the backscattered electric field is collected for a scenario such that the airplane is 16 km away from the radar and moving in the direction that makes a 30° angle with the radar's line-of-sight axis. The target has a translation speed of $v_t = 1$ m/s while rotating with an angular speed of $\omega_t = 0.24$ rad/s.

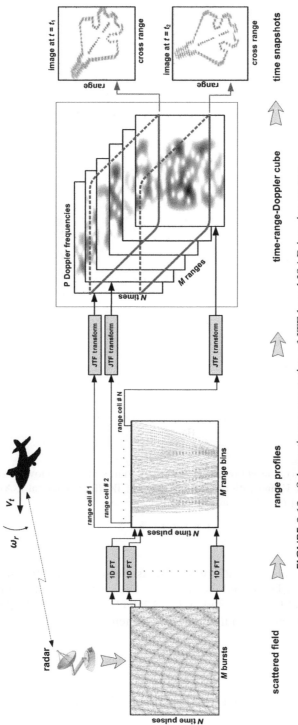

FIGURE 8.13 Schematic representation of JTF-based ISAR imaging system.

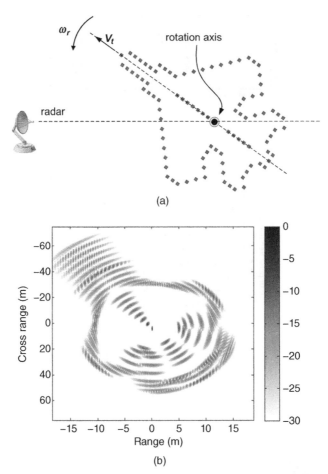

FIGURE 8.14 (a) A target (consists of perfect point scatterers) moving with a translational speed of V_t and with an angular speed of W_t, (b) Traditional ISAR image of the moving target ($v_t = 1$ m/s, $\omega_r = 0.24$ rad/s).

A total of 128 pulses in each of 512 bursts are selected for the SFCW radar simulation of backscattered electric field. The center frequency and the frequency bandwidth are selected as $f_c = 3.256$ GHz and $B = 512$ MHz, respectively. The corresponding pulse duration is then

$$T_p = \frac{(\# \, of \, pulses - 1)}{B} = 248.05 \; ns. \tag{8.34}$$

The PRF is chosen as 20 KHz. So, the PRI or the time between the two consecutive bursts is then

$$\text{PRI} = \frac{1}{\text{PRF}} = 50 \text{ μs}. \tag{8.35}$$

First, the traditional ISAR image is obtained via applying 2D IFT to the back-scattered field. The resulting image is plotted in Figure 8.14b, where the image suffers from blurring effect due to fast rotation rate of the target. Since the target's angular and translational location are different for the first and the last pulse of the radar, the maneuvering effect can also be seen in the ISAR image as the aircraft's image is smeared. This is pretty analogous to any optical imaging system: When the object point is moving fast, it occupies several pixels in the image during the time the lens stays open. Therefore, the resulting picture of the fast-moving object becomes blurred.

As suggested in the above algorithm, the time-dependent ISAR images of the target can be formed, thanks to the JTF-based ISAR imaging process that can take time snapshots of the scene. Therefore, we applied the above methodology to the backscattered field data from the target while it was moving. During the implementation of the algorithm, a Gabor-wavelet transform that uses a Gaussian blur function [38] is employed as the JTF tool. At the end of the algorithm steps, a corresponding 3D time-range-cross-range cube is obtained for the simulated moving target. In Figure 8.15, time snapshots of the 3D time-range-cross-range cube are plotted for the selected nine different time instants. Each subfigure corresponds to a particular ISAR image for a particular time instant. As time progresses, the movement of the fighter's radar image is clearly observed by looking from the first ISAR image to the last one.

8.3.4 Algorithm for JTF-Based Translational and Rotational MOCOMP

Here, we will briefly present an algorithm that compensates the motion errors in two steps. In the first step, the motion parameters are estimated using the well-known *matching pursuit (MP) technique* and compensate the translational motion errors. In the second step, we apply a JTF tool to compensate the rotational motion errors.

Assuming that the target has both translational and rotational motion components, the time-domain backscattered signal at the radar receiver can be approximated as given in Equation 8.32. In the first step of the algorithm, translational motion parameters such as translation radial speed and translation radial acceleration are estimated and then compensated for afterward. We start the algorithm by rewriting the formula in Equation 8.32 as the following:

$$g(x,t) = \sum_{k=1}^{K} A_k(x, y_k) \cdot \exp\left\{-j4\pi \frac{f_o}{c} x\right\} \cdot \varphi^t(R_o, v_t, a_t, t) \cdot \varphi^r(\omega_r, \alpha_r, x, y_k, t),$$

$$\tag{8.36}$$

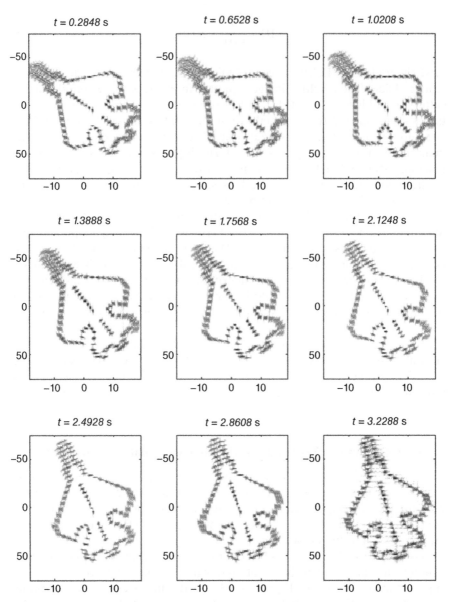

FIGURE 8.15 2D range-cross range ISAR images for different time snapshots (images are retrieved from the 3D time-range-cross-range cube).

where φ^t and φ^r are the phase terms that contain both the translational and the rotational motion parameters as

$$\varphi^t(v_t, a_t, t) = \exp\left\{-j4\pi\frac{f_o}{c}\left[\left(R_o + v_t t + \frac{1}{2}a_t t^2\right) + \cdots\right]\right\}$$

$$\varphi^r(\omega_r, \alpha_r, x, y_k, t) = \exp\left\{-j4\pi\frac{f_o}{c}\left[\omega_r y_k t + \frac{1}{2}(-\omega_r^2 x + \alpha_r y_k)t^2 + \cdots\right]\right\}.$$

(8.37)

As the first goal, translational MOCOMP has to be accomplished. We start with the algorithm by estimating the translational motion parameters with the help of an MP-type searching routine. In general, an MP algorithm predicts a suboptimal solution to the problem of a signal with known basis functions of unknown parameters. The MP algorithm is commonly used in time-frequency analysis [39, 40] and also in radar imaging problems [15, 41].

By utilizing the MP algorithm, the unknown translational motion parameters of v_t and a_t are estimated with an iterative search by projecting the received signal with every motion parameter basis on a 2D space. The parameters that give the largest projection are assumed to be estimated values, as the search can be modeled as

$$A = \max_{(v_t, a_t)}\left\{\langle g(x, t) \cdot \varphi^t(v_t, a_t, t)\rangle\right\},$$

(8.38)

where the inner product is defined as

$$\langle g \cdot \varphi^t \rangle = \int_{v_t}\int_{a_t} g(x, t) \cdot \left(\varphi^t(v_t, a_t, t)\right)^* d(a_t)d(v_t).$$

(8.39)

After the end of this iterative search, the parameter values of v_t^{est} and a_t^{est} that maximize the projection of the received signal onto 2D $v_t - a_t$ space are considered to be the resultant estimated values. Then, the translational MOCOMP is finalized by multiplying the original received signal with the phase correction term as

$$s(x, t) = g(x, t) \cdot \varphi^t\left(v_t^{est}, a_t^{est}, t\right)^*.$$

(8.40)

Once the translational motion errors are mitigated, only the rotational motion effects stay in the received signal. In the second part of the algorithm, therefore, the compensation for rotational motion errors is done. For this goal, the JTFT-based MOCOMP routine as explained in Section 8.3.3.2 can be applied.

8.3.4.1   A Numerical Example   We will now demonstrate an example that simulates the above algorithm. The target is assumed to consist of a set of perfect point scatterers that have equal scattering amplitudes as shown in Figure 8.7. The target's initial location is $R_o = 1.3$ km away from the radar and

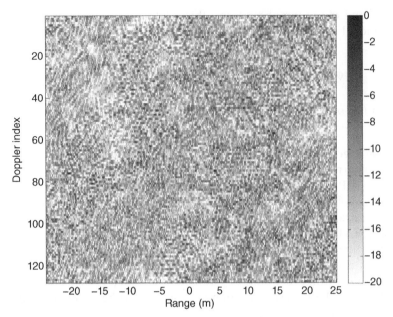

FIGURE 8.16   Conventional ISAR image of the airplane target with translational and rotational motion (no compensation).

moving with a radial translational velocity of $v_t = 35$ m/s and with a radial translational acceleration of $a_t = -1.9$ m/s^2. The target's angular velocity of the target is $\varphi_r = 0.15$ rad/s (8.5944°/s). The radar's starting frequency is $f_o = 3$ GHz and the total bandwidth is $B = 384$ MHz. The radar transmitter sends out 128 modulated pulses in each of 512 bursts. PRF is chosen to be 20 KHz.

Without applying any compensation routine, the regular range-Doppler ISAR image of the target is formed using the traditional imaging methodology. The corresponding range-Doppler ISAR image is shown in Figure 8.16. As clearly seen from the figure, the image is highly distorted, defocused, and blurred due to high velocity values in both the translational and the rotational directions. To observe the frequency (or phase) shifts among the received time pulses, the spectrogram for consecutive time pulses is shown in Figure 8.17. This spectrogram clearly demonstrates the severe frequency shifts between the pulses. Because of the translational acceleration, the nonlinearity of the frequency shifts is also observed. After compensating for the errors associated with target's motion, these shifts are expected to be minimal.

In the first step of the algorithm that was explained in Section 8.3.3.4, the translational motion parameters together with target's initial distance R_o were estimated using an MP-type search routine. After this exhaustive iterative search, the correct values of $R_o^{est} = 1.3$ km, $v_t^{est} = 35$ m/s, and $a_t^{est} = -1.9$ m/s^2 were successfully found. Figure 8.18 demonstrates the 2D search space of MP

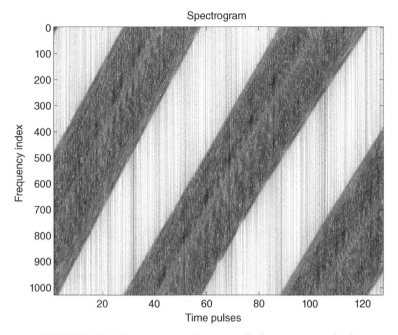

FIGURE 8.17   Spectrogram of range cells (no compensation).

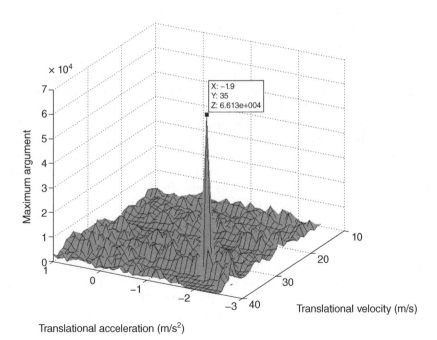

FIGURE 8.18   2D Matching pursuit search space for the translational velocity and the translational acceleration.

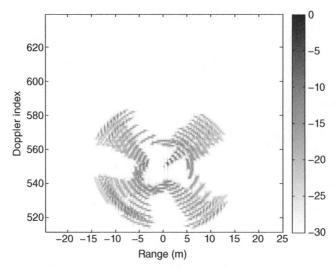

FIGURE 8.19 ISAR image of the airplane target after translational motion compensation.

in the translational radial velocity and translational radial acceleration. As seen from the figure, the argument in MP search makes a maximum when v_t^{est} becomes equal to 35 m/s and a_t^{est} equals to -1.9 m/s^2.

After the translational motion parameters of the target were predicted, the translational MOCOMP was finalized by employing the formula in Equation 8.40. Then, the ISAR image corresponding to the modified electric field is plotted in Figure 8.19. This figure clearly demonstrates the success of the translational MOCOMP such that only the rotational motion-based defocusing is noted in the ISAR image. The spectrogram of the time pulses in the modified received signal is also plotted in Figure 8.20 to investigate the frequency shifts between the consecutive time pulses. As seen from this spectrogram, although severe frequency shifts mainly due to the target translational velocity were mitigated, there still exists some fluctuation in the phase of the modified received signal due to rotational motion errors.

In the second part of the algorithm, the errors associated with target's rotational motion are compensated. For this goal, a Gabor-wavelet transform that uses a *Gaussian blur function* [38] is employed as the JTF tool to compensate the rotational motion effects in the ISAR image; the resultant image is given in Figure 8.21 where all the phase errors due to the translational and the rotational motion of the target were eliminated. The resultant motion-free ISAR image is very well focused and the scattering centers around the target are well localized. The final check is also performed by looking at the spectrogram of the compensated received signal as plotted in Figure 8.22 where the frequencies (or the contents of the phases) of time pulses are well aligned.

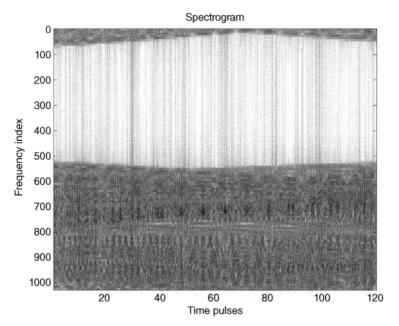

FIGURE 8.20 Spectrogram of time pulses (after translational motion compensation).

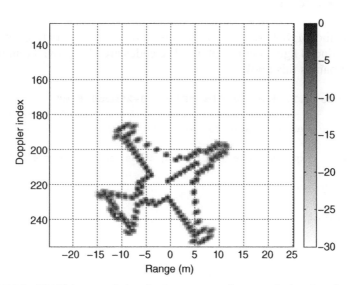

FIGURE 8.21 ISAR image of the airplane target after translational and rotational motion compensation.

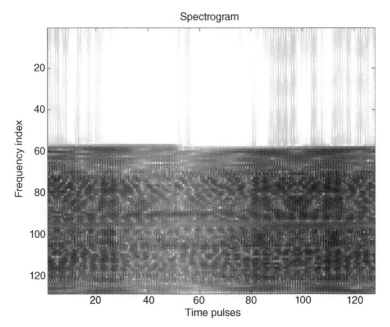

FIGURE 8.22   Spectrogram of time pulses (after translational and rotational motion compensation).

8.4   MATLAB CODES

Below are the Matlab source codes that were used to generate all of the Matlab-produced figures in Chapter 8. The codes are also provided in the CD that accompanies this book.

Matlab code 8.1: Matlab file "Figure8-2thru8-6.m"

```
%-------------------------------------------------------------
% This code can be used to generate Figures 8.2 - 8.6
%-------------------------------------------------------------
% This file requires the following files to be present in the
same
% directory:
%
% Fighter.mat
clear all
close all
clc
%---Radar parameters------------------------------------------
pulses = 128; % # no of pulses
burst = 128; % # no of bursts
c = 3.0e8; % speed of EM wave [m/s]
f0 = 10e9;   % Starting frequency of SFR radar system [Hz]
```

```
bw = 128e6; % Frequency bandwidth [Hz]
T1 = (pulses-1)/bw; % Pulse duration [s]
PRF = 20e3; % Pulse repetition frequency [Hz]
T2 = 1/PRF; % Pulse repetition interval [s]
dr = c/(2*bw); % range resolution [m]

%---Target parameters-----------------------------------------
W = 0.03; % Angular velocity [rad/s]
Vr = 70.0; % radial translational velocity of EM wave [m/s]
ar = 0.1; % radial accelation of EM wave [m/s^2]
R0 = 16e3; % target's initial distance from radar [m]
theta0 = 0; % Initial angle of target's wrt target [degree]

%---Figure 8.2------------------------------------------------
%load the coordinates of the scattering centers on the
fighter
load Fighter
h = plot(-Xc,Yc,'o', 'MarkerSize',8,'MarkerFaceColor', [0, 0,
1]);grid;
set(gca,'FontName', 'Arial', 'FontSize',12,'FontWeight',
'Bold');
axis([-35 35 -30 30])
xlabel('X [m]'); ylabel('Y [m]');
%Scattering centers in cylindirical coordinates
[theta,r]=cart2pol(Xc,Yc);
theta=theta+theta0*0.017455329; %add initial angle

i = 1:pulses*burst;
T = T1/2+2*R0/c+(i-1)*T2;%calculate time vector
Rvr = Vr*T+(0.5*ar)*(T.^2);%Range Displacement due to radial
vel. & acc.
Tetw = W*T;% Rotational Displacement due to angular vel.

i = 1:pulses;
df = (i-1)*1/T1; % Frequency incrementation between pulses
k = (4*pi*(f0+df))/c;
k_fac=ones(burst,1)*k;

%Calculate back-scattered E-field
 Es(burst,pulses) = 0.0;
 for scat = 1:1:length(Xc);
 arg = (Tetw - theta(scat) );
 rngterm = R0 + Rvr - r(scat)*sin(arg);
 range = reshape(rngterm,pulses,burst);
 range = range.';
 phase = k_fac.* range;
 Ess = exp(j*phase);
 Es = Es+Ess;
 end
 Es = Es.';
```

```
%---Figure 8.3--------------------------------------------
%Form ISAR Image (no compansation)
X = -dr*((pulses)/2-1):dr:dr*pulses/2;Y=X/2;
ISAR = abs(fftshift(fft2((Es))));
h = figure;
matplot2(X,1:pulses,ISAR,20);
colormap(1-gray); colorbar;
set(gca,'FontName', 'Arial', 'FontSize',12,'FontWeight',
'Bold');
xlabel('Range [m]');
ylabel('Doppler index');


%-Cross-Correlation Algorithm Starts here------------------
RP=(ifft(Es)).';% Form Range Profiles


for l=1:burst; % Cross-correlation between RPn & RPref
 cr(l,:) = abs(ifft(fft(abs(RP(1,:))).*
fft(abs(conj(RP(l,:))))));
 pk(l) = find((max(cr(l,:))== cr(l,:)));%Find max. ind. (range
shift) range)
end


Spk = smooth((0:pulses-1),pk,0.8,'rlowess');%smoothing the
delays
RangeShifts = dr*pk;% range shifts
SmRangeShifts = dr*Spk;% range shifts


RangeDif = SmRangeShifts(2:pulses)-SmRangeShifts(1:pulses-
1);%range differences
RangeDifAv = mean(RangeDif);% average range differences


T_burst = T(pulses+1)-T(1); % time between the bursts
Vr_Dif = (-RangeDif/T_burst);% estimated radial velocity from
each RP
Vr_av = (RangeDifAv /T_burst);% estimated radial velocity
(average)


%---Figure 8.4--------------------------------------------
h = figure;
plot(i,RangeShifts,'LineWidth',2);hold
plot(i,SmRangeShifts,'-.k.','MarkerSize',4);hold
axis tight
legend('RP shifts','Smoothed RP shifts');
set(gca,'FontName', 'Arial', 'FontSize',12,'FontWeight',
'Bold');
xlabel('range profile index');
%---Figure 8.5--------------------------------------------
```

```
h = figure;
subplot(211);plot(RangeDif,'LineWidth',2);
axis([1 burst -.75 -.25 ])
set(gca,'FontName', 'Arial', 'FontSize',10,'FontWeight',
'Bold');
xlabel('Range profile index');
ylabel('Range differences [m] ')

subplot(212);
plot(Vr_Dif,'LineWidth',2);
axis([1 burst Vr-5 Vr+5 ])
set(gca,'FontName', 'Arial', 'FontSize',10,'FontWeight',
'Bold');
xlabel('Range profile index');
ylabel('Radial speed [m/s] ')
text(15,74,['Actual Speed = ',num2str(Vr),' m/s , Est.
average speed = ',num2str(-Vr_av),' m/s']);
% Compansating the phase
f = (f0+df);% frequency vector
T = reshape(T,pulses,burst); %prepare time matrix
F = f.'*ones(1,burst); %prepare frequency matrix
Es_comp = Es.*exp((j*4*pi*F/c).*(Vr_av*T));%Phase of E-field
is compansated

%---Figure 8.6-----------------------------------------------
win = hanning(pulses)*hanning(burst).'; %prepare window
ISAR = abs((fft2((Es_comp.*win)))); % form the image
ISAR2 = ISAR(:,28:128);
ISAR2(:,102:128)=ISAR(:,1:27);
h = figure;
matplot2(Y,1:pulses,ISAR2,20); % motion compansated ISAR
image
colormap(1-gray);colorbar;
set(gca,'FontName', 'Arial', 'FontSize',12,'FontWeight',
'Bold');
xlabel('Range [m]'); ylabel('Doppler index');
title('Motion compansated ISAR image')
```

Matlab code 8.2: Matlab file "Figure8-7thru8-12.m"
```
%------------------------------------------------------------
% This code can be used to generate Figures 8.7 thru 8.12
%------------------------------------------------------------
% This file requires the following files to be present in the
same
% directory:
%
% fighter2.mat
clear all
close all
```

```
clc
%---Radar parameters--------------------------------------
pulses = 128; % # no of pulses
burst = 128; % # no of bursts
c = 3.0e8; % speed of EM wave [m/s]
f0 = 8e9; % Starting frequency of SFR radar system [Hz]
bw = 384e6; % Frequency bandwidth [Hz]
T1 = (pulses-1)/bw; % Pulse duration [s]
PRF = 14.5e3; % Pulse repetition frequency [Hz]
T2 = 1/PRF; % Pulse repetition interval [s]
dr = c/(2*bw); % slant range resolution [m]

%---Target parameters-------------------------------------
W = 0.06; % Angular velocity [rad/s]
Vr = 4.0; % radial translational velocity of EM wave [m/s]
ar = 0.6; % radial accelation of EM wave [m/s^2]
R0 =.5e3; % target's initial distance from radar [m]
theta0 = 125; % Initial angle of target's wrt target
[degree]

%---Figure 8.7--------------------------------------------
%load the coordinates of the scattering centers on the
fighter
load fighter2

h = plot(Xc,Yc,'o', 'MarkerSize',8,'MarkerFaceColor', [0, 0,
1]);
set(gca,'FontName', 'Arial', 'FontSize',12,'FontWeight',
'Bold');
axis([-20 20 -20 20])
xlabel('X [m]');
ylabel('Y [m]');
%Scattering centers in cylindirical coordinates
[theta,r]=cart2pol(Xc,Yc);
theta=theta+theta0*0.017455329; %add initial angle

i = 1:pulses*burst;
T = T1/2+2*R0/c+(i-1)*T2;%calculate time vector
Rvr = Vr*T+(0.5*ar)*(T.^2);%Range Displacement due to radial
vel. & acc.
Tetw = W*T;% Rotational Displacement due to angular vel.

i = 1:pulses;
df = (i-1)*1/T1; % Frequency incrementation between pulses
k = (4*pi*(f0+df))/c;
k_fac=ones(burst,1)*k;

%Calculate back-scattered E-field
 Es(burst,pulses)=0.0;
```

```
for scat=1:1:length(Xc);
arg = (Tetw - theta(scat) );
rngterm = R0 + Rvr - r(scat)*sin(arg);
range = reshape(rngterm,pulses,burst);
range = range.';
phase = k_fac.* range;
Ess = exp(-j*phase);
Es = Es+Ess;
end
Es = Es.';

%---Figure 8.8 --------------------------------------------------
%Form ISAR Image (no compansation)
X = -dr*((pulses)/2-1):dr:dr*pulses/2;Y=X/2;
ISAR = abs(fftshift(fft2((Es))));
h = figure;
matplot2(X,1:pulses,ISAR,20);
colormap(1-gray); colorbar;
set(gca,'FontName', 'Arial', 'FontSize',12,'FontWeight',
'Bold');
xlabel('Range [m]'); ylabel('Doppler index');

%---Figure 8.9 --------------------------------------------------
% JTF Representation of range cell
EsMp = reshape(Es,1,pulses*burst);
S = spectrogram(EsMp,128,64,120);
[a,b] = size(S);
h = figure;
matplot2((1:a),(1:b),abs(S),50);
set(gca,'FontName', 'Arial', 'FontSize',12,'FontWeight',
'Bold');
colormap(1-gray);
xlabel('time pulses');
ylabel('frequency index');
title('Spectrogram');
%Prepare time and frequency vectors
f = (f0+df);% frequency vector
T = reshape(T,pulses,burst); %prepare time matrix
F = f.'*ones(1,burst); %prepare frequency matrix

% Searching the motion parameters via min. entropy method
syc=1;
V = -15:.2:15;
A = -0.4:.01:1;
m = 0;
for Vest = V;
m = m+1;
n = 0;
for iv = A;
```

```
n = n+1;
VI(syc,1:2) = [Vest,iv];
S = exp((j*4*pi*F/c).*(Vest*T+(0.5*iv)*(T.^2)));
Scheck = Es.*S;
ISAR = abs(fftshift(fft2((Scheck))));
SumU = sum(sum(ISAR));
I = (ISAR/SumU);
Emat = I.*log10(I);
EI(m,n) = -(sum(sum(Emat)));
syc = syc+1;
end
end


[dummy,mm] = min(min(EI.')); %Find index for estimated
velocity
[dummy,nn] = min(min(EI)); %Find index for estimated
acceleration
%---Figure 8_10 -------------------------------------------------
h =surfc(A,V,EI);
colormap(gray)
set(gca,'FontName', 'Arial', 'FontSize',12,'FontWeight',
'Bold');
ylabel('Translational velocity [m/s]');
xlabel('Translational acceleration [m/s^2]');
zlabel ('Entropy value')
saveas(h,'Figure9-10.png','png');


% Form the mathing phase for compensation
Sconj = exp((j*4*pi*F/c).*(V(mm)*T+(0.5*A(nn)*(T.^2))));
% Compansate
S_Duz = Es.*Sconj;


%---Figure 8.11 -------------------------------------------------
% ISAR after compensation
h = figure;
matplot2(X,burst,abs(fftshift(fft2(S_Duz))),20);
colormap(1-gray);
colorbar;%grid;
set(gca,'FontName', 'Arial', 'FontSize',12,'FontWeight',
'Bold');
xlabel('Range [m]');
ylabel('Doppler index');


%---Figure 8.12 -------------------------------------------------
% Check the compensation using via JTF Representation of
range cells
EsMp = reshape(S_Duz,1,pulses*burst);
S = spectrogram(EsMp,128,64,120);
[a,b] = size(S);
```

```
h = figure;
matplot2((1:a),(1:b),abs(S),50);
colormap(1-gray);
set(gca,'FontName', 'Arial', 'FontSize',12,'FontWeight',
'Bold');
xlabel('time pulses');
ylabel('frequency index');
title('Spectrogram');
```

Matlab code 8.3: Matlab file "Figure8-14.m"
```
%------------------------------------------------------------
% This code can be used to generate Figure 8.14
%------------------------------------------------------------
% This file requires the following files to be present in the
same
% directory:
%
% scat_field.mat
clear all
close all

%---Load the Scattered Field ------------------------------
load scat_field

% Npulse = 128; % number of pulses in one burst
% Nburst = 512; % number of bursts
% f1 = 3e9; % starting frequency for the EM wave
% BWf = 512e6; % bandwidth of the EM wave
% T1 = (Npulse-1)/BWf; % pulse duration
% PRF = 20e3; % Pulse Repetition Frequency
% PRI = 1/PRF; % Pulse Repetition Interval
% W = 0.16; % angular velocity [rad/s]
% Vr = 1.0; % radial velocity [m/s]
% ar = 0.0; % acceleration [m/s2]

% c = 3.0e8; % speed of the EM wave

%---Figure 8.14(a)-------------------------------------------
plot(-Xc,Yc,'square', 'MarkerSize',5,'MarkerFaceColor', [0,
0, 1]);
set(gca,'FontName', 'Arial', 'FontSize',14,'FontWeight','Bold
');
xlabel('X [m]');
ylabel('Y [m]');
axis([min(-Xc)*1.1 max(-Xc)*1.1 min(Yc)*1.1 max(Yc)*1.1])

%---Figure 8.14(b)-------------------------------------------
%---Form Classical ISAR Image ------------------------------
w=hanning(Npulse)*hanning(Nburst)';
```

```
Es=Es.*w;
Es(Npulse*4,Nburst*4)=0;
ISAR=abs(fftshift(ifft2((Es))));
figure;matplot2(XX,YY,ISAR,30);
colorbar; colormap(1-gray);
set(gca,'FontName', 'Arial', 'FontSize',14,'FontWeight','Bold
');
xlabel('Range [m]');
ylabel('X-Range [m]');
```

Matlab code 8.4: Matlab file "Figure8-15.m"
```
%------------------------------------------------------------
% This code can be used to generate Figure 8.15
%------------------------------------------------------------
% This file requires the following files to be present in the
same
% directory:
%
% scat_field.mat
clear all
close all

%---Load the Scattered Field ------------------------------
load scat_field

% Npulse = 128; % number of pulses in one burst
% Nburst = 512; % number of bursts
% f1 = 3e9; % starting frequency for the EM wave
% BWf = 512e6; % bandwidth of the EM wave
% T1 = (Npulse-1)/BWf; % pulse duration
% PRF = 20e3; % Pulse Repetition Frequency
% PRI = 1/PRF; % Pulse Repetition Interval
% W = 0.16; % angular velocity [rad/s]
% Vr = 1.0; % radial velocity [m/s]
% ar = 0.0; % acceleration [m/s2]
c = 3.0e8; % speed of the EM wave
N=1;
T = 2*R0/c+(((1:Npulse*Nburst)-1)*PRI;
tst = PRI*Npulse; % dwell time
Nt=T(1:Npulse:Npulse*Nburst);

Es_IFFT = ifft(Es)'; % take 1D IFFT of field
%----Prepare JTF filter function-----------------------
n=0;figure
for frame=90:115:1100; % select time frames
 n=n+1; % counter for plotting
 fp=145;
 tp=((frame-1)*tst)/2 % window center for each frame
```

```
%----Prepare JTF filter function (Gabor function with
Gaussian Blur)
 Alpha_p=(0.04); % Blurring coefficient
 for i=1:Npulse
 part1=1/sqrt(2*pi*(Alpha_p)^2); % normalized term
 part2=exp(-((Nt-tp).^2)/(2*Alpha_p)); % Gaussian window term
 part3=exp((-j*2*pi*fp*(Nt-tp))/N); % Harmonic function
 GaborWavelet(i,1:Nburst)=part1*part2.*part3;
 fp=fp+1/(Npulse*tst);
 end;
% --- Wavelet Transform
 St =fftshift(GaborWavelet*Es_IFFT);
 subplot(3,3,n);matplot2(XX,YY,(St.'),25); colormap(1-gray);
 set(gca,'FontName', 'Arial', 'FontSize',10,'FontWeight','Bol
d');
 title(['t = ',num2str(tp),' s'],'FontAngle','Italic');
end
```

Matlab code 8.5: Matlab file "Figure8-16thru8-22.m"

```
%-----------------------------------------------------------
% This code can be used to generate Figures 8.16 thru 8.22
%-----------------------------------------------------------
% This file requires the following files to be present in the
same
% directory:
%
% Fighter3.mat
clear all
close all
clc
%---Radar parameters----------------------------------------
pulses = 128; % # no of pulses
burst = 512; % # no of bursts
c = 3.0e8; % speed of EM wave [m/s]
f0 = 3e9; % Starting frequency of SFR radar system [Hz]
bw = 384e6; % Frequency bandwidth [Hz]
T1 = (pulses-1)/bw; % Pulse duration [s]
PRF = 20e3; % Pulse repetition frequency [Hz]
T2 = 1/PRF; % Pulse repetition interval [s]
theta0 = 0; % Initial angle of target's wrt target
[degree]
W = 0.15; % Angular velocity [rad/s]
Vr = 35.0; % radial translational velocity of EM wave [m/s]
ar = -1.9; % radial accelation of EM wave [m/s^2]
R0 = 1.3e3; % target's initial distance from radar [m]
dr = c/(2*bw); % range resolution [m]
theta0 = -30; % Look angle of the target
```

```
%---Figure 8.16 -----------------------------------------
%load the coordinates of the scattering centers on the
fighter
load Fighter3
h = plot(-Xc,Yc,'o', 'MarkerSize',8,'MarkerFaceColor', [0, 0,
1]);
grid on;
set(gca,'FontName', 'Arial', 'FontSize',12,'FontWeight',
'Bold');
axis([-20 20 -20 20])
xlabel('X [m]'); ylabel('Y [m]');

%Scattering centers in cylindirical coordinates
[theta,r] = cart2pol(Xc,Yc);
theta = theta+theta0*0.017455329; %add initial angle

i = 1:pulses*burst;
T = T1/2+2*R0/c+(i-1)*T2;%calculate time vector
Rvr = Vr*T+(0.5*ar)*(T.^2);%Range Displacement due to radial
vel. & acc.
Tetw = W*T;% Rotational Displacement due to angular vel.

i = 1:pulses;
df = (i-1)*1/T1; % Frequency incrementation between pulses
k = (4*pi*(f0+df))/c;
k_fac = ones(burst,1)*k;

%Calculate back-scattered E-field
 Es(burst,pulses) = 0.0;
 for scat = 1:1:length(Xc);
 arg = (Tetw - theta(scat) );
 rngterm = R0 + Rvr - r(scat)*sin(arg);
 range = reshape(rngterm,pulses,burst);
 range = range.';
 phase = k_fac.* range;
 Ess = exp(-j*phase);
 Es = Es+Ess;
 end
 Es = Es.';

%---Figure 8.17 -----------------------------------------
%Form ISAR Image (no compansation)
X = -dr*((pulses)/2-1):dr:dr*pulses/2;Y=X/2;
ISAR = abs((fft2((Es))));
h = figure;
matplot2(X,1:pulses,ISAR,20);
colormap(1-gray);
colorbar;
set(gca,'FontName', 'Arial', 'FontSize',12,'FontWeight',
'Bold');
```

```
xlabel('Range [m]'); ylabel('Doppler index');

%---Figure 8.18 -----------------------------------------------
% JTF Representation of range cell
EsMp = reshape(Es,1,pulses*burst);
S = spectrogram(EsMp,128,64,128);
[a,b] = size(S);
h = figure;
matplot2((1:a),(1:b),abs(S),50);
set(gca,'FontName', 'Arial', 'FontSize',12,'FontWeight',
'Bold');
colormap(1-gray);
xlabel('time pulses');
ylabel('frequency index');
title('Spectrogram');

%Prepare time and frequency vectors
f = (f0+df);% frequency vector
T = reshape(T,pulses,burst); %prepare time matrix
F = f.'*ones(1,burst); %prepare frequency matrix

% Searching the motion parameters via Matching
Pursuit
syc = 1;
RR = 1e3:1e2:2e3;
V = 10:40;
A = -2.5:.1:1;
m = 0;
clear EI

for Vest = V;
 m = m+1;
 n = 0;
 for iv = A;
 n = n+1;
 p = 0;
 for Rest = RR;
 p = p+1;
 VI(syc,1:2) = [Vest,iv];
 S = exp((j*4*pi*F/c).*(Rest+Vest*T+(0.5*iv)*(T.^2)));
 Scheck = Es.*S;
 SumU = sum(sum(Scheck));
 EI(m,n,p) = abs(SumU);
 end
end
end

[dummy,pp] = max(max(max((EI)))); %Find index for estimated
velocity
```

```
[dummy,nn] = max(max((EI(:,:,pp)))); %Find index for
estimated velocity
[dummy,mm] = max(EI(:,nn,pp)); %Find index for estimated
acceleration
%---Figure 8.19 -------------------------------------------
figure;
h = surfc(A,V,EI(:,:,pp));
colormap(gray)
set(gca,'FontName', 'Arial', 'FontSize',12,'FontWeight',
'Bold');
ylabel('Translational velocity [m/s]');
xlabel('Translational acceleration [m/s^2]');
zlabel ('maximum argument')
%saveas(h,'Figure9-16.png','png');

% Form the mathing phase for compensation
Sconj = exp((j*4*pi*F/c).*(V(mm)*T+(0.5*A(nn)*(T.^2))));
% Compansate
S_Duz = Es.*Sconj;

%---Figure 8.20 -------------------------------------------
% ISAR after compensation
h = figure;
matplot(X,burst,abs(fftshift(fft2(S_Duz))),30);
colormap(1-gray);
colorbar;%grid;
set(gca,'FontName', 'Arial', 'FontSize',12,'FontWeight',
'Bold');
xlabel('Range [m]'); ylabel('Doppler index');
%---Figure 8.21 -------------------------------------------
% Check the compensation using via JTF Representation of
range cells
Sconjres = reshape(S_Duz,1,pulses*burst);
S = spectrogram(Sconjres,128,64,120);
[a,b] = size(S);
h =figure;
matplot2((1:a),(1:b),abs(S),50);
colormap(1-gray);
set(gca,'FontName', 'Arial', 'FontSize',12,'FontWeight',
'Bold');
xlabel('time pulses'); ylabel('frequency index');
title('Spectrogram');

%---This part for Rotational motion compensation-----------------
Ese = S_Duz;
win = hamming(pulses)* hamming(burst).';% Prepare Window
Esew = Ese.*win; % Apply window to the E-field
Es_IFFT = (ifft(Esew)).'; % Range profiles
```

```matlab
i = 1:pulses*burst;
T = T1/2+2*R0/c+(i-1)*T2;

%** Apply Gaussian Blur Filter via Gabor Function
N = 1; % Sampling #
tst = T2*pulses; % dwell time
t = T(1:pulses:pulses*burst); % time vector for bursts

fp = 160;
Alpha_p = 0.04; % Blurring coefficient
t_istenen = 100; % tp=1 sec, T=2.1845 sec
tp = ((t_istenen-1)*tst)/2; % Center of Gaussian window

% % Gabor function and Gaussian Blur function
 parca1 = 1/sqrt(2*pi*(Alpha_p)^2); % normalized term
 parca2 = exp(-((t-tp).^2)/(2*Alpha_p)); % Gaussian window
term
for i=1:pulses
 parca3 = exp((-j*2*pi*fp*(t-tp))/N); % Harmonic function
 GaborWavelet(i,1:burst) = parca1*parca2.*parca3;% Gabor
Wavelet func.
 fp=fp+1/(pulses*tst);
end;

% %** Wavelet Transform
St_Img = fftshift(GaborWavelet*Es_IFFT); % shift image to
the center

h = figure;
matplot2(X,pulses,(St_Img.'),30);
colorbar;
set(gca,'FontName', 'Arial', 'FontSize',12,'FontWeight',
'Bold');
colormap(1-gray);
grid;
xlabel('Range [m]');
ylabel('Doppler index');

%---Figure 8.22 -----------------------------------------------
EMp = reshape(St_Img,1,128*128);
S = spectrogram(EMp,256,120);
h = figure;
matplot2((1:pulses),(1:pulses),abs(S.'),60);
colormap(1-gray);
set(gca,'FontName', 'Arial', 'FontSize',12,'FontWeight',
'Bold');
xlabel('time pulses'); ylabel('frequency index');
title('Spectrogram');
```

REFERENCES

1 W. Haiqing, D. Grenier, G. Y. Delisle, and D.-G. Fang. Translational motion compensation in ISAR image processing. *IEEE Transactions on Image Processing* 4(11) (1995) 1561–1571.

2 P. H. Eichel, D. C. Ghiglia, and C. V. Jakowatz, Jr. Speckle processing method for synthetic-aperture-radar phase correction. *Optics Letters* 14(1) (1989) 1–3.

3 C.-C. Chen and H. C. Andrews. Target-motion-induced radar imaging. *IEEE Transactions on Aerospace and Electronic Systems* AES-16(1) (1980) 2–14.

4 Z. D. Zhu and X. Q. Wu. Range-Doppler imaging and multiple scatter-point location, *Journal of Nanjing Aeronautical Institute*, 23 (1991) 62–69.

5 T. Itoh, H. Sueda, and Y. Watanabe. Motion compensation for ISAR via centroid tracking. *IEEE Transactions on Aerospace and Electronic Systems* 32(3) (1996) 1191–1197.

6 J. S. Cho, D. J. Kim, and D. J. Park. Robust centroid target tracker based on new distance features in cluttered image sequences. *IEICE Transactions on Information and Systems* E83-D(12) (2000) 2142–2151.

7 G. Y. Wang and Z. Bao. The minimum entropy criterion of range alignment in ISAR motion compensation. Proceedings of Radar Conference, 236–239, 1999.

8 L. Xi, L. Guosui, and J. Ni. Autofocusing of ISAR image based on entropy minimisation. *IEEE Transactions on Aerospace and Electronic Systems* 35 (1999) 1240–1252.

9 T. M. Calloway and G. W. Donohoe. Subaperture autofocus for synthetic aperture radar. *IEEE Transactions on Aerospace and Electronic Systems* 30(2) (1994) 617–621.

10 D. E. Wahl. Phase gradient autofocus—A robust for high resolution SAR phase correction. *IEEE Transactions on Aerospace and Electronic Systems* AES-30 (1994) 827–835.

11 T. Küçükkılıç. *ISAR Imaging and Motion Compensation*, M.S. thesis, Middle East Tech. Univ., 2006.

12 B. Haywood and R. J. Evans. Motion compensation for ISAR imaging. Proceedings of Australian Symposium on Signal Processing and Applications, 1989, pp. 112–117.

13 I. S. Choi, B.-L. Cho, and H.-T. Kim. ISAR motion compensation using evolutionary adaptive wavelet transform. *IEE Proceedings—Radar Sonar and Navigation* 150(4) (2003) 229–233.

14 V. C. Chen and H. Ling. *Time-frequency transforms for radar imaging and signal processing*. Artech House, Norwood, MA, 2002.

15 Y. Wang, H. Ling, and V. C. Chen. ISAR motion compensation via adaptive joint time-frequency technique. *IEEE Transactions on Aerospace and Electronic Systems* 34(2) (1998) 670–677.

16 V. C. Chen and S. Qian. Joint time-frequency transform for radar range-Doppler imaging. *IEEE Transactions on Aerospace and Electronic Systems* 34(2) (1998) 486–499.

17   J. C. Kirk. Motion compensation for synthetic aperture radar. *IEEE Transactions on Aerospace and Electronic Systems* 11 (1975) 338–348.

18   J. Walker. Range-Doppler imaging of rotating objects. *IEEE Transactions on Aerospace and Electronic Systems* 16 (1980) 23–52.

19   H. Wu, et al. Translational motion compensation in ISAR image processing. *IEEE Transactions on Image Processing* 14(11) (1995) 1561–1571.

20   R. Xu, Z. Cao, and Y. Liu. Motion compensation for ISAR and noise effect. *IEEE Aerospace and Electronic Systems Magazine* 5(6) (1990) 20–22.

21   F. Berizzi, M. Martorella, B. Haywood, E. Dalle Mese, and S. Bruscoli. A survey on ISAR autofocusing techniques. International Conference on Image Processing (ICIP 2004), vol. 1: 9–12, 2004.

22   M. Martorella, B. Haywood, F. Berizzi, and E. Dalle Mese. Performance analysis of an ISAR contrast-based autofocusing algorithm using real data. Proceedings of the International Radar Conference, 30–35, 2003.

23   F. Wenxian, L. Shaohong, and H. Wen. Motion compensation for spotlight SAR mode imaging. Proceedings of the CIE Interntional Conference on Radar, 938–942, 2001.

24   J. Yu and J. Yang. Motion compensation of ISAR imaging for high-speed moving target. IEEE International Symposium on Knowledge Acquisition and Modeling Workshop, 124–127, 2008.

25   J. M. Munoz-Ferreras, J. Calvo-Gallego, F. Perez-Martinez, A. Blanco-del-Campo, A. Asensio-Lopez, and B. P. Dorta-Naranjo. Motion compensation for ISAR based on the shift-and-convolution algorithm. IEEE Conference on Radar, 24–27, 2006.

26   B. D. Steinberg. Microwave imaging of aircraft. *Proceedings of the IEEE* 76(12) (1988) 1578–1592.

27   W. G. Carrara, R. S. Goodman, and R. M. Majevski. *Spotlight synthetic aperture radar: Signal processing algorithms*. Artech House, Norwood, MA, 1995.

28   S. Werness, W. Carrara, L. Joyce, and D. Franczak. Moving target imaging algorithm for SAR data. *IEEE Transactions on Aerospace and Electronic Systems* AES-26 (1990) 57–67.

29   W. S. Cleveland. Robust locally weighted regression and smoothing scatterplots. *Journal of the American Statistical Association* 74(368) (1979) 829–836.

30   C. E. Shannon. Prediction and entropy of printed English. *The Bell System Technical Journal* 30 (1951) 50–64.

31   S. Y. Shin and N. H. Myung. The application of motion compensation of ISAR image for a moving target in radar target recognition. *Microwave and Optical Technology Letters* 50(6) (2008) 1673–1678.

32   A. Moghaddar and E. K. Walton. Time-frequency distribution analysis of scattering from waveguide cavities. *IEEE Transactions on Antennas and Propagation* 41 (1993) 677–680.

33   C. Ozdemir and H. Ling. Joint time-frequency interpretation of scattering phenomenology in dielectric-coated wires. *IEEE Transactions on Antennas and Propagation* 45(8) (1997) 1259–1264.

34   L. C. Trintinalia and H. Ling. Extraction of waveguide scattering features using joint time-frequency ISAR. *IEEE Microwave and Guided Wave Letters* 6(1) (1996) 10–12.

35 K.-T. Kim, I.-S. Choi, and H.-T. Kim. Efficient radar target classification using adaptive joint time-frequency processing. *IEEE Transactions on Antennas and Propagation* 48(12) (2000) 1789–1801.

36 H. Kim and H. Ling. Wavelet analysis of radar echo from finite-sized targets. *IEEE Transactions on Antennas and Propagation* 41 (1993) 200–207.

37 X.-G. Xia, G. Wang, and V. C. Chen. Quantitative SNR analysis for ISAR imaging using joint time-frequency analysis-Short time Fourier transform. *IEEE Transactions on Aerospace and Electronic Systems* 38(2) (2002) 649–659.

38 http://en.wikipedia.org/wiki/Gaussian_blur

39 S. G. Mallat and Z. Zhang. Matching pursuit with time-frequency dictionaries. *IEEE Transactions on Signal Processing* 41 (1993) 3397–3415.

40 P. J. Franaszczuk, G. K. Bergey, P. J. Durka, and H. M. Eisenberg. Time-frequency analysis using the matching pursuit algorithm applied to seizures originating from the mesial temporal lobe. *Electroencephalography and Clinical Neurophysiology* 106(6) (1998) 513–521.

41 T. Su, C. Özdemir, and H. Ling. On extracting the radiation center representation of antenna radiation patterns on a complex platform. *Microwave and Optical Technology Letters* 26(1) (2000) 4–7.

CHAPTER NINE

Some Imaging Applications Based on Inverse Synthetic Aperture Radar

In this chapter, some of the recent imaging applications based on inverse synthetic aperture radar (ISAR) imaging methodology are presented. As the conventional ISAR imaging algorithm is based on the assumption that the transmitter and the receiver are situated at the far field of the target, we will show in this chapter that it is also possible to form radar images of an object of interest if the transmitter or both the transmitter and the receiver are at the near field of the target.

As thoroughly discussed in the previous chapters, an ISAR image displays the amplitudes and the locations of the different scattering mechanisms from the target (or the platform). If we look at the ISAR image from the physical cause-and-effect point of view, an ISAR image can be regarded as the far-field electromagnetic (EM) wave interaction between the transmitter and the receiver over the target. In a similar manner, it may also be possible to form the radar image for other types of interactions such as near-field-to-far-field interaction or near-field-to-near-field interaction. The radar images of such scenarios can be successfully imaged using an analogy with the conventional ISAR imaging concept.

In this chapter, two of such radar imaging algorithms, namely, antenna synthetic aperture radar (ASAR) and antenna coupling synthetic aperture radar (ACSAR), will be presented. Another interesting application is forming the image from the EM scattering of subsurface objects or discontinuities. This

Inverse Synthetic Aperture Radar Imaging with MATLAB Algorithms, First Edition.
Caner Özdemir.
© 2012 John Wiley & Sons, Inc. Published 2012 by John Wiley & Sons, Inc.

problem is also called in the radar community as ground-penetrating radar (GPR) imaging. As we shall see in Section 9.3, the use of synthetic aperture radar (SAR)/ISAR approaches in GPR imaging problems provides well-focused images of buried objects.

9.1   IMAGING ANTENNA-PLATFORM SCATTERING: ASAR

It is well known that antenna characteristics can be dramatically altered by the platform that supports the antenna structure (Fig. 9.1). It becomes there-fore important to study platform effects and explore ways to decrease or diminish such effects. When the antenna is mounted on a platform, a very important problem is formulating the "inverse" algorithm of spatially pin-pointing where on the platform the undesirable scattering is.

The approach for pinpointing the locations of the platform scattering is based on the ISAR concept. In fact, the antenna-platform scattering is imaged by extending the ISAR imaging concept to the antenna radiation problem. By collecting the multifrequency, multiangle radiation data from an antenna mounted on a complex platform, it will be demonstrated that an ASAR image of the platform (or target) that displays the dominant secondary radiation off the platform can be generated [1, 2]. Contrary to conventional ISAR imaging, a key complication of the ASAR imaging scenario is that the antenna is located in the near field of the platform, as illustrated in Figure 9.2.

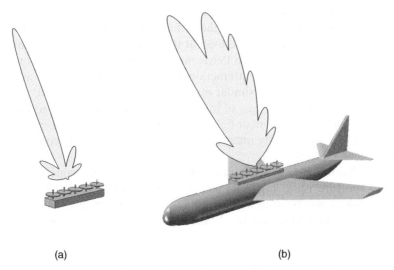

(a)   (b)

FIGURE 9.1   (a) The pattern of a typical stand-alone antenna. (b) The pattern is altered by the complex platform that holds the antenna.

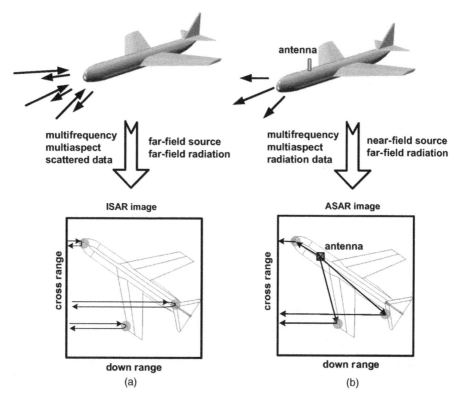

FIGURE 9.2   (a) The conventional ISAR scenario. (b) The ASAR scenario.

9.1.1   The ASAR Imaging Algorithm

The formulation of the ASAR algorithm starts with the consideration of the geometry given in Figure 9.3. Without loss of generality, the transmitter antenna is assumed to be located at the origin, and the scattered electric field is collected around the $-x$ direction. Our aim is to image any point scatterer, $P(x_0, y_0, z_0)$, on the platform. In addition to the direct radiation from the antenna, the secondary radiation from the antenna-platform interaction (i.e., the scattering from the platform) at the point $P(x_0, y_0, z_0)$ can be approximated as

$$E^s \cong A \cdot e^{-jk \cdot r_0} \cdot e^{-j\vec{k} \cdot \vec{r}_0}$$
$$= A \cdot e^{-jk \cdot r_0} \cdot e^{-j(k_x \cdot x_0 \hat{x} + k_y \cdot y_0 \hat{y} + k_z \cdot z_0 \hat{z})}, \tag{9.1}$$

where A is the strength of the scattered signal and $r_0 = (x_0^2 + y_0^2 + z_0^2)^{1/2}$ is the path traveled by the radiated signal from the antenna to the scatterer at point P, and k is the free-space wave number. In Equation 9.1, the first exponential accounts for the phase lag from the antenna to point P, and the second

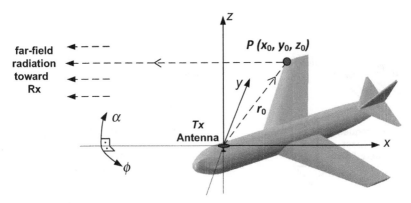

FIGURE 9.3   The path of the radiated signal which is scattered off a point on the platform.

exponential accounts for the additional phase delay from the scatterer to the observation direction relative to the direct radiation from the antenna. Therefore, the phase center of the problem is again selected as the origin as has been done in ISAR formulation.

In Equation 9.1, the spatial frequencies in x, y, and z directions are defined as

$$\vec{k}_x = k\sin\theta\cos\varnothing \cdot \hat{x} = k\cos\alpha\cos\varnothing \cdot \hat{x}$$
$$\vec{k}_y = k\sin\theta\sin\varnothing \cdot \hat{y} = k\cos\alpha\sin\varnothing \cdot \hat{y} \qquad (9.2)$$
$$\vec{k}_z = k\cos\theta \cdot \hat{z} = k\sin\alpha \cdot \hat{z}.$$

Here, α is the angle from the x–y plane toward the z-axis, and $\varnothing$ is the angle defined from $-x$ to $-y$ as depicted in Figure 9.3. Notice that $\alpha = 90 - \theta$ where θ is defined in the spherical coordinate system as the angle from the z-axis toward the x–y plane. Then, the scattered electric field is given as

$$E^s(k, \alpha, \varnothing) = A \cdot e^{-jk\cdot r_0} \cdot e^{-jk(\cos\alpha\cos\varnothing \cdot x_0 + \cos\alpha\sin\varnothing \cdot y_0 + \sin\alpha \cdot z_0)}. \qquad (9.3)$$

Next, we apply the small-bandwidth, small-angle assumptions, that is, the observation angles α and $\varnothing$ are small, and the radiation data are collected within a small frequency bandwidth. Then, the spatial frequencies k_x, k_y, and k_z are approximated as

$$k_x = k \cdot \cos\alpha \cdot \cos\varnothing \cong k$$
$$k_y = k \cdot \cos\alpha \cdot \sin\varnothing \cong k_c \cdot \varnothing \qquad (9.4)$$
$$k_z = k \cdot \sin\alpha \cong k_c \cdot \alpha,$$

where k_c is the wave number at the center frequency. With these assumptions, the scattered field formulation is, then, reduced to

$$E^s(k, \alpha, \emptyset) = A \cdot e^{-jk \cdot (r_0 + x_0)} \cdot e^{-jk_c \emptyset \cdot y_0} \cdot e^{-jk_c \alpha \cdot z_0}. \tag{9.5}$$

In the above equation, it is obvious that there are Fourier transform (FT) relationships between k and $(r + x)$, $(k_c \emptyset)$ and y, and $(k_c \alpha)$ and z, respectively. The idea behind the ASAR algorithm is that if the three-dimensional (3D) inverse Fourier transform (IFT) of the scattered field is taken with respect to k, $(k_c \emptyset)$, and $(k_c \alpha)$, it is possible to extract the information about the location and strength of the point scatterer on the platform. For simplicity, we define a new variable of $u = r + x$ to be used during the analysis of ASAR [1, 2]. If the 3D IFT of Equation 9.3 is taken with respect to k and $(r + x)$, $k_c \emptyset$ and y, we get

$$
\begin{aligned}
E^s(u, y, z) &= \mathcal{F}_3^{-1}\{E^s(k, \alpha, \emptyset)\} \\
&= \mathcal{F}_3^{-1}\{A \cdot e^{-jk \cdot (r_0 + x_0)} \cdot e^{-jk_c \emptyset \cdot y_0} \cdot e^{-jk_c \alpha \cdot z_0}\} \\
&= \mathcal{F}_3^{-1}\{A \cdot e^{-jk \cdot u_0} \cdot e^{-jk_c \emptyset \cdot y_0} \cdot e^{-jk_c \alpha \cdot z_0}\} \\
&= A \cdot \delta(u - u_0, y - y_0, z - z_0) \\
&\triangleq ASAR(u, y, z).
\end{aligned} \tag{9.6}
$$

Therefore, by taking the 3D IFT of the multifrequency, multiaspect secondary radiation data, the point scatterer will manifest itself as a peak in the ASAR image at (u_0, y_0, z_0) with the correct scattering amplitude of A. In practice, the exact spot size of the peak will of course not be infinitesimal but will inversely depend on the frequency bandwidth and the look-angle width.

Note that an additional tricky situation exists as the ASAR image has not yet been formed in the real coordinate system of (x, y, z). To do this, a transformation should be applied to pass the ASAR image from the (u, y, z) domain to the desired (x, y, z) domain. This step is performed by the following transformation [2, 3]:

$$x = \frac{u^2 - y^2 - z^2}{2u} \tag{9.7}$$

The effect of the u-to-x transformation is illustrated in Figure 9.4. In Figure 9.4a, the uniform grid of $u - y$ for $z = 0$ is plotted. As is obvious from Equation 9.7, transformation from u to x is not linear; therefore, the resulting x–y grid after the transformation becomes nonuniform as shown in Figure 9.4b. The points $x > 0$ correspond to the region behind the antenna (i.e., away from the observation region of interest). In this region the grid is nearly uniform. However, the x–y grid becomes highly distorted for the $x < 0$ region that corresponds to the region in front of the antenna toward the observation direction. Therefore, as will be demonstrated while presenting

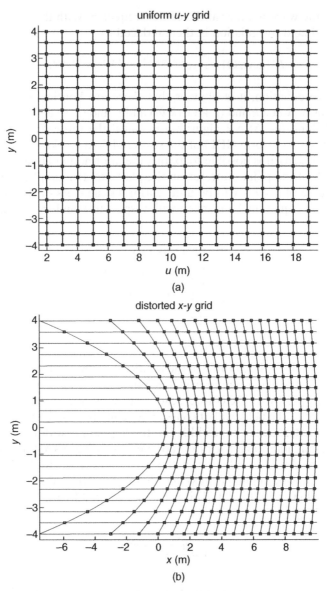

FIGURE 9.4   An example showing the effect of the u-to-x transformation: (a) the uniform u–y grid for $z = 0$; (b) the corresponding distorted x–y grid.

ASAR examples, the scattering mechanisms from the region in front of the antenna will experience distortion due to this transformation.

If the ASAR imaging algorithm is summarized, the following are the key three steps:

1 Collect the multifrequency, multiaspect radiated electric field data $E^s(k, \alpha, \emptyset)$ from the antenna-platform pair,

2. Take the 3D IFT of $E^s(k, \alpha, \emptyset)$ to generate $ASAR(u, y, z)$.

3. Use the u-to-x transformation formula in Equation 9.7 to produce $ASAR(x, y, z)$.

Once the ASAR image is constructed, the dominant scattering locations on the platform at the observation angles of interest should then be manifested as peaks in the ASAR image. Before demonstrating ASAR images using examples, several comments about the features of ASAR imaging are listed below:

1. Using the analogy to the ISAR imaging concept, the x direction is referred to as the *down range* or *range* direction. The y and the z axes are two orthogonal *cross-range* directions (see Fig. 9.3).

2. From Equation 9.5, it is obvious to note that the resolution in x comes from frequency diversity, while the resolutions in y and z directions are facilitated by the two orthogonal angular diversities.

3. Since the ASAR concept is based on one-way radiation, unlike the two-way propagation as in the case of ISAR imaging, it requires twice the bandwidth in frequency and angles to have the same resolution in ISAR imagery. For instance, to achieve 15 cm of down-range resolution requires 1 GHz of bandwidth in ISAR, but 2 GHz of bandwidth in ASAR.

4. Contrary to ISAR, an additional transformation from the u–y plane to the x–y plane is necessary to unwrap the nonlinear phase term arising from the propagation delay from the antenna to the point scatterer on the platform. This results in an additional image distortion which can be severe along the $-x$ direction. Physically, this phenomenon can be explained by the fact that it is not possible to resolve point scatterers along the $-x$-axis, as all these mechanisms have the same path length. Although the image distortion can be partially alleviated by oversampling in the u-domain, it cannot be completely circumvented.

5. The ASAR algorithm, just like the ISAR algorithm, is based on a single-bounce assumption. Multiple bounce mechanisms will not be correctly mapped in the ASAR image to the actual locations of the platform scattering. However, as was shown in Reference 3, a higher order scattering mechanism will simply be delayed in the down-range direction but will be mapped to the last bounce point on the platform along the two cross-range directions.

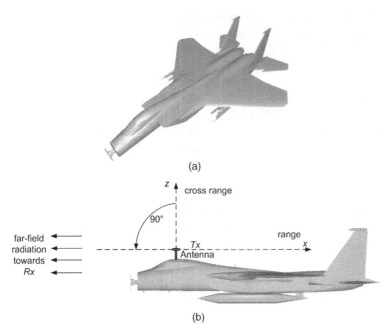

FIGURE 9.5   (a) Aircraft's CAD model used for ASAR imaging. (b) The simulation geometry for the aircraft–antenna pair. The monopole antenna is on the top of the cockpit.

9.1.2   Numerical Example for ASAR Imagery

An example of the ASAR imaging algorithm is demonstrated via an antenna mounted on a complex airplane model. The computer-aided design (CAD) model of the airplane consists of approximately 8000 triangular patches and is assumed to be a perfect electric conductor (see Fig. 9.5). A monopole antenna whose length is electrically small ($\sim\lambda/10$) is used as the transmitting source and positioned above the cockpit. The EM field calculation of radiation from the airplane–antenna pair is carried out by a physical optics-shooting and bouncing ray (PO-SBR)-based simulator [4]. During the one-way radiation calculation of the scenario, the radiated field data are collected between 9.45 GHz and 10.55 GHz for a total of 128 discrete frequencies. While changing the frequencies, the data are collected at the far field on a two-dimensional (2D) angular grid of different azimuth and elevation angles. The collection angles in azimuth range from −1.67° to 1.67° for a total of 32 equally spaced points. The elevation angle variation is between −0.85° and 0.85° for 8 discrete angle points. Therefore, for each frequency point, the radiated data are collected for the total of $32 \times 8 = 256$ aspect points. In the computed data, only the scattered field from the aircraft platform is included; the direct (or primary) radiation due to the antenna in the absence of the platform is not considered.

Next, the 3D ASAR image is generated using the algorithm discussed earlier. A 3D Hanning window and two-times zero-padding procedure are used prior to the IFT operation in the algorithm. Since the ASAR image is 3D, the 2D ASAR slices of the fighter model are presented on the x–y plane for different z values as given in Figure 9.6. A number of distinct secondary point scatterers on the wings and the tail of the aircraft are observed. Because the antenna is located near the cockpit, the strongest platform scattering occurs from the specular points around the nose of the fighter. However, as expected, the region from antenna to observation region around the $-x$ direction is distorted due to the nonlinear feature of the u-to-x transformation. The 2D projected ASAR image in the y–z plane is also provided in Figure 9.7.

The distortion problem in ASAR imaging can be overcome with the help of the scattering center concept [3].

9.2 IMAGING PLATFORM COUPLING BETWEEN ANTENNAS: ACSAR

While ASAR can successfully image the secondary radiators off the platform, an equally important problem is to image the platform interaction between multiple antennas mounted on the platform. The EM coupling between antennas on a complex platform is actually important for the antenna designer and the electromagnetic compatibility (EMC)/electromagnetic interference (EMI) engineer. It is, therefore, useful to identify the areas on the platform where the strong coupling between the different antennas occurs. For this purpose, the ACSAR imaging algorithm has been developed to pinpoint the dominant scattering locations on the platform that give rise to antenna interactions from the coupling data between the antennas [3, 5, 6].

In approaching this problem, the ISAR concept is utilized. While the ASAR imaging algorithm has been derived in the previous section, the ISAR concept has been extended to the near-field antenna radiation problem. In the ACSAR case, ISAR concept is further extended to the near-field antenna coupling problem in order to generate the ACSAR image of the platform.

In this section we will show that an ACSAR image that displays the dominant scattering locations on the platform between the antennas can be formed by collecting the multifrequency, multispatial coupling data. To achieve the required spatial diversity, the data should be collected on a 2D grid at the receiver site. The differences between the ISAR scenario and the ACSAR scenario are explained in Figure 9.8. While the transmitter and receiver antennas are at the far field of the target for the ISAR problem, these antennas are in the near field of the platform in the ACSAR case.

In Figure 9.9, a conceptual comparison of different imaging algorithms is pointed out by showing the EM radiation situations. For the ISAR case, both the transmitter and the receiver are at the far field of the target. In extending the ISAR concept to the ASAR concept, the transmitting antenna is located

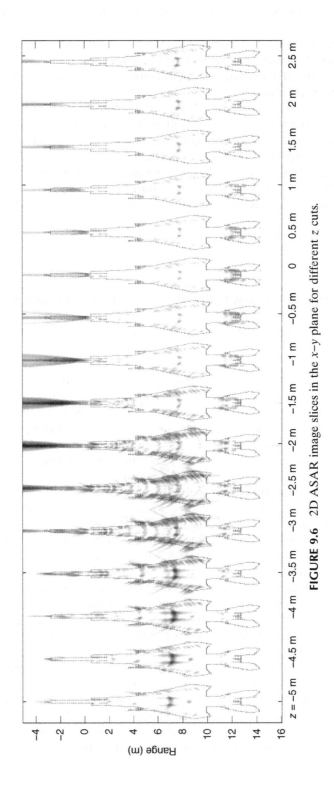

FIGURE 9.6 2D ASAR image slices in the x–y plane for different z cuts.

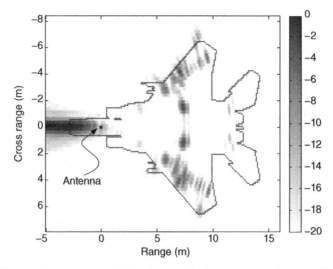

FIGURE 9.7   2D projected ASAR image obtained by summing up all z slices.

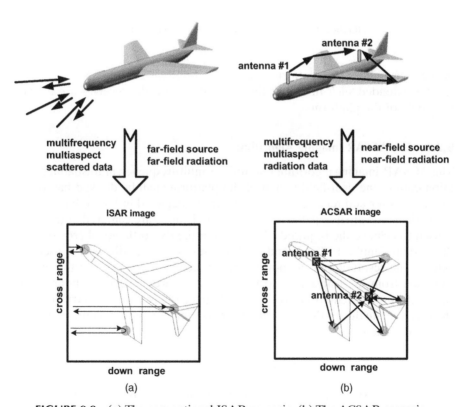

FIGURE 9.8   (a) The conventional ISAR scenario. (b) The ACSAR scenario.

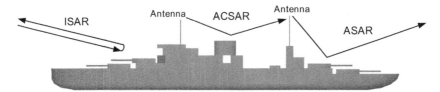

FIGURE 9.9   Conceptual comparison of different radar imaging scenarios.

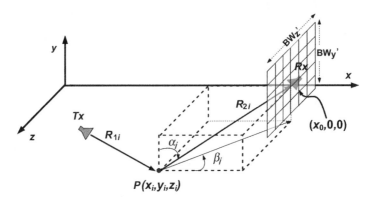

FIGURE 9.10   The geometry for ACSAR imaging.

in the near field of the platform. For the ACSAR case, the ISAR concept is further extended such that both the transmitter and the receiver are at the near field of the platform.

9.2.1   The ACSAR Imaging Algorithm

The ACSAR imaging algorithm that utilizes multifrequency, multispatial radiation data in the near-field region of the platform shall be derived based on the transmitter and receiver setup geometry as depicted in Figure 9.10.

At the receiver site, spatial diversity on a 2D aperture centered at $(x_o, 0, 0)$ is used to achieve the required resolutions in the two orthogonal cross-range dimensions. Similar to ISAR and ASAR cases, frequency diversity is used to achieve the resolution in the down-range direction. In addition to the direct radiation from the transmitter to the receiver, there are contributions from the platform interactions between the antennas. The scattered electric field at the receiver site due to the scattering from a point $P(x_o, y_o, z_o)$ on the platform can be written as

$$E^s \cong A \cdot e^{-jk \cdot R_{1o}} \cdot e^{-jk \cdot R_2}, \qquad (9.8)$$

where A is the strength of the scattered field, R_{1o} is the path length from the transmitter antenna to point P, R_2 is the path length from P to the receiver, and k is the free-space wave number. Next, two approximations will be made to the above equation to arrive at a Fourier-based imaging algorithm [5]. The first assumption, commonly used in ISAR imaging, is that the radiation data are collected within a certain frequency bandwidth that is small compared to the center frequency of operation; this is the small bandwidth approximation. It is further assumed that the size of the aperture at the receiver site is small compared to path length R_2 [3, 5]. Combining these two assumptions, the phase lag in the second phase term can be approximated as follows:

$$-jkR_2 \cong -jkR_{2i} - jk_o(ycos\alpha_i + zsin\alpha_i sin\beta_i), \tag{9.9}$$

where α_i and β_i are defined in Figure 9.10 and R_{2i} is the path length from P to the center of the 2D aperture. Therefore, the scattered electric field can then be approximated by

$$E^s(k, y, z) \cong A_i \cdot e^{-jk \cdot (R_{1i}+R_{2i})} \cdot e^{-jk_o \cdot ycos\alpha_i} \cdot e^{-jk_o \cdot zsin\alpha_i sin\beta_i}. \tag{9.10}$$

In the above equation, an FT relationship exists between the variables (k, y, z) and $(R_i = R_{1i} + R_{2i}, u_i = k_o \cdot \cos\alpha_i, v_i = k_o \cdot \sin\alpha_i \sin\beta_i)$. By taking the 3D IFT of the scattered field with respect to k, y, and z, one can obtain a 3D ACSAR image of the platform as demonstrated below:

$$
\begin{aligned}
ACSAR(R, u, v) &\cong \mathcal{F}_3^{-1}\{E^s(k, y, z)\} \\
&= \mathcal{F}_3^{-1}\{A_i \cdot e^{-jk \cdot R_i} \cdot e^{-jk_o \cdot u_i} \cdot e^{-jk_o \cdot v_i}\} \\
&= A_i \cdot \delta(R - R_i) \cdot \delta(u - u_i) \cdot \delta(v - v_i).
\end{aligned}
\tag{9.11}
$$

Therefore, by inverse Fourier transforming the multifrequency, multispatial coupling data, the point scatterer P will manifest itself as a peak in the image at (R_i, u_i, v_i) with amplitude A_i. In practice, since the frequency bandwidth and the aperture size are not infinite, the actual spot size of the point scatterer will be inversely proportional to the frequency bandwidth and the aperture size. Note that the ACSAR image has not yet been constructed in the original (x, y, z) coordinates. However, it is straightforward to transform the ACSAR image from the (R, u, v) to the (x, y, z) coordinates by the following transformation formulas [3, 5]:

$$
\begin{aligned}
x &= \frac{x_o^2 + R^2 - 2Rx_o\sqrt{1+c}}{2(x_o - R\sqrt{1+c})} \\
y &= -\frac{x_o - x}{tan\alpha \cdot cos\beta} \\
z &= tan\beta \cdot (x_o - x).
\end{aligned}
\tag{9.12}
$$

In the above equation, the constant c is given by

$$c = tan^2\beta + \frac{sec^2\beta}{tan^2\alpha}. \tag{9.13}$$

As is obvious from Equations 9.12 and 9.13, the transformation from (R, u, v) to (x, y, z) is not linear. Therefore, there will be some distortion effects in the resultant $ACSAR(x, y, z)$ image. This issue will be addressed in the next section while demonstrating a numerical example.

In brief, the ACSAR imaging algorithm can be summarized as the following three basic steps:

1. Collect the multifrequency, multispatial EM coupling data $E^s(k, y, z)$.
2. Take the 3D IFT of $E^s(k, y, z)$ to form the $ACSAR(R, u, v)$.
3. Use the transformation formulas in Equation 9.12 to generate the $ACSAR(x, y, z)$ image in the platform coordinates.

In the resulting ACSAR image, the dominant scattering points on the platform due to platform interactions between the transmitter and the receiver will be manifested as peaks.

9.2.2   Numerical Example for ACSAR

An example of the above imaging algorithm is generated using a test object whose CAD geometry is shown in Figure 9.11. This test object contains a number of shapes on top of the platform including a closed cylinder, an open cylinder, a corner reflector, and a step region. The platform is assumed to be perfectly conducting.

A half-wave dipole at 10 GHz is used as the transmitter placed at the origin. A total of 40 discrete frequencies are computed within the bandwidth of 252.3 MHz at the center frequency of 10 GHz. The EM scattering simulation is carried out by a modified version PO-SBR code [4]. The scattered field is collected on an aperture centered at (50 m, 0, 0). There are $40 \times 32 = 1280$ spatial points on the aperture, ranging from −1.03 m to 0.97 m along the y direction for a total of 40 discrete points and from −0.29 m to 0.27 m along the z direction for a total of 32 uniformly spaced points. In the computed data, only the scattered field from the platform is considered, and the primary radiation due to the antenna is not included.

Using the algorithm presented above, the 3D ACSAR image is generated for this scenario. While applying the algorithm, a 3D Hanning window is used prior to the FFT operation to suppress the sidelobes in the image. For display purposes, the 3D ACSAR image is projected onto the 2D $R–u$, $R–v$, and $u–v$ planes as shown in Figure 9.12. As shown from the figure, there are many point scatterers that show up at different locations in the (R, u, v) domain. To view the ACSAR image in the platform coordinates of (x, y, z), the transformations

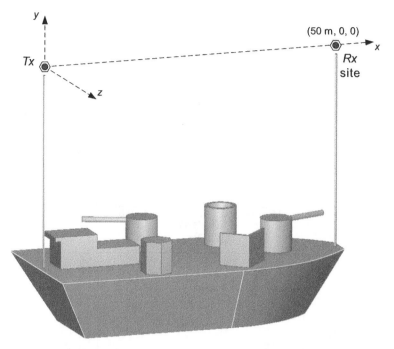

FIGURE 9.11   The geometry for generating the ACSAR image of a ship-like platform.

in Equation 9.12 should be employed. Then, the 3D $ACSAR(x, y, z)$ is obtained. Again, for display purposes, the image is projected onto the 2D y–z, x–y and x–z planes as shown in Figure 9.13. It is observed that the dominant coupling mechanism happens to be around the top of the ship that is roughly midway between the transmitter and the receiver site. Other scattering mechanisms show up around the surrounding region of the midway between the antennas as they experience specular-like EM scattering. The ACSAR image in Figure 9.13 shows that the scattering from other structures is also visible.

9.3   IMAGING SCATTERING FROM SUBSURFACE OBJECTS: GPR-SAR

Detecting objects buried beneath the ground or situated interior to a visually opaque medium has been an interesting topic for many researchers across different disciplines [7–10]. In achieving this goal, different subsurface detection and imaging techniques have been developed. Among those, GPR is one effective tool based on the processing of EM scattering of beneath objects [7, 11–13]. A typical GPR image yields information on the spatial position and the reflectivity of buried objects.

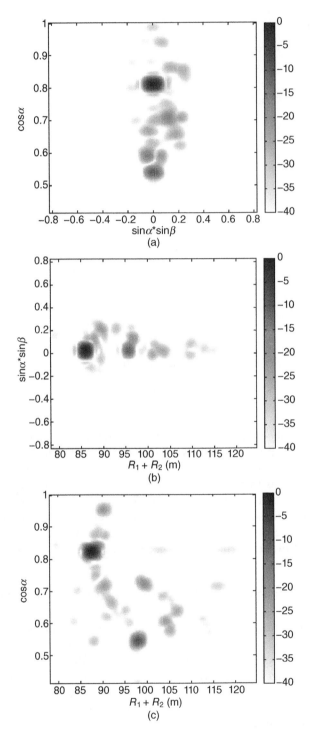

FIGURE 9.12 2D projected ACSAR images of a ship platform in (b) $R-u$ plane, (b) $R-v$ plane, and (c) $u-v$ plane.

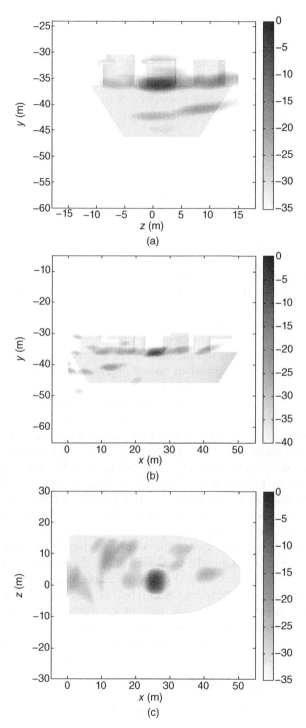

FIGURE 9.13 2D projected ACSAR images of a ship platform: (a) projected y–z image, (b) projected x–y image, (c) projected x–z image.

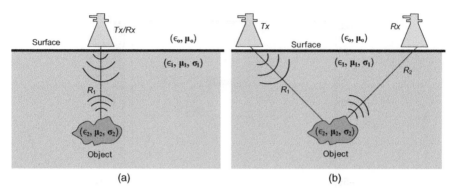

(a)   (b)

FIGURE 9.14 Geometry for GPR problem (a) monostatic scenario, (b) bistatic scenario.

9.3.1 The GPR Problem

GPR scenarios for the bistatic and the monostatic configurations are shown in Figure 9.14. The main idea in GPR is to detect the discontinuities beneath the surface as the EM wave travels inside the opaque medium. EM scattering/reflection occurs when the wave intercepts an object or intersection that has different EM properties, that is, characteristic impedance ($\eta = \sqrt{\epsilon / \mu}$), than the surrounding medium. If the medium is lossy (σ is finite), the penetration depth decreases, and detectibility of buried objects becomes more difficult as the EM wave attenuates more while traveling.

In terms of the GPR operation and the recording method of the GPR data, measurement techniques called A-scan, B-scan, and C-scan are commonly used by researchers [7]. A-scan process provides an amplitude–time record of a single measurement over the target (see Fig. 9.15a). The measurement data can be obtained by either sending a time-domain pulse and recording the reflecting wave or collecting the backscattered field data over frequencies within a finite bandwidth. Of course an IFT operation has to be applied to have the data in time (or range) domain for the latter case. A-scan measurement, in fact, can also be regarded as the range (or depth) profile of the target.

In Figure 9.15, a typical example of A-scan is given. This measured GPR image is constructed for a metal pipe buried under the sand environment [14]. The dielectric constant of the sand is almost stable at 2.1 for the measured frequencies between 4.8 GHz and 8.5 GHz. The resultant A-scan image in Figure 9.15b is constructed by inverse Fourier transforming the frequency diverse backscattered field data from a network analyzer-based measurement system. The first big return is from the ground surface, which is unavoidable and considered to be one of the main problems in GPR. The second return is from the pipe buried about 25 cm from the surface.

A usual GPR system collects reflectivity of the ground and beneath objects while the radar is moving on top of the ground. For this situation, series of A-scan measurements is repeated along a line of synthetic aperture. The

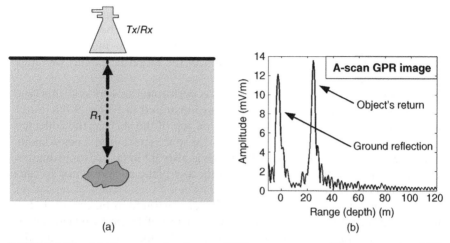

FIGURE 9.15   (a) Geometry for A-scan GPR measurement, (b) an example of 1D A-scan image for a buried pipe.

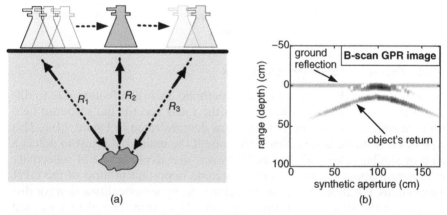

FIGURE 9.16   (a) Geometry for B-scan GPR measurement, (b) an example of 2D B-scan image for a buried pipe.

resulting data set is called the B-scan measurement as illustrated in Figure 9.16a. For the monostatic arrangement of the radar sensor, a single point scatterer appears as the hyperbola in the space-time GPR image when the radar sensor moves over the surface.

An example of a B-scan GPR image is illustrated for the same buried pipe in the same sand environment. For the same frequencies between 4.8 GHz and 8.5 GHz, the B-scan measurement is performed for along the synthetic aperture of 171 cm for a set of 58 discrete A-scans [14]. The resultant measured B-scan is demonstrated in Figure 9.16b. The ground reflection from the surface shows up as a straight line in the space-time (or space-range) B-scan GPR image. The return from the buried pipe appears as a hyperbola in the image since the path of the EM wave varies as the GPR sensor moves along the surface (see Fig. 9.16a). The actual location of the target is actually the apex of this hyperbola.

When a series of B-scan measurements is performed side by side, this type of GPR measurement is called a C-scan as illustrated in Figure 9.17a. Since the data are collected over a 2D aperture on top of the surface, the collected data set is 3D either in the space-space-time or space-space-frequency domain. Therefore, the resultant C-scan GPR image is also 3D in (x, y, z) coordinates. Usually, 2D slices of the C-scan image in the $x-y$ plane are presented for different values of depth, z.

An example of C-scan GPR imagery is demonstrated in Figure 9.17b. For the same sand environment, 2D synthetic aperture of 15 cm by 60 cm is constructed for a total of $15 \cdot 30$ discrete data collection grid to measure the backscattered field data from a bottled water target buried about 45 cm below the sand's surface [14]. For each data set, the frequency is varied from 4.8 GHz to 8.5 GHz. A slice of the resultant 3D C-scan GPR image for the depth of 44.64 cm is shown in Figure 9.17b. As is obvious from the image, C-scan GPR image has the capability of displaying the target's outline, which is, of course, very important for diagnostic applications.

9.3.2 Focused GPR Images Using SAR

A usual B-scan GPR image has the hyperbolic distortion feature due to different travel times of the EM wave as the radar records the scattered field while moving above the buried object, as demonstrated in Figure 9.16a. This type of GPR image is sometimes adequate if the main goal is just to detect a pipe or similar canonical objects. However, size, depth, and EM reflectivity information about the buried object become important in most of the GPR applications. For these situations, therefore, the hyperbolic diffraction (or dispersion) in the space-time GPR image should be transformed to a focused pattern that shows the object's true location and size together with its reflectivity. The process of translating the hyperbolic diffraction or any other kind of dispersion to a focused image is often called *migration* (or *focusing*).

Many algorithms have been adopted by various researchers for the goal of obtaining a focused subsurface image of buried objects [8–25]. Kirchhoff wave-equation [15] and frequency-wave number $(\omega - k)$-based [16, 17] migration techniques are widely accepted and applied. The similarities between the acoustic and EM wave equations have led to the use of the same processing techniques for GPR image processing as in the case of acoustic imaging [17–19]. Among these, the wavenumber domain-focusing techniques were originally developed in seismic imaging applications [18] and have been widely implemented and also adapted to SAR imaging [20–23]. The wavenumber domain algorithms have been developed by various groups in SAR commu-

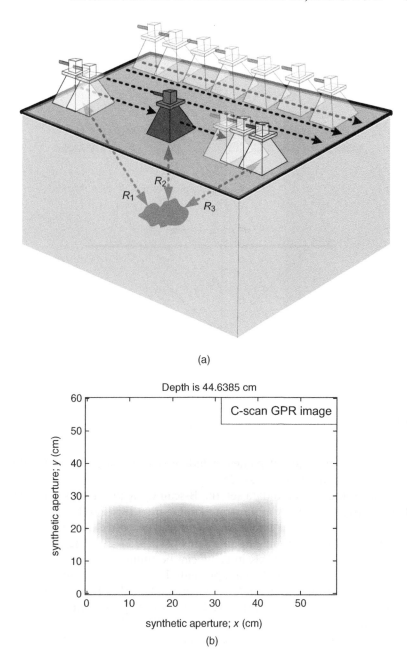

(a)

(b)

FIGURE 9.17   (a) Geometry for C-scan GPR measurement, (b) an example of one 2D slice of a 3D C-scan image for a buried bottled water.

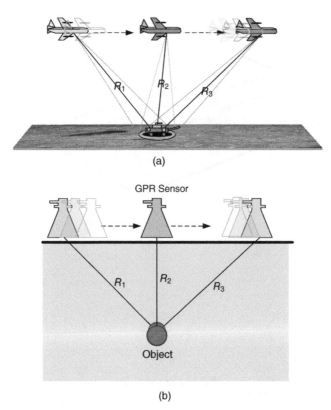

(a)

(b)

FIGURE 9.18   Geometry for (a) the stripmap SAR problem, (b) the B-scan GPR problem.

nity, and called by different names including seismic migration [20, 23] and $\omega - k$ (or $f - k$) migration [13, 24, 25].

In terms of the collected data set, the B-scan GPR problem is very similar to stripmap SAR geometry as illustrated in Figure 9.18b [26]. The geometrical similarities between these two have led to the use of SAR-based focusing techniques in subsurface imaging problems [24, 25, 27–29]. The detailed formulation for the stripmap SAR imaging can be found in Reference 28. Dual algorithms for the GPR problem are presented by several researchers [13, 27]. Here, a $\omega - k$ migration domain SAR imaging algorithm based on the plane wave decomposition of spherical wave fronts will be presented.

The geometrical layout of a typical B-scan GPR problem is shown in Figure 9.18b as the 2D scattered electric field $E^s(x, \omega)$ is recorded for different synthetic aperture points and frequencies. Assuming that the propagation medium is homogeneous and the antenna is close to the ground surface, the frequency domain backscattered field from a point scatterer at r distance from the antenna will have the form of

$$E^s(\omega) = A \cdot e^{-j2\omega r/v_m}, \tag{9.14}$$

where $\omega = 2\pi f$, A is the strength of the scattered from the point target, and v_m is the velocity of the EM wave inside the medium. The number "2" in the exponential accounts for the two-way propagation between the radar and the scatterer. This static measurement at a single spatial location is just an A-scan. In the case of B-scan setup, however, the 2D GPR data are obtained by collecting a series of A-scan measurement along the synthetic aperture axis, say X. For an arbitrary measurement point x_n on the synthetic aperture path, the distance, r, from the point target at (x_i, z_i) to the sensor is equal to

$$r = \sqrt{(x_n - x_i)^2 + z_i^2} \quad n = 1, 2, \ldots, N, \qquad (9.15)$$

where N specifies the total number of A-scan measurements and the z-axis represents the depth direction. If there exist M point scatterers within the subsurface environment, the total 2D backscattered field is the sum of all scattered fields from each point scatterer along the B-scan path as

$$E^s(x, \omega) = \sum_{i=1}^{M} A_i \cdot e^{-j2\frac{\omega}{v_m}\sqrt{(x-x_i)^2 + z_i^2}}. \qquad (9.16)$$

Here, A_i stands for the scattered field strength from the ith point scatterer. Taking the 1D FT of Equation 9.16 along the x direction to get the field in the spatial frequency (k_x) domain,

$$E^s(k_x, \omega) = \sum_{i=1}^{M} A_i \cdot \int_{-\infty}^{\infty} e^{-j2\frac{\omega}{v_m}\sqrt{(x-x_i)^2 + z_i^2}} e^{jk_x x} dx. \qquad (9.17)$$

By utilizing the *exploding source model* [19], the wave velocity v_m is replaced with $c/2$. Then, applying the *principle of stationary phase* [30], the above integral can be approximately solved as

$$E^s(k_x, \omega) \cong \frac{e^{-j\pi/4}}{\sqrt{4k^2 - k_x^2}} \sum_{i=1}^{M} A_i \cdot e^{-j\left(k_x \cdot x_i + \sqrt{4k^2 - k_x^2} \cdot z_i\right)}, \qquad (9.18)$$

where the ratio $e^{-j\pi/4}/\sqrt{4k^2 - k_x^2}$ is the complex amplitude term and has a constant phase. Therefore, it can be neglected for image displaying purposes. Thus, Equation 9.18 can be normalized to give

$$\bar{E}^s(k_x, \omega) = \sum_{i=1}^{M} A_i \cdot e^{-j\left(k_x \cdot x_i + \sqrt{4k^2 - k_x^2} \cdot z_i\right)}. \qquad (9.19)$$

Here, $\sqrt{4k^2 - k_x^2}$ is nothing but the wave number in z domain, namely, k_z. With this construct, we can map the data from the (k_x, ω) domain to the (k_x, k_z) domain by applying the nonlinear transformation of $k_z = \sqrt{4k^2 - k_x^2}$. Since the mapped data will not be on a uniformly spaced grid any longer, the data should also be interpolated. After applying the required mapping and interpolation schemes, the GPR data will be in the form of

$$\tilde{E}^s(k_x, k_z) = \sum_{i=1}^{M} A_i \cdot e^{-j(k_x \cdot x_i + k_z \cdot z_i)}, \tag{9.20}$$

where $\tilde{E}^s$ is the mapped and interpolated data on an equally spaced rectangular grid of k_x and k_z so that 2D inverse fast Fourier transform (IFFT) can be utilized. If the 2D IFT of Equation 9.20 is taken with respect to k_x and k_z, we get

$$e^s(x, z) = \sum_{i=1}^{M} A_i \cdot \iint_{-\infty}^{\infty} e^{-j(k_x \cdot x_i + k_z \cdot z_i)} \cdot e^{j(k_x x + k_z z)} dk_x dk_z, \tag{9.21}$$

which results in

$$e^s(x, z) = \sum_{i=1}^{M} A_i \cdot \delta(x - x_i, z - z_i). \tag{9.22}$$

In this equation, $\delta(x, z)$ is the 2D impulse (Dirac delta) function which perfectly pinpoints the locations of M point scatterers. In reality, the data are collected within a certain bandwidth of frequencies and synthetic aperture length which means that the limits of the integrals in Equation 9.21 are finite in reality. Therefore, the impulse functions degrade to *sinc* functions for practical applications. A flow chart representation of the algorithm steps is given in Figure 9.19 for clear understanding of the SAR-based focusing algorithm. To summarize,

1. Collect the scattered field data either in the time domain to have $E^s(x, t)$ or in the frequency domain to have $E^s(x, \omega)$.
2. Take the 2D IFT of $E^s(x, t)$ or 1D FT of $E^s(x, \omega)$ to transform the data into the wave number-frequency domain as $E^s(k_x, \omega)$ and normalize to get $\bar{E}^s(k_x, \omega)$.
3. Map the data from $k_x - \omega$ domain to $k_x - k_z$ domain and interpolate to have the data on a uniformly spaced rectangular grid as $\tilde{E}^s(k_x, k_z)$.
4. Take the 2D IFT of $\tilde{E}^s(k_x, k_z)$ to get the final focused image in Cartesian coordinates as $e^s(x, z)$.

An example of SAR-based $\omega - k$ migration imaging is demonstrated in Figure 9.20 [14]. Two metal pipes were buried flat around $z = 30$ cm and $z = 40$ cm in a sand pool whose geometry is seen in Figure 9.20a. Using stepped frequency continuous wave (SFCW) radar system, B-scan GPR measurements were collected along the straight path while the frequency is varied from 4.0 to 7.1 GHz. In Figure 9.20b, the raw GPR image obtained by taking 1D IFT of the measured spatial-frequency data is displayed. Obviously, the image is distorted in the synthetic aperture domain due to the well-known hyperbolic behavior.

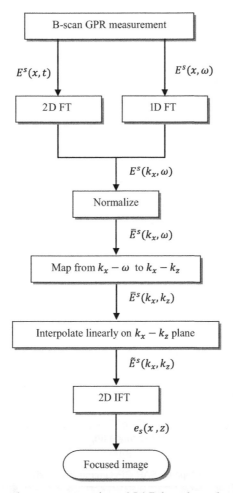

FIGURE 9.19   Flow chart representation of SAR based $\omega - k$ migration algorithm.

Since the pipes are close to each other, the tails of these hyperbolas interact. In Figure 9.20c, the focused GPR image after applying the SAR-based $\omega - k$ migration algorithm is presented.

9.3.3   Applying ACSAR Concept to the GPR Problem

With the bistatic configuration in Figure 9.14b, the data collection setup in the bistatic GPR is very similar to one in ACSAR imaging scenario. Noting this similarity, it is possible to make use of the ACSAR approach in imaging buried objects [31]. The only difference is that the 2D receiver grid will lie parallel to or on top of the surface as depicted in Figure 9.21. This modification only yields a few changes in the coordinate transformation formulas, as listed in Reference 31.

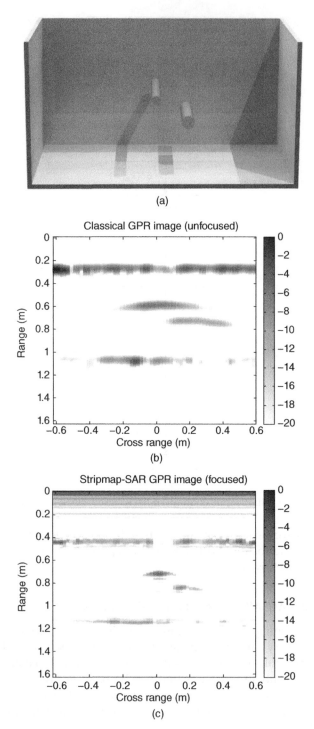

FIGURE 9.20 (a) Configuration of two closely placed pipes buried in sand medium (b) classical B-scan GPR image, (c) focused image after SAR based $\omega - k$ migration algorithm.

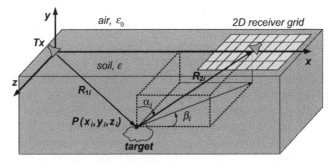

FIGURE 9.21 Applying ACSAR approach to the GPR problem.

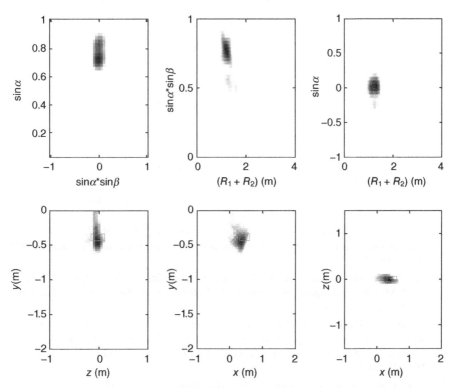

FIGURE 9.22 2D projected ACSAR images of buried bottle beneath the sand surface: Upper figures show the 2D projected radar images in (u, v), (R, v), and (R, u) domains. Lower figures shows the 2D projected radar images (y, z), (x, y), and (x, z) planes.

As an example, a measured GPR image of plastic object full of water is shown in Figure 9.22. The object is buried 38 cm inside the sand medium. The transmitter and the receiver antennas were placed on top of the sand surface. The center of the receiving grid was 1 m away from the transmitter antenna. With the help of a network analyzer, scattered field data are collected on the

10-by-10 2D spatial grid whose size is 52 cm (in the x direction) and 14.7 cm (in the z direction). The frequency was altered from 5.06 to 5.90 GHz for 25 evenly sampled points. After applying the Fourier-based ACSAR imaging routine to the 3D frequency-spatial data, the image was first formed in the distance-angle domain, that is, the (R, u, v) domain, as shown in Figure 9.22 (upper). To get the final GPR image in the (x, y, z) domain, we applied the required coordinate transformation formulas given in Reference 31. Then, a 3D image of the subsurface region was obtained. Figure 9.22 (lower) shows the 2D projected GPR images in the (x, y, z) domain after applying the coordinate transformation. The outline of the buried bottled water is also displayed as the reference. The ACSAR approach to the GPR problem successfully points out the location of the buried object. As seen from the radar images from Figure 9.22, the images are distorted due to nonlinearity of the transformation from (R, u, v) to (x, y, z) as expected.

REFERENCES

1 C. Ozdemir, L. C. Trintinalia, and H. Ling. Antenna synthetic aperture radar (ASAR) image formation. Antennas and Propagation Society International Symposium, 1997. IEEE., 1997 Digest, vol. 4, no., pp. 2601–2604 vol. 4, 13–18 Jul 1997.

2 C. Özdemir, R. Bhalla, L. C. Trintinalia, and H. Ling. ASAR—Antenna synthetic aperture radar imaging. *IEEE Transactions on Antennas and Propagation* 46(12) (1998) 1845–1852.

3 C. Özdemir. *Synthetic aperture radar algorithms for imaging antenna-platform scattering*, Ph.D. Dissertation, the Univ. of Texas at Austin, 1998.

4 H. Ling, R. Chou, and S. W. Lee. Shooting and bouncing rays: Calculation the RCS of an arbitrary shaped cavity. *IEEE Transactions on Antennas and Propagation* 37 (1989) 194–205.

5 C. Özdemir and H. Ling. ACSAR—Antenna coupling synthetic aperture radar (ACSAR) imaging algorithm. *Journal of Electromagnetic Waves and Applications* 13(3) (1999) 285–306.

6 C. Özdemir. Platform effect reduction between antennas using antenna coupling synthetic aperture radar (ACSAR) imaging concept. International Conference on Electrical and Electronics Engineering—ELECO'2001, Bursa, vol. Electronic, 214–217, 2001.

7 D. J. Daniels. *Surface-penetrating radar*. IEE Press, London, 1996.

8 L. Peters, Jr. D. J. Daniels, and J. D. Young. Ground penetrating radar as a subsurface environmental sensing tool. *Proceedings of the IEEE* 82(12) (1994) 1802–1822.

9 S. Vitebskiy, L. Carin, M. A. Ressler, and F. H. Le. Ultrawide-band, short pulse ground-penetrating radar: Simulation and measurement. *IEEE Transactions on Geoscience and Remote Sensing* 35 (1997) 762–772.

10 L. Carin, N. Geng, M. McClure, J. Sichina, and L. Nguyen. Ultra-wide-band synthetic-aperture radar for mine-field detection. *IEEE Transactions on Antennas and Propagation* 41 (1999) 18–33.

REFERENCES **373**

11 K. Gu, G. Wang, and J. Li. Migration based SAR imaging for ground penetrating radar systems. *IEE Proceedings—Radar Sonar and Navigation* 151(5) (2004) 317–325.

12 J. Song, Q. H. Liu, P. Torrione, and L. Collins. Two-dimensional and three-dimensional NUFFT migration method for landmine detection using ground-penetrating radar. *IEEE Transactions on Geoscience and Remote Sensing* 44(6) (2006) 1462–1469.

13 C. Gilmore, I. Jeffrey, and J. LoVetri. Derivation and comparison of SAR and frequency-wavenumber migration within a common inverse scalar wave problem formulation. *IEEE Transactions on Geoscience and Remote Sensing* 44 (2006) 1454–1461.

14 C. Özdemir. *Yeni Bir "Yere Nüfuz Eden Radar (YNR)" Algoritması için Deney Düzeneğinin Oluşturulması, Saha Uygulamaları ve 3 Boyutlu Gerçek YNR Görüntülerinin Elde Edilmesi (in Turkish)*, TÜBİTAK Project no. EEEAG-104E085, Tech. Report, 2005.

15 W. A. Schneider. Integral formulation for migration in two and three dimensions. *Geophysics* 43 (1978) 49–76.

16 J. Gazdag. Wave equation migration with the phase-shift method. *Geophysics* 43 (1978) 1342–1351.

17 R. H. Stolt. Migration by Fourier transform. *Geophysics* 43 (1978) 23–48.

18 E. Baysal, D. D. Kosloff, and J. W. C. Sherwood. Reverse time migration. *Geophysics* 48 (1983) 1514–1524.

19 C. J. Leuschen and R. G. Plumb. A matched-filter-based reverse-time migration algorithm for ground-penetrating radar data. *IEEE Transactions on Geoscience and Remote Sensing* 39 (2001) 929–936.

20 C. Cafforio, C. Prati, and F. Rocca. Full resolution focusing of Seasat SAR images in the frequency-wave number domain. *Journal of Robotic Systems* 12 (1991) 491–510.

21 C. Cafforio, C. Prati, and F. Rocca. SAR data focusing using seismic migration techniques. *IEEE Transactions on Aerospace and Electronic Systems* 27 (1991) 194–207.

22 A. S. Milman. SAR imaging using the w-k migration. *International Journal of Remote Sensing* 14 (1993) 1965–1979.

23 H. J. Callow, M. P. Hayes, and P. T. Gough. Wavenumber domain reconstruction of SAR/SAS imagery using single transmitter and multiple-receiver geometry. *Electronics Letters* 38 (2002) 336–337.

24 A. Gunawardena and D. Longstaff. Wave equation formulation of synthetic aperture radar (SAR) algorithms in the time-space domain. *IEEE Transactions on Geoscience and Remote Sensing* 36 (1998) 1995–1999.

25 Z. Anxue, J. Yansheng, W. Wenbing, and W. Cheng. Experimental studies on GPR velocity estimation and imaging method using migration in frequency-wavenumber domain. Proceedings ISAPE Beijing China 2000, 468–473.

26 M. Soumekh. *Synthetic aperture radar signal processing: With Matlab algorithms.* John Wiley & Sons, New York, 1999.

27 E. Yigit, S. Demirci, C. Ozdemir, and A. Kavak. A synthetic aperture radar-based focusing algorithm for B-scan ground penetrating radar imagery. *Microwave and Optical Technology Letters* 49 (2007) 2534–2540.

28   M. Soumekh. A system model and inversion for synthetic aperture radar imaging. *IEEE Transactions on Image Processing* 1 (1992) 64–76.

29   V. Kovalenko, A. Yarovoy, and L. P. Ligthart. *A SAR-based algorithm for imaging of landmines with GPR*, International Workshop on Imaging Systems and Techniques (IST 2006), Minori, Italy, 2006, 65–70.

30   W. C. Chew. *Waves and fields in inhomogeneous media*, 2nd ed. IEEE Press, New York, 1995.

31   C. Özdemir, S. Lim, and H. Ling. A synthetic aperture algorithm for ground-penetrating radar imaging. *Microwave and Optical Technology Letters* 42(5) (2004) 412–414.

Appendix

In this appendix, we present some useful Matlab functions that are used by the Matlab codes given in the chapters of this book.

Matlab code A.1: Matlab file "stft.m"

```
function [B,T,F]=stft(Y,f,BW,r,d)
%STFT Calculates the Short Time Fourier Transform of Vector
Y

% Inputs:
% Y : signal in the frequency domain
% f : vector of frequencies [Hz]
% BW : bandwidth (same unit as F) of the sliding window
% f : desired dinamic range of the display
% d : additional delay [s] (if desired)
%
% Outputs:
% B : stft of vector Y
% F : frequency vector [GHz]
% T : frequency vector [ns]

% The window used is a Kaiser window with beta=6.0

df = f(2)-f(1); % frequency resolution
Ws = round(BW/df);
if (Ws<2),
  Ws=2;
end;
```

Inverse Synthetic Aperture Radar Imaging with MATLAB Algorithms, First Edition.
Caner Özdemir.
© 2012 John Wiley & Sons, Inc. Published 2012 by John Wiley & Sons, Inc.

```
W = kaiser(Ws,6); % window is a Kaiser with beta=6.0
N = max(size(Y)); % find length of Y

% Spectrogram
[B,T,F] = specgram((Y.*exp(-j*2*pi*f*d))',N,1/df,W,
Ws-1);
F = F+((Ws-1)/2)*df+f(1); % set frequency axis
T = T-d; % set time axis

% Treshold the image to the dynamic range
bmax = max(max(abs(B)));
ra = bmax/(10^(r/20));
B = B.*(abs(B)>=ra)+ra*ones(size(B)).*(abs(B)<ra);

% Display the STFT
colormap(jet(256)); %set colormap
imagesc(T*10^(9),F*10^(-9),20*log10(abs(B')/bmax))
axis xy; % change origin location
xlabel('Time [ns]'),
ylabel('Frequency [GHz]');
```

Matlab code A.2: Matlab file "cevir2.m"

```
function out=cevir2(a,nx,ny)
% This function converts a 1D vector to a 2D matrix
% Inputs:
% a : 1D vector of length (nx*ny)
% nx : column length of the output matrix
% ny : row length of the output matrix

% Output:
% out : 2D matrix of size nx by ny


for p = 1:nx;
  out(p,1:ny) = a(1,(p-1)*ny+1:ny*p);
end;
```

Matlab code A.3: Matlab file "shft.m"

```
function [out]=shft(A,n)
%This function shifts (circularly) the vector A with
% an amount of n

% Inputs:
% A : the vector to be shifted
% n : shift amount

% Output:
% out : shifted vector

out = A(1-n+2:1);
out(n:1) = A(1:1-n+1);
```

Matlab code A.4: Matlab file "matplot.m"

```
function [p]=matplot(X,Y,A,r)
% This function displays a matrix within the
% dynamic range of r

% Inputs:
% A : the matrix
% r : dynamic range of the display [dB]
% X : x-label Vector
% Y : y-label vector

% Output:
% p : matrix thresholded to r(dB)

b = max(max(abs(A))); %find max value of A
ra = b/(10^(r/20)); % make it to dB

% treshold A to the dynamic range of r[dB]
p = A.*(abs(A>=ra)+ra*ones(size(A)).*(abs(A)<ra);
pp = 20*log10(abs(p)/b);

colormap(jet(256))
imagesc(X,Y,pp)
axis xy; % change the location of origin
```

Matlab code A.5: Matlab file "matplot2.m"_____

```
function [p]=matplot(X,Y,A,r)
% This function displays a matrix within the
% dynamic range of r

% This function is similar to matplot.m except the location
of the origin

% Inputs:
% A : the matrix
% r : dynamic range of the display [dB]
% X : x-label Vector
% Y : y-label vector

% Output:
% p : matrix thresholded to r(dB)

b = max(max(abs(A))); %find max value of A
ra = b/(10^(r/20)); % make it to dB

% treshold A to the dynamic range of r[dB]
p = A.*(abs(A>=ra)+ra*ones(size(A)).*(abs(A)<ra);
pp = 20*log10(abs(p)/b);

colormap(jet(256))
imagesc(X,Y,pp)
```

Index

WILEY SERIES IN MICROWAVE AND OPTICAL ENGINEERING

KAI CHANG, Editor
Texas A&M University

Printed and bound by CPI Group (UK) Ltd, Croydon, CR0 4YY

16/04/2025

14658424-0001